Rethinking Development Geographies

Development as a concept is notoriously imprecise ... an-
ing of this fiercely contested term have had profo ... aces
across the globe.

Rethinking Development Geographies offers a st ... eog-
raphy and development. In doing so it sets out to ex ... tice.
The book highlights the geopolitical nature of dev ... r. It
also reflects critically on the historical engagement ... and
the 'South'. The dominant economic and politic ... ives
of major institutions are explored here. The inter ... are
highlighted through an examination of local, na ... s of
development.

The text provides an accessible introduction to ... obal
development. Informative diagrams, cartoons and ... obal
geographies of economic and political change *R* ... in a
concern with people and places, the 'view from b ... th'.

Marcus Power is a Lecturer in Human Geograp ... y of
Bristol.

RETHINKING DEVELOPMENT **GEOGRAPHIES**

MARCUS POWER

LONDON AND NEW YORK

First published 2003
by Routledge
2 Park Square, Milton Park, Abingdon, Oxon, OX14 4RN

Simultaneously published in the USA and Canada
by Routledge
270 Madison Ave, New York, NY10016

Reprinted 2004

Routledge is an imprint of the Taylor & Francis Group

Transferred to Digital Printing 2007

Typeset in Bell Gothic and Times New Roman
by RefineCatch Limited, Bungay, Suffolk
Printed and bound in Great Britain by
Bell & Bain Ltd, Glasgow

British Library Cataloguing in Publication Data
A catalogue record for this book is available from the British Library

Library of Congress Cataloging in Publication Data
Power, Marcus
Rethinking development geographies/Marcus Power.
p. cm.
Simultaneously published in the USA and Canada.
Includes bibliographical references and index.
1. Economic geography. 2. Economic development.
3. Developing countries–Economic conditions.
4. Globalisation. 5. Economic history. 1. Title.
HF1027 .M23 2003
338.9–dc21 2002155680
ISBN-10: 0-415-25078-1 (hbk)
ISBN-10: 0-415-25079-X (pbk)
ISBN-13: 978-0-415-25078-8 (hbk)
ISBN-13: 978-0-415-25079-5 (pbk)

FOR AIDAN AND SAMUEL LUEBCKE

Contents

Acknowledgements x
Copyright acknowledgements xi

Chapter 1 Introduction: what is 'development'? 1
People, places and progress: thinking spatially about development
Mental maps and imagined geographies of difference
Development and the geography of power
Re/thinking geographies of development: a guide to what follows

Chapter 2 Illuminating the dark side of development 20
Introduction: an incomplete and contested picture of progress
The global 'protest caravan'
Decolonising the mindset of development
The political and psychological satisfaction of 'helping' the poor
The (geo)politics of aid
Conclusions: geopolitics and the 'apothecary' of development remedies

Chapter 3 Geographers and 'the Tropics' 45
Introduction: Orientalism and tropical geographies
Geographies of colonial modernity
From tropical to development geography (via imperialism)
From 'pious Eurocentrism' to geographies of 'anti-development'

Conclusions: geo-writings and the tropical worlds of development

Chapter 4 Development thinking and the mystical 'kingdom of abundance' 71

Introduction: knowledge and the era of modernity
The Enlightenment and the theorisation of development as progress
The 'Sinatra Doctrine' and the Cold War
Dependency: just another narration of tradition vs. modernity?
Imagining a post-development era
Conclusions: deconstructions and reconstructions of development

Chapter 5 Thirdworldism and the imagination of global development 95

Introduction: a critical geopolitics of Third World development
'The West' and 'the rest'
The concept of the three worlds
Lost for wor(l)ds?: worlding and otherness
Conclusions: colonialism, decolonisation and the pursuit of development

Chapter 6 Postcolonial geographies of development 119

Introduction: what is postcolonialism?
Postcolonialism and national belonging
Language, culture and identity
Partnership or trusteeship?
Conclusions: postcolonialism and the voices of the poor: can anyone hear us?

Chapter 7 Globalisation, government and power 143

Introduction: globalisation and the non-western world
Neoliberalism and 'actually existing' globalisation
Globalisation and the myth of free trade
Globalisation and the post-Washington apothecary
Conclusions: governance, geopolitics and development: towards a new kind of multilateralism?

Chapter 8 The dissemination of development 169

Introduction: development as a 'global moral imperative'?
Obscuring the *causes* of poverty: the poverty process acronyms
Poverty, livelihoods discourse and social capital

Deconstructing the 'development gateway'
Conclusions: solutions in search of problems?

Chapter 9 'Theorising back': views from the South and the globalisation of resistance 194
Opening a space in the history and geography of international development
Empowerment from above or below?
Geopolitics from below and the crises of state-centred developmentalism
'We do not need you to save our forests'
Neoliberalism, democracy and resistance in post-apartheid South Africa
Conclusions: a global fabric of struggle?

Chapter 10 Conclusions: resisting the temptations of remedies, mirages and fairy-tales 219
Towards 2015
Geographies of neoliberalisation
Imagining a post-development era
Decolonisation and development (geography): beyond trusteeship

Glossary 236
Bibliography 238
Index 259

Acknowledgements

This book is a product of many different ideas and sources of inspiration. My greatest debt is to my parents Ann and Maurice and to my sisters Pauline and Jennifer for all the love and support they have offered to me throughout this project and throughout my academic career. I also owe a particular debt of gratitude to James Sidaway for all his valuable and insightful suggestions on possible directions in which to take this book. I would also like to thank James for his friendship over the years, and his continued support and encouragement.

As well as those mentioned above, I would also like to thank the following for their influence and help: Andrew Crampton, Giles Mohan, Claire Mercer, Rachel Slater, Rob Kitchin, Dave Griffith, Alix Wood, Adrian Smith, Rob Potter, David Simon, Paul Waley, Dave Clarke, Adrian Bailey, Frank Cudjoe, Felix Driver, Peter Bazimya, Reg Cline-Cole, Andrew Mould and Melanie Attridge. Thanks also to the anonymous referees for their helpful comments on earlier drafts of this book.

Copyright Acknowledgements

The author and publishers would like to thank the following for granting permission to reproduce images in this work:

Adbusters Media Foundation for Figures 1.5, 9.2, 9.11, 9.13 and 10.3.

Alex Hofford for Figures 2.4, 5.5, 5.8.

Alix Wood for Figures 5.2, 5.6, 6.3.

Andrzej Krauze for Figure 10.5.

Atlantic Syndication for Figure 10.6, Beattie.

Cagle Cartoons Inc for Figure 6.7 cartoon by B. Farrington, 16 April 2002; Figure 6.9 cartoon by D. Cagle, 6 September 2001; Figure 7.5, M. Lane; Figure 8.3, cartoon by D. Cagle, 11 May 2001; Figure 10.4, cartoon by M. Keefe, 26 March 2002; Figure 10.9, cartoon by M. Lane, 3 April 2002.

Cartoonist and Writers Syndicate (CWS) for Figure 1.1, Oliver Schopf; Figure 5.3, Palomo; Figure 5.7, Paresh; Figure 5.9, Hajjaj; Figure 7.4, Ammer; Figure 7.6, Miel, Figure 7.8, Bado, Figure 7.9, Wonsoo; Figure 7.10, Gable; Figure 8.2, Phore; Figure 8.10, Gable; Figure 10.1, Zetterling.

Dave Brown and *The Independent* (2001) for Figure 1.4 and Figure 6.8.

Figure 4.4 © David Griffith and Rob Kitchen.

Environment & Urbanization for Figure 1.9, originally published on the front cover of Vol. 12, No.2, October 2000; and Figure 4.2, originally published on the front cover of Vol. 13, No. 1, April 2001.

François Houtart for Figure 1.8

Frelimo/Forum Mulher for Figure 6.1 and Figure 9.3.

Globalise Resistance for Figure 1.12 and Figure 9.12.

Humana People to People for Figure 4.3.

ID21 for Figure 9.4, cartoon image by Maddocks from *Insights* magazine, issue 38 (November 2001).

Jonathon Shapiro for Figure 7.1, cartoon which appeared in *The Sowetan* on 3 May 2001.

Lily Kong for Figure 3.4, Figure 7.7.

Michael Chanan for Figure 6.4, film still courtesy of ICAIC.

NBA for Figures 7.3, 9.6, 9.7, 9.8, 9.9, 9.10, 10.8.

New Internationalist Publications for Figure 4.1.

Oxfam Publishing, 274 Banbury Road, Oxford, OX2 7DZ for Figure 2.9 and Figure 9.5. Reproduced with permission.

Figure 2.7 © Chappatte, originally published in *International Herald Tribune* – www.globecartoon.com.

Shaun Askew for www.brettonwoodsproject.org for Figure 7.2 and Figure 8.5.

Taylor & Francis for Figure 1.10 from *Geography of the Third World* by Dickenson *et al.*, Routledge, 1996, and Figures 4.5 and 5.1 from Critical Geopolitics by Gearoid O'Tuathail.

The Associated Press Ltd and *The Economist* for Figure 2.8.

The Straits Times for Figure 10.2.

THINK AGAIN/protestgraphics.org 2002 for Figures 1.6, 2.3, 2.5, 2.6, 6.6, and D. Alwan/Protest Graphics for Figure 10.10.

United Nations Department of Public Information for Figures 1.2, 1.3, 1.11, 3.5, 4.6, 6.2, 8.1, 8.4, 8.6, 8.7, 8.8, 8.9, 10.7.

World Bank for Figure 1.7 and Figure 2.2.

Every effort has been made to contact copyright holders for their permissions to reprint material in this book. The publishers would be grateful to hear from any copyright holder who is not here acknowledged and will undertake to rectify any errors or omissions in future editions of this book.

1 Introduction

what is 'development'?

> [D]evelopment is a continuous intellectual project as well as an ongoing material process.
>
> (Apter, 1987: 7)

At the beginning of the twenty-first century, a number of high-profile events and processes have illustrated that the challenges of global 'development' are becoming an increasingly important part of international relations and world politics. After the terrorist attacks on the United States in September 2001, many people were quick to point out that poverty and inequality between nations was becoming *the* most important issue in building world peace and international political stability. In the context of the global war on terrorism that has followed 9/11 and the international concern to rebuild and reconstruct Afghanistan and Iraq, issues of poverty and development have again taken centre-stage. Furthermore, in the spring of 2002 at a United Nations (UN) meeting in Monterrey, Mexico, world leaders made many promises to deliver more aid and assistance to poor countries and to open their markets to trade with the 'lesser developed world'. At the time of writing, the World Summit on sustainable development is meeting in Johannesburg (South Africa) to discuss current trends in global production and consumption and the social and environmental strains that threaten to 'derail development efforts and erode living standards' (Wolfensohn, 2002: 21). What is particularly interesting about these and similar world gatherings is that they bring together nations and peoples with often vastly different and incredibly varied levels of social and economic resources for 'development' to discuss common approaches and to devise collective solutions. The United States and Uganda, for example, are characterised by marked differences in living standards, with very different cultures and histories as well as often divergent perspectives on how to change the world economy and how to devise policies that improve economic growth opportunities for all peoples. In seeking to understand these kinds of differences between the 'rich' and 'poor' nations of the world, geography and geographical analysis have a particularly important role to play.

One of the major difficulties in finding common approaches, policies and solutions to these challenges is that the idea of 'development' is difficult to define, since the term has a whole variety of meanings in different times and places. In Malaysia, for example, debates about the nature of development and its importance to national 'progress' and social change have taken on a very different complexion when compared to a country such as South Africa or Sri Lanka. Even within such countries, the meanings and definitions of 'development' vary substantially across national territory and between different social groups or are, in a way, 'place-specific'. As if to complicate our study of 'development' and its geographies even further, it might be said that the term actually has no clear and unequivocal meaning and is in a sense truly the stuff of myth, mystique and mirage. Little consensus exists around the meaning of this heavily contested term yet most if not all leaders of the world's many nation-states and international organisations claim to be pursuing this objective in some way. This book seeks to show that, by contrast, the strength of the term comes directly from its power to seduce, to please, to fascinate, to set dreaming, but also from its power to deceive and to turn away from the truth (Rist, 1997: 1). Development is nearly always seen as something

Figure 1.1 The World Summit for Sustainable Development as 'seen' on TV by President Bush
Source: Oliver Schopf and CWS

that is possible, if only people or countries follow through a series of stages or prescribed instructions. It is also often assumed that an organic and inherently 'natural' process of evolution is somehow at work, one that is both progressive and forward-moving. The assumption here is that from little acorns do giant oaks grow. Development also often means simply 'more': whatever we might have some of today we might or should have more of tomorrow (Wallerstein, 1994). In this book we will be delving further beneath the surface of these different definitions in an attempt to show how they have often been 'normative' (saying what should be done) or instrumental (serving as an instrument or means), usually pointing to things that are *lacking* or *deficient* (e.g. knowledge) or things that need to be *intensified* (e.g. democracy). As we shall see, at the core of development studies and also development geography there has been this very normative preoccupation with the poor and with what are often referred to as the 'marginalized and exploited people of the South' (Schuurman, 2001: 9).

The term 'development' often refers simply to vague notions of 'good change', an unquestionably positive phrase that in everyday parlance is practically synonymous with 'progress' and is viewed typically in terms of increased living standards, better health and well-being and other forms of common good which are seen to benefit society at large. The core definition of development in the United Nations Development Programme's annual *Human Development Reports* revolves around the idea of an 'enlargement of people's choices'. The term is used regularly in a wide variety of senses however, which embody 'competing political aims and social values and contrasting theories of social change' (Thomas, 2000: 23). Thus conflicting and contradictory views about the relationships between development and capitalism can be enfolded within popular assumptions about positive change and how to bring about poverty eradication.

A number of the major development agencies have recently tried to demonstrate that both 'developing' and 'developed' societies can be rank ordered according to common measures of progress and change such as in the case of the UN's Human Development Index (HDI). This tried to escape the limitations of previous measures by combining four variables for each country: income, life expectancy, human liberty and level of education. The definition of poverty (like that of development) remains hotly contested however, while conditions of poverty clearly vary between different areas of the world, as does the way poverty is experienced, making it more difficult to reach a 'universal' definition that can be agreed upon. Poverty is of course a relational issue which refers to life chances and experiences which are uneven socially and spatially, but debates about poverty have generally focused on those groups that are 'deprived' and lacking in social power, assets or capital rather than focusing

Figure 1.2 Fifty years of successful UN operations?
Source: UN Department of Public Information

Figure 1.3 To what extent does the United Nations create a better world?
Source: UN Department of Public Information

Figure 1.4 Western political leaders like Tony Blair have a very positive view of globalisation
Source: © Dave Brown 2001, *The Independent*

on the associated issues of wealth and consumption. Moreover, poverty is clearly far from being a new concern, but 'development' *is* an idea that has acquired particular meanings over the past two centuries, especially since the end of the Second World War.

The vision of development articulated by many global agencies however is still very much based around the idea of a linear progression, indicated as different degrees of departure from already established western ideals and experiences. As in the case of Gross National Product (GNP) per capita, the HDI does in many senses (albeit at a very crude level) point to growing gaps between 'rich' and 'poor' areas and peoples of the world, but their significance is almost cheapened by the failures of these 'generalised' indicators. Too often these 'gaps' are taken as a given in the study of development and rarely is this sufficiently problematised or questioned in studies and representations of poverty which may therefore even be responsible for deepening, reproducing and widening these gaps (Rist, 1997). None the less, even though the data give only a shallow and sometimes misleading impression, it is still possible (and necessary) to theorise about different levels of economic development in the twenty-first century (Peet with Hartwick, 1999: 9).

In many western societies, poverty is now often considered in relative terms, where the wider social exclusion of the poor is sometimes taken into consideration. This notion of poverty is also relevant in non-western contexts where a wide range of social 'resources' may be available to an individual other than monetary income (Sen, 1983, 1985). None the less, it is argued here that despite some of these changes, the measures adopted by many global development organisations remain powerful examples of precisely how *not* to go about counting the poor (Reddy and Hogge, 2002). This book seeks to show that people experience impoverishment in and through their particular places and localities. This also needs to be considered as an important dimension of poverty in a global and inclusive definition, which does not seek to 'objectify the poor', to declare them a 'problem' and to invent a solution, but rather to convey the variety of complex daily struggles that poverty involves on a number of geographical scales, from the local to the global. What is particularly intriguing here is that despite nearly six decades of 'global' strategies of development, few if any of the international agencies know exactly how many poor people there are in the world and only very recently have these agencies actually begun to ask: How do poor people *themselves* view issues of poverty, development and well-being? (Narayan *et al.*, 2002).

As we shall see, the notion of development thus represents a great variety of definitions, theories and approaches, but it can also be considered something of a hosting metaphor that contains within it 'crucial connecting assumptions between growth and democracy' (Apter, 1987: 7). In the chapters that follow we will be exploring the ways in which development also often serves as a *heuristic* device, as something to stimulate thought or discussion, a way to propose strategies, theories and ideologies of how to overcome and eradicate poverty. Before we set out on this journey it is worth remembering that development is one of the most complex words in the English language and its varied intellectual history or genealogy can confuse and limit its contemporary usage in myriad ways (Williams, 1976).

PEOPLE, PLACES AND PROGRESS: THINKING SPATIALLY ABOUT DEVELOPMENT

> This is the power of development: the power to transform old worlds, the power to imagine new ones.
>
> (Crush, 1995: 2)

One of the most intriguing and useful discussions of the meaning and significance of development was written in 1997 by Majid Rahnema, who once served as a minister in the government of Iran in the 1960s and later headed a UN development programme in Mali before engaging in debates about 'alternative' development in his later career:

> I remember very well that, for example, in my own country, dozens of researchers have set out to find an equivalent term in Persian. There were many concepts that spoke of "good life", "solidarity", "prosperity", "blossoming", "the happiness of living together", and "the beneficial flow of life". These words were far from hollow, fossilized or pompous.
>
> (Rahnema, 1997a: 6)

Rahnema is referring to the political and economic climate of the 1960s when Iranians searched for an equivalent term in Persian. 'Dozens' of researchers have since set out to understand how the idea of development could be made relevant to their particular historical and geographical contexts. In particular, Rahnema recalls how, in the case of the Arab and Persian world, the term frequently acquired and was imbued with these positive characteristics while debates around the meaning of 'development' were far from empty or 'hollow': 'they stirred people's hearts, spoke to them, resonant as they were of everything that gave meaning to their dreams and the well-being of a happy community' (Rahnema, 1997a: 6).

Thus, while the Middle East does not usually figure prominently in debates about global development, Rahnema shows how the meanings of development in the Iranian context also stirred people's hearts, dreams and imagination. This raises fundamental questions about the relations between people and places that we need to examine further. Debates about development have resonated with people and communities in a variety of places and spaces around the world over a long period of time.

One objective of this book is to formulate a view of development which focuses on the relationships to households and communities and not only (as often happens) on 'formal' institutions such as the state, the transnational corporation, the international development agencies or non-governmental organisations (NGOs). This is a very difficult balance to maintain consistently, since development operates across so many spatial scales and involves different degrees of formality simultaneously.

Geographers have often focused on deprived and poor areas and have raised questions about different regions and inequalities, adopting a range of different approaches over the past six decades. What we will focus attention on in this book is the interactions between different spaces and places and the ways in which these interactions have been stretched and extended at different times and through different processes. In this sense, our consideration of space, place and scale can offer alternative organising principles around which to think about development. In particular this book is aimed at intermediate and advanced level undergraduate students studying development, and on one level it sets out to offer a thematic and historical critique of development, arguing that we thus need to understand the variety of relationships between people and places by examining historical relationships at the *global* level. On another level, it is argued that such an exploration needs to be grounded, to be seen as rooted in the 'everyday' practices, movements and behaviours of individual people based in particular places.

It is worth remembering that there is also no easy definition of place here, since we are referring to such a complex diversity of peoples, cultures and localities. Places are not bounded however, but are 'open' and linked to a 'space of flows' and they can also be seen as something that is 'made' by the media or understood as a context for social and political relations (Holloway and Hubbard, 2001). People and places have all too often been disregarded in the rush to form grand theories of economic or political development and this imbalance needs to be urgently rectified. It is useful to think then about how people have created and transformed the places and spaces of development and have perceived and imagined particular kinds of localities or historical experiences in their conceptions of those worlds deemed worthy and in need of 'developing'. In this sense, space can be theorised in material, relational, imaginary, abstract and metaphorical forms, all of which need to be considered in thinking critically about the *spatiality of development*. In a whole variety of ways, development has always been about spatial imaginaries that operate at the local, national and global scale. Thus it is useful to focus on the extent to which there is a translation of the meaning and objectives of development theories and practices into a variety of seemingly very different geographical contexts. As Rahnema explains about the use of equivalent terms in Persian and other contexts and settings, these words seemed almost impossible to translate, as they represented what appeared to be 'totally different perceptions of what a good life is' (Rahnema, 1997a: 7).

This then is a central and fundamental issue in the process of rethinking our approaches to the study of development's geographies: how to translate these 'totally' different and supposedly incompatible conceptions of what a good life is or of what 'good change' is necessary to help bring this about for the majority. For students of 'development', learning to understand and appreciate social and economic differences between one area of the world and another is a difficult and value-laden process, particularly complicated in the context of contemporary globalisation. In a study of student travellers to the Third World, Desforges (1998) illustrates, for example, the links between travel and identity for many students who sought to 'collect places' which offered authentic individual knowledges and personal experiences that could be gained through journeying to other worlds. This involved a kind of framing of Third World peoples and places as different and also the assumption that it was possible to 'collect' experiences of Third World places. In this way, travel may be understood as one way in which 'youth identities "stretch out" beyond the local to draw in places from around the globe' (Desforges, 1998: 176). This sense of identities being 'stretched out' and places being drawn into the global is a crucial theme in thinking about geography and development today.

Most recently, debates about 'globalisation' and development (which have often been abstract and lacking in clarity) have come to provide a key means through which many people seek to understand transitions among human societies at the start of the third millennium. Much has been made of the impact of globalisation, the consequent 'annihilation of space by time' and the increased ease of travel and communication that characterise the contemporary world. There has also been a tendency to speak of the 'end of geography', envisioning a time when all places will have similar social and cultural characteristics as global corporations, and as media agencies spread similar products and images across the globe (Holloway and Hubbard, 2001). In this respect, the nations of Africa, Central and Latin America and much of Asia – collectively often referred to as the 'Global South' – are seen to face great challenges and to be confronted by a number of opportunities as a consequence of 'globalisation'. In this book, we will be examining the dynamic relations between territories, power and identity in the 'Global South' (which is said to include well over two-thirds of the world's nation-states). In particular we will be examining the possibility of what some observers have referred to as 'deterritorialisation' or the displacement of identities, peoples and meanings that is said to result from globalisation in a postmodern world system. It is important then to examine the way in which territory is seen to lose significance as a result of globalisation and the 'unstoppable' juggernaut of global changes in information, knowledge and technology. What is often not clear in these debates is what role there is for the people of the 'South' to shape their own destinies and to what extent borders and geographical divides are becoming less important in a global context.

The media in particular sometimes simplify the complexities of development and globalisation and present poverty as universal or globalisation as novel. The media also relay images and information about inequality within and between worlds almost every day but these global inequalities have often been misrepresented, while the means of conducting an assessment or evaluation of this unevenness has also been widely disputed. The ways in which the peoples and places of development have been represented is crucial here. If people construct places, so too people are constructed by places and thus there is a reciprocity and relationality between peoples and places that geographers are beginning to turn their attention to (Holloway and Hubbard, 2001). In a whole variety of important ways the media play a key role in building a spatial imagination of the globe and globalisation, enframing the way that Third World others are viewed in western societies.

MENTAL MAPS AND IMAGINED GEOGRAPHIES OF DIFFERENCE

> By whom is development being done? To whom? These are potent questions. They should particularly be asked when 'solutions' are put forward that begin 'We should . . .', without making clear who 'We' are and what interests 'We' represent.
>
> (Allen and Thomas, 2001: 4)

One of the focal points of this book is the contention that representations of the 'Third World' and of the 'West' imagine places which are imbued with certain sorts of moral, cultural and socio-political attributes. These important representations also have a significant impact on the spatial scales at which development policy takes place and ultimately also determine the kinds of resistance to and contestation of these policies which are possible. This book also raises questions about the principal sources and means we use to help understand and make sense of 'tropical areas', the 'Third World' or 'developing areas'. These might include, for example, the mass media (e.g. television, radio, cinema and the written press), school, college and university education, and even friends and relatives. These images of other worlds are built up from our cognition and perception of the environment and summarise each individual's knowledge of their surroundings. They are not necessarily 'false' or 'wrong' but partial because they are not always based on first-hand experience and are thus in some ways 'fictions' (Holloway and Hubbard, 2001). These are simplified and distorted images, which can relate to different kinds of places and spatial scales (rooms, houses, neighbourhoods, cities, regions, nations and the entire globe). Imagining the geography of development through mental maps, we create maps in our minds which have meaning for us, and come to form something of a 'mental atlas' where each page can be referred to in a particular situation (Holloway and Hubbard, 2001). They allow us to make sense of broad and varied parts of the world with which we may yet have little or any direct contact either through experience or learning. These mental maps can be called 'imagined geographies'. They represent individual and collective imaginations of the world and refer to the way in which we imagine the geography of Africa, Asia or Latin America and understand the 'development' of these regions. One of the major objectives of this book is to encourage readers to think critically about how these maps are drawn and how they affect perceptions of political, socio-economic or cultural differences between world regions, peoples and places. What kinds of interaction do we have as consumers, for example, with these other worlds, and how is this interaction mediated at different spatial scales? Geographers are increasingly raising important questions about the ways in which spaces, places and scales are being reconfigured to the exclusion of certain groups (such as women, 'the poor' or people with disabilities) and in the reconstruction of new kinds of communities, which transcend the boundaries of particular places and localities (see Massey, 1992).

Imagined geographies conjure up for us in our own mind's eye a view of the world and may be drawn from a variety of sources, meaning that they are not necessarily unique to us as individuals. Many people will have the same or remarkably similar mental maps and imagined geographies of the world since in many cases they will have been

Figure 1.5 Sweatshops and the production of consumer goods have attracted more and more attention in recent years
Source: Image courtesy of www.adbusters.org (2001)

formed using the same sources. The views and projections of the globe that emerge outside Europe and North America are important here (Chaliand and Rageau, 1985) as we need also to think about the way in which the world may be viewed differently by, for example, Chinese, Russian, African and Muslim peoples. Eurocentrism thus has its own imagined geographies of the world (Blaut, 1993), an ordering of the global map, which begins with Europeans and has European spaces at its centre (Hall, 1992). In Eurocentric or Americocentric representations of the world and 'modern' development, Europe and North America are often depicted as the highest stages of civilisation and global progress. Despite the continuance of 'persistent poverty' in the USA (Glasmeier, 2002) throughout the twentieth century and up to the present, America is still seen as the richest and most 'developed' economy on the planet. Many North American geographies of development remain centred on Canada or the United States of America, with these states seen as pivotal axes in the global development industry.

A central theme explored in this book concerns the power and capacity of these pivotal axes and of global development organisations such as the UN agencies and the World Bank to put forward dominant spatial imaginations of other peoples and places and to provide 'truths' about successful growth, miracle economies and supposedly 'miraculous' economic recovery. In so doing, we are not trying to create a simple 'pro' or 'anti' debate about such countries and organisations or the more general themes of 'development' here but rather to understand the power of these states, ideas and institutions in particular places and at certain times in world history. Development in some conceptions may be understood as a dogmatic belief, almost a 'religion' (Rist, 1997) involving a series of practices that, taken together, are riddled with contradictions, manifest in a variety of place-specific ways. From their pulpits, global development agencies preach a number of creeds and doctrines of good change while forcing countries to seek their blessing and baptism on the global alter of development. Throughout this book we will be examining the relevance and value of these creeds and doctrines, exploring the possibility that under the guise of 'attacking poverty' some of these global institutions are actually attacking the poor (Cammack, 2002). Development arguably operates in a way as a kind of global 'industry', one with various technical services and its own 'pharmacy' of prescriptions for poor and needy countries or indebted nations 'thirsty' for credit. According to World Bank President James Wolfensohn, fighting poverty is at the very centre of the work done by these global development agencies, since the '4.8 billion people who are our ultimate clients deserve nothing less' (quoted in Cammack, 2002: 125). Why does this organisation choose to locate its headquarters in Washington, DC if its 'ultimate clients' are concentrated in other localities?

Development is partly also quite a religious enterprise, where objectives are pursued with missionary zeal and a pious devotion to particular theologies and beliefs. Various images and stories about distant places and peoples emerge from these institutions, and their constructions and imaginations of other worlds. In this way it is important to remember that development is simultaneously both an intellectual project and a material process (Apter, 1987) and that 'development' represents a broad 'fund' of knowledge about the modern world and progress within it which different individuals and institutions draw upon at different historical conjunctures to articulate their plans for global, local or national societies. In noting that development is profoundly material (and about more than telling stories) we are recognising that the term is more than just a 'plastic' word and that in its name:

> schools and clinics are built, exports encouraged, wells dug, roads laid, children vaccinated, funds collected, plans established, national budgets revised, reports drafted, experts hired, strategies concocted, the international community mobilised, dams constructed, forests exploited, high-yield plants invented, trade liberalised, technology imported, factories opened, wage-jobs multiplied, spy satellites launched.
>
> (Rist, 1997: 11)

To this we might add that the 'fund' of knowledge upon which global development draws is also very spatial in character, often centred upon the experiences of certain countries in particular political and economic spaces. Thus the prescribed remedies for development 'problems' and the problems of 'developing countries' usually issue from certain types of places and localities and must be seen as shaped by certain spaces of the world economy and certain recollections of world history.

As mentioned above, one key measure that has been used to depict patterns of global unevenness is Gross National Product (GNP) per capita. A growing GNP became a kind of myth in development circles, or 'a dogma, a shibboleth in economic reasoning and a golden calf and centre of economic worship' in the 1950s and 1960s (Weisskopf, 1964: 14). UN agencies are fond of monetary criteria like these which categorise the countries of the world (according to groups of income, literacy, calorie supply, life-expectancy or volume of export-trade) but these indicators often mask more than they reveal about impoverishment and inequality (Power, 2000; Matsu, 2002). Average and aggregated measures such as these are often meaningless in terms of 'representing the real situation' on the ground or in particular places (Peet with Hartwick, 1999: 8). To focus on a growing GNP at the expense of other ways of thinking about development (and how to

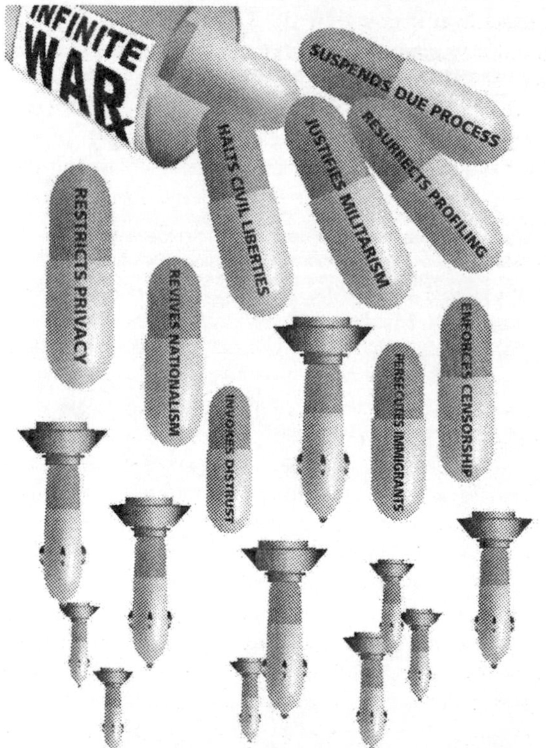

Figure 1.6 The idea that there are miracle cures to the problems of the South has been an enduring one
Source: © Protest Graphics

measure it) has thus arguably weakened and undermined the search for new and alternative ways of thinking about poverty and inequality. As we shall see in Chapters 7, 8 and 9, the contemporary global development industry is currently dominated by an ideology known as 'neoliberalism', which has become the predominant approach to thinking about and practising development today. Manufactured in Chicago in the 1970s by philosopher-economists such as Milton Friedman, neoliberalism has since become almost a religion in itself, which preaches restraint and the prominence of the market, becoming a kind of ideological 'thought virus' (Beck, 2000: 122) made in Europe and North America. Neoliberalism is problematic in that it promotes and normalises an economic growth-first strategy where social and welfare concerns come later (Peck and Tickell, 2002). It also naturalises an image of international markets as fair and efficient (when the reality is altogether different) and privileges 'lean government', economic deregulation and the removal of state subsidies to marginal and disadvantaged communities. More importantly, in many ways neoliberalism forecloses alternative paths of development, narrowing the ideological space in which it is possible to think outside the 'development box'. Of particular interest here is that behind neoliberalism there is the core assumption that the economy should dictate its rules to society rather than the other way around.

The global capitalist economy, it could be argued, actively *produces* inequality and uneven development in a kind of 'zero-sum' game where for every winner there is a loser, for every place that comes to share in global wealth another is pushed out of the equation:

> A big part of the 'economic development', i.e. the wealth, of the rich countries *is* wealth imported from the poor countries. The world economic system *generates* inequality and it runs on inequality. . . . It is a fraud to hold up the image of the world's rich as a condition available to all.
>
> (Lummis, 1992: 46–47, emphasis in original)

If the global economy is a pyramid and everyone is encouraged to stand on top, how is this to be arranged? If China, India and Brazil were able to truly 'catch up', is it possible that other 'developed' countries would have to decline as a locus of development, consumption or accumulation? Development equality and the idea of catching up with the wealthy through economic activity has dominated development thinking for many years but seems increasingly less likely at the beginning of the twenty-first century in the context of growing evidence of widening and deepening patterns of global inequality. If we look at the representation of the 'champagne glass' of global social inequality produced by the United Nations Development

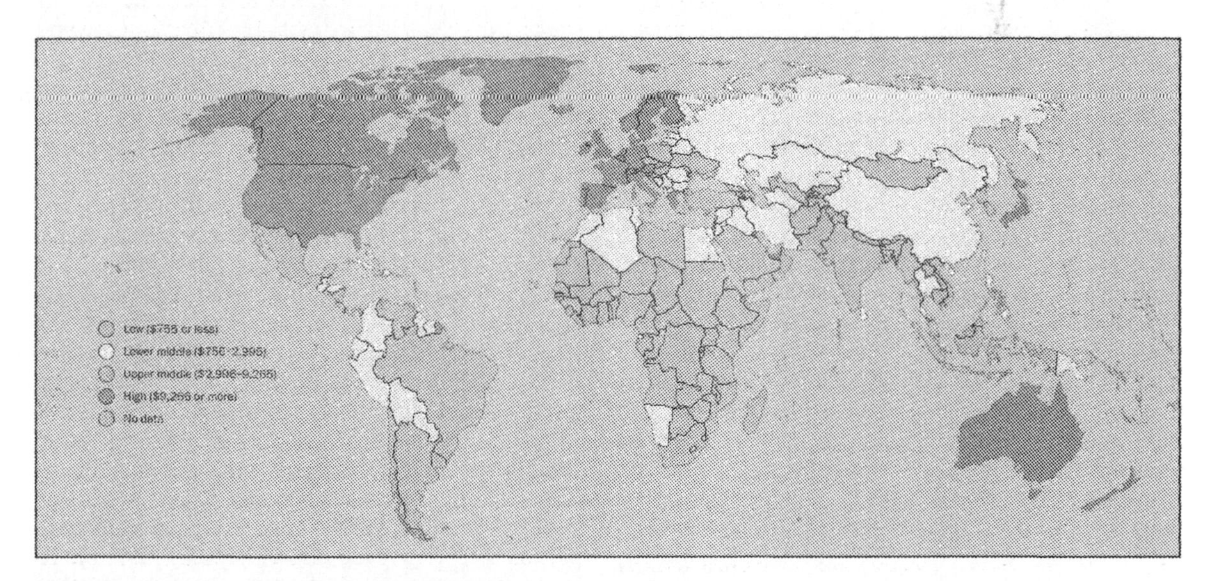

Figure 1.7 A mapping of the world according to World Bank income data
Source: World Bank

Programme (UNDP) in 1992 we can see how entrenched these sorts of inequality actually are (see Figure 1.8). Central to the notion of (neoliberal) development is the idea that 'latecomers' must catch up with already existing and 'advanced' systems and practices, but with such marked and entrenched differences and inequalities, how is this possible? For much of the postwar era of development, implicit in the representation of the developed world is the idea that blueprints can be created, that progress can be copied, mimicked, transferred or trickled down to 'LDCs' (to use another annoying locution that arrogantly distances non-western societies as 'less developed'). 'Development' has therefore often served as a 'lighthouse' (Sachs, 1992) or as a 'lodestar' and an 'illusion' (Wallerstein, 1994) into which several different movements, governments and institutions have invested faith and meaning. Like the term 'Third World', development partly represents a geographical imagining and the representation of a better world. It also incorporates a belief in the idea of correctable inequalities and injustices between nations, states and regions and (usually) within the existing global framework of economic structures. The idea of the Third World also gives enormous power to western development institutions to shape popular perceptions of Africa, Asia, Latin America and the Caribbean (Escobar, 1995). In this sense, the idea of three worlds has structured many political scriptings of old and new 'world orders'. It has also shaped many of the ways in which people have imagined complex geographies of global change.

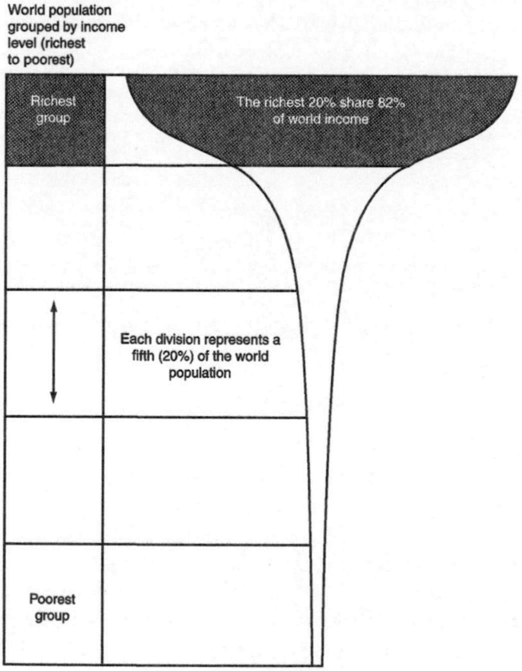

World population grouped by income level (richest to poorest)

Richest group

The richest 20% share 82% of world income

Each division represents a fifth (20%) of the world population

Poorest group

The poorest 20% share 1.4% of world income

Figure 1.8 The UNDP's favoured 'champagne glass' of world income distribution
Source: Houtart (2001)

DEVELOPMENT AND THE GEOGRAPHY OF POWER

For Rahnema (1992: 158) the term 'global poverty' is in some ways 'an entirely new and modern construct', while attempts to comparatively measure inequalities between nations and to eradicate them can be considered to be partly a product and invention of the postwar era (after the Second World War). From this modern construct comes a complex and widely significant politics of naming and labelling whole areas of the globe as 'poor' and the invention of global solutions to problems of impoverishment in the South. This is partly a reference to the end of colonial rule in some areas of the non-western world in the second half of the twentieth century and the way in which 'development' became a means of enframing or constructing the new 'challenges' and 'opportunities' of self-government that lay ahead for the newly independent countries. One example of this idea comes from the struggle for independence in Ghana in the late 1940s and early

Figure 1.9 Cartoon illustrating the dominance of northern streams of environmentalism
Source: *Environment & Urbanization* (2000)

1950s, when Kwame Nkrumah (who later went on to become the country's first leader after independence from Britain in 1957) regularly called upon Ghanaians to organise themselves so as to challenge their colonial masters. On a nationwide tour in 1950, Nkrumah pointed out: 'We prefer self-government in danger, to servitude in tranquillity. Forward ever, backward never!' (Nkrumah, 1961: 56).

Many nation-states in the South sought to deepen their victory over colonial rule by embracing 'development' as a national framework for building upon independence. As we shall see, both as an idea and as an arena of practice, development was borne partly out of the ashes of colonialism and the growing concern with 'decolonisation' in the non-western world. Former US President Harry Truman, for example, in a famous speech of 1949, spoke of this emerging 'underdeveloped' world, representing it as a 'handicap and threat both to them [the underdeveloped and poor] and the more prosperous areas' (Truman, 1949). As we shall see, the issue of how to conceptualise 'poverty' is crucial in any geographical interpretation of development and so it is important to examine the idea of the poor as a threat to established social or political orders and additionally to look at the way in which 'poor people' have been historically (as well as geographically) constructed as an 'object of humanitarian concern' (Allen and Thomas, 2001: 16). The idea of development as 'forwardness' (as opposed to 'backwardness') quickly took centre-stage in the decades that followed as part of the war on communism and in some of the global political dramas of the Cold War. There is no space to explore the nature of the Cold War here but this is none the less a fundamental cornerstone of the discussion that follows in this book (particularly in the first four chapters). As a crucially important global social and political conflict between capitalism and communism, the Cold War gave rise to something of a new world order in ways which continue to have a bearing on the contemporary geography of power in the world today. President Truman went on to explain the need for 'modern, scientific and technical knowledge' and announced the beginning of a 'bold new program' to resolve 'underdevelopment' and poverty in 'backward' areas (Truman, 1949).

One of the most interesting aspects of this speech was its *normative* dimensions, or the fact that it declared what *ought* to be done, what the 'underdeveloped world' ought to do to change its political and economic status before the 'West'. Thus in some ways development was emerging as a key principle upon which the United States would seek to build its own global empire after 1945. Since then 'development' has become:

> an amoeba-like concept, shapeless but ineradicable [which] spreads everywhere because it connotes the best of intentions [creating] a common ground in which right and left, elites and grassroots, fight their battles.
>
> (Sachs, 1992: 4)

Ideological interpretations of development have varied but they have always claimed 'the best of intentions'. The idea of development is often discussed in relation to 'developing countries' (also a value-laden term) but is still very relevant in societies that proclaim themselves as 'developed'. Can any society claim to be fully 'developed' when it forms part of a world that allows around 800 million people to suffer from hunger and malnutrition every year? According to Jean Ziegler, a UN rights expert, global hunger today remains a 'silent genocide' (Zeigler, quoted in Chomsky, 2001). Furthermore, in 1998 a high-ranking US official reportedly claimed that a child born in New York stood a smaller chance of living to age 5 or learning to read than a child born in Shanghai (quoted in Sogge, 2002: 27). Thus North and South, First World and Third World are not as separate as they might appear, since, as some observers have pointed out, there are simultaneous and intertwined processes of 'Thirdworldisation' in the 'First World' and processes of 'Firstworldisation' in the 'Third World'. None of this is particularly new when it is remembered that terms such as 'First World'/'Third World' and 'North'/'South' were themselves constituted by colonial encounters of various kinds, but it is important to remember the complex geographies of migration in the contemporary world which mean that living standards in the 'North' are not universally high and neither are they universally low in the 'South'. In Chapters 5 and 6, for example, we will be examining the history of the notion of 'three worlds' and of 'East'/'West', exploring the possibility of alternative representations.

This book is therefore very much about questions of 'race' and the possibility of anti-racist representations of other cultures and economies, highlighting the reality that, as Gilroy (2000) has argued, racial hierarchies and categories continue to

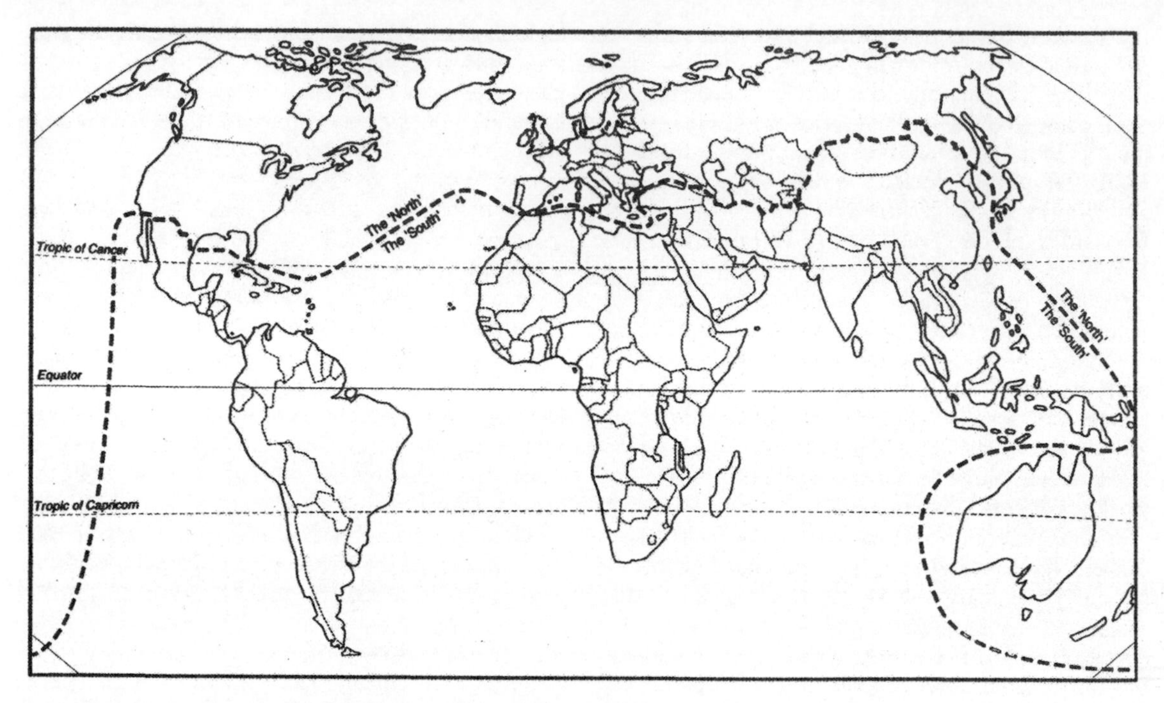

Figure 1.10 A map of the world indicating important North/South distinctions
Source: From Dickenson *et al.* (Routledge 1996)

structure the ways in which differences between peoples and places are approached. In this sense, the world of global development still remains entrenched in camps whether they be national, geopolitical, racial or cultural. In choosing a title like *Rethinking Development Geographies* this is precisely what I want to signal: that it is important and necessary to challenge the enduring influence of racial thinking and racialised categories bequeathed partly as a legacy of the days of Empire. In this book we will be seeking to complicate such categories and labels and to flag up the continued importance of whiteness, racism and racial formations in thinking about, mapping and practising development today.

This book argues that development has to be considered therefore in large part as related to a 'geopolitics' of race, where 'developed' countries take an interest in and consider the needs of poor countries in ways that often issue directly from their own preoccupations and strategic political, cultural and economic objectives and perspectives. One useful example of this is a report by the US Central Intelligence Agency (CIA) written in 2000 and entitled 'Global Trends 2015', where it is argued that the changing global economy will create many economic 'winners', warning that it 'will not lift all boats' and will therefore lead to further political conflicts and instabilities:

> Regions, countries, and groups feeling left behind will face deepening economic stagnation, political instability, and cultural alienation. They will foster political, ethnic, ideological and religious extremism, along with the violence that accompanies it.
>
> (CIA, 2000: 7)

The involvement of the US Central Intelligence Agency (CIA) in this concern for the 'global trends' of development is interesting here. The notion of developing countries or 'emerging markets' lagging or falling behind in some way runs throughout the report as does the image of these countries as relatively unchanged, with 'brittle' or 'frail' economies and societies. Although written before the terrorist attacks of 11 September 2001, the report identifies key 'drivers' and trends that will shape the world between 2000 and 2015. These include future conflict, natural resources and the environment, national and international governance, the global economy and globalisation, and finally the preponderant role of the United States itself. This

BOX 1.1

Consumers, corporations and branding

Brands are in the dock, accused of all sorts of mischief, from threatening our health and destroying our environment to corrupting our children.

(*The Economist*, 8 September 2001: 10).

A number of recently published books such as *No Logo: Taking Aim at the Brand Bullies* (Klein, 2000) and *The World is Not For Sale* (Duffuor and Bové, 2001) have been influential in highlighting important issues of consumption. In the new global economy, brands generate a huge portion of a company's value and can be a major source of profits – companies are thus switching to branding lifestyles, images and aspirations rather than actual products. Production is sometimes 'shipped out to the Third World' and as a result 'people, countries and companies are all racing to turn themselves into brands' (*The Economist*, 8 September 2001). Brands are a conduit through which influence flows between companies and consumers (these are not unidirectional flows however). Corporations also try to appeal to emotions, to construct a branded personality for products to make bonds with consumers and will construct themselves as having a 'social responsibility' if they think it will help market their goods. Global development agencies are big fans of these kinds of brand corporations (e.g. Coca-Cola), arguing that they pay the best wages, have the best working conditions and more generally represent positive examples of corporations operating in economies 'open' to trade and investment as in Asia (and unlike the supposedly 'enclosed' spaces of Africa). It is important therefore to examine consumption patterns nationally and internationally and to link these to the corresponding prevalence of poverty in other parts of the world, to see 'development' in one region as connected to the simultaneous 'underdevelopment' of another.

Running almost parallel with debates in the 'global North' in the 1970s and 1980s about consumption has been the emergence of environmental consciousness in the international arena of development. Key stages in this process were the publication of the Brundtland World Commission on Environment and Development (WCED, 1987) and the Rio Earth Summit of 1992 that considered the possibility of direct linkages between poverty and environmental degradation. The notion of 'sustainable development' is also relatively recent, although the term is exceptionally vague and appeals to mainstream development institutions that believe in evolutionary rather than revolutionary change. The concept of sustainability has its roots in northern streams of environmentalism (Redclift, 1987; Adams, 1990; Elliott, 1994) and does not always reflect perceptions of environmental crises held by people of the 'global South'. Some commentators have even suggested that 'sustainable development' is incompatible with neoliberal economics that favour the principles of the market since there is much evidence to suggest that such perspectives have aggravated conditions of poverty and other causes of environmental degradation. According to Chossudovsky (1997) the World Bank has claimed to have shifted policy focus towards 'sustainable development' and 'poverty alleviation', but neoliberal policy prescriptions go on relatively untransformed. In 1992 the environment re-emerged on the international stage with the first Earth Summit in Rio de Janeiro. Like many other UN gatherings, Rio was big on rhetoric but weak on actual commitment to change, and consensus was difficult to achieve between world regions with very different priorities. None the less, with a staff of over 500,000 and a total budget of US$1.35 billion, the UN does have enormous power to shape international conceptions of development for the better. The 2002 Earth Summit (Rio+10) held in Johannesburg in August and September 2002 also did not come cheap and may well be no more effective at pioneering new and alternative long-term perspectives on how to link debates about consumption, the environment and development than its predecessors (except through some of the social movements that have mobilised around it).

book seeks to explore many of these themes, especially the particular (geo)political imaginations of development that agencies such as the CIA produce. What we need to ask here is why is it so common to see 'poor countries' as tradition-bound, unchanging, brittle, weak and frail and in what sense are such representations a legacy of the past? In this sense, the CIA is engaging in a kind of writing of the global spaces of development, constructing ways of knowing the 'developing world' as well as inventing and legitimating forms of intervention in the political and economic affairs of other countries. The actual contribution of the CIA to this process is not especially important here; rather our concern is with the broader question of writing about global spaces of development (and interventions therein), and with an examination of the links between development theories and practices and their relationship to the changing nature of international political relations over time.

It is thus important to remember that the industry, business and 'machinery' of global development (and the flow of representations that this creates) is dynamic since its languages, strategies and practices have been shaped over a long period. Development is therefore not just a European creation (since as we shall see 'Europe' itself is a product of the 'Third World') but must also be considered as a 'reflection of the responses, reactions and resistance of the people who are its object' (Crush, 1995: 8). The key point here is not that development was invented by European or even North American countries for and on behalf of non-western countries, but rather that development is also partly the product of people's resistance, particularly those groups constructed as the 'objects' of intervention and policy-making. None the less, there has been a persistent tendency for development professionals and institutions to devise mechanisms and procedures that make societies fit pre-existing models which embrace the functions of 'modernity' and the 'modern' world, rather than each society being seen as *able to make its own history and geography*, to tell its own stories of progress (in conditions of its own choosing) and to represent its own particular political and cultural traditions (Escobar, 1995).

Throughout this book reference will be made to the 'discourses of development', and what we are talking about here is a kind of flow of ideas that are connected to one another, a flow of representations and conversations about global development and society over a period of time, even a century or more. The terms 'discourse' and 'discursive practices' usually refer to verbal and written communications but have come to include the 'ensemble of social practices through which the world is made meaningful and intelligible to oneself and others' (Johnson *et al.*, 1994: 136). Thus particular kinds of knowledges of world development are (re)produced through the representations and practices of discourses. Geographers have increasingly emphasised the variation of discourses over space and time and have shown how spaces themselves (such as that of

Figure 1.11 The UN has always been concerned with the possibility of international peace and stability
Source: UN Department of Public Information

Figure 1.12 A call to attack the World Economic Forum meeting in January 2002
Source: Globalise Resistance

BOX 1.2

A satirical take on the ability of world leaders to understand 'poverty'

At a UN meeting in Monterrey (Mexico) in March 2002 world leaders assembled to form what has since become known as the 'Monterrey consensus', envisaging more foreign aid and increased foreign investment as the way forward for poor countries. Some US$12 billion was pledged at the conference towards this end. Few new commitments were actually made however, and instead a range of stipulations and conditions were established that recipient countries would be forced to meet. In a satirical article about how rich western leaders understand the lives of poor people, an organisation known as SatireWire focused on the Monterrey conference in a quite interesting way. At the UN conference to address world poverty, the article reports that dozens of leaders from the planet's wealthiest nations concluded they were 'totally in awe' of the 1.2 billion people who reportedly live on less than US$1 a day. The following are some satirical quotations from comments that world leaders might have made in conversation with each other about the nature of poverty:

> 'Lord knows I couldn't do it,' US President George W. Bush told a roundtable of colleagues on Friday. 'The mortgage on my ranch alone is $5500 a month.'
>
> 'I think the hardest part would be keeping it up,' added South African Finance Minister Trevor Manuel. 'Every now and then I'd have to sneak out for a nice meal with my wife, or maybe a movie.'
>
> 'How often do you see a really poor person splurge for a steak dinner, or an upgrade to first class on an airliner?' said French President Jacques Chirac.
>
> 'I tell you, these people are incredible.' [President] Bush said, after a momentary pause: 'I'd like to propose that the world doesn't just have a billion poor people, it's got a billion heroes'.

However, some attending the International Conference on Financing for Development respectfully suggested that the leaders, while apparently no longer indifferent, were missing the point.

> 'The poor don't have mortgages. They don't have houses,' said UN Secretary General Kofi Annan. He also noted that 'Most days, they don't even have food.'
>
> 'Oh heavens no, I'd have to eat,' replied World Trade Organization chief Michael Moore. 'I could maybe skimp on dessert, but that's it.'

As the meeting adjourned a spokesman for the World Development Movement announced that almost half the Earth's population – 2.8 billion people – live on less than US$2 a day. The attendees stood in unison to cheer until it was explained to them that these people wanted money.

The full article from which these quotations are drawn may be found at http://www.SatireWire.com/

the imaginary 'Third World') are discursively produced (McDowell and Sharp, 1999). This is very much a postmodern approach to the study of development and, as we shall see, is not without its own problems and limitations.

The book draws in particular on the notion of discourse articulated in the work of Michel Foucault (1926–1984) who worked as Professor of the History of Thought at the College de France. Foucault attempted to question some of the basic ideas which people normally take to be permanent truths about human nature and societal change, and his writings continue to loom large over any geographical discussion of power (Philo, 1992) but also over contemporary debates about the power of development. Although Foucault's work is focused on a particular series of historical moments and the experiences of European countries and social relations, these ideas arguably have a much wider relevance in any examination of power relations in the Third World (Escobar, 1985).

In his work, Foucault (1979, 1980) was not writing explicitly about the idea of development, but these important writings did offer new ways of thinking that challenged people's assumptions about prisons, the police, insurance, gay rights, welfare and the care of the mentally ill. This work also had an important focus on the process of transition to a 'modern' western society and explored how this was based around new forms of control dispersed through complex networks of power and knowledge and a number of 'technologies of domination'. Perhaps most usefully, these interventions focused on people as subjects and the question of how individuals were encouraged by certain kinds of institution to behave in particular ways, to reach certain social 'standards' and to observe certain rules or to adopt particular ideas and ideologies of modern social order, progress and development. This approach is quite complex and challenging but it does capture something of the diffuse and subtle nature of power relations between peoples and places and of the diverse nature of important resistances to those relations of power and domination.

In a related sense, as Hart (2001) suggests, a distinction can also usefully be made between 'big D' Development and 'little d' development. The former refers to a post-Second World War project of intervention in the Third World that emerged during the time of decolonisation and the Cold War. This would suggest that it is necessary to explore the 'disciplinary power' of development, or the extent to which, as Foucault's work suggests, people become good, docile, governable citizens in the course of their society's modernisation. The phrase 'little d' development, however, points to the development of capitalism as a 'geographically uneven, profoundly contradictory set of historical processes' (Hart, 2001: 650). D/development can thus be viewed simultaneously as both a project and a process, with overt and covert forms, shot through with contradiction and unevenness. Both senses are deployed throughout this book and this is also combined with a concern for the *deconstruction of development*, following the work of Jacques Derrida, born in El-biar, Algeria in 1930. Deconstruction refers to a specific kind of analysis which Derrida applied to literature, linguistics, philosophy, law and architecture, and suggests new ways of thinking about texts, languages and reading. Pointing to a number of problematic assumptions in the nature of texts and highlighting the multiple layers of meaning at work in language, Derrida's writings are useful in thinking about development in that they can help us to understand the unavoidable tensions between the ideals of the developers and the developed, and the absence of clarity and coherence in philosophies of progress. Again, this work emerges from a particular context but has major relevance as a possible framework and methodology for critiquing development and exploring alternative ways of thinking about relations between society and space. In this sense, the approach adopted here follows work by a number of researchers who have discussed the possibility of combining productively the interventions of Foucault and Derrida in the study of international relations and development (Doty, 1996; Power, 1998; Manzo, 1999; Yapa, 2002). This combined approach allows us to unpack the conflicting meanings and contradictions of development, and its narrations of progress and social change in other worlds. It also calls for an opening up of the process of reading which allows for multiple interpretations, not all of which are intentional, that may be found within the languages and margins of any textual writing.

RE/THINKING GEOGRAPHIES OF DEVELOPMENT: A GUIDE TO WHAT FOLLOWS

> The day will come when nations will be judged not by their military or economic strength, not by the splendour of the their capital cities and buildings but by the well-being of their peoples.
>
> (UNICEF, 2000: 1)

Development is often presented as a collective task that (despite the relative lack of success globally) appears to be justified beyond all dispute, as inherently 'good' for people as apple pie. Thus we should not take it for granted that development is always positive, necessary or associated with progressive values and changes. All these questions are up for debate in what follows in this book. We will be trying to peel back these outer layers of self-evidence and this image of inherently positive values, to try to understand how the concern for development has been constructed and embedded within a particular geography, history and culture. In addition, our perspective on the images and practices of development depends in large part on whether we adopt the viewpoint of the 'developer' or those of the 'developed' (Rist, 1997). In this way the book seeks to show that there are a whole variety of perspectives involved here and argues that as a result, it is clearly a good time to take a fresh look at the geographies of development, to go back to the drawing-board in order to interrogate further the geographical imagination of the local, national and global development mosaic. A rethinking of development geographies is particularly necessary because the lines that have so far divided North and South are now present within every nation-state and

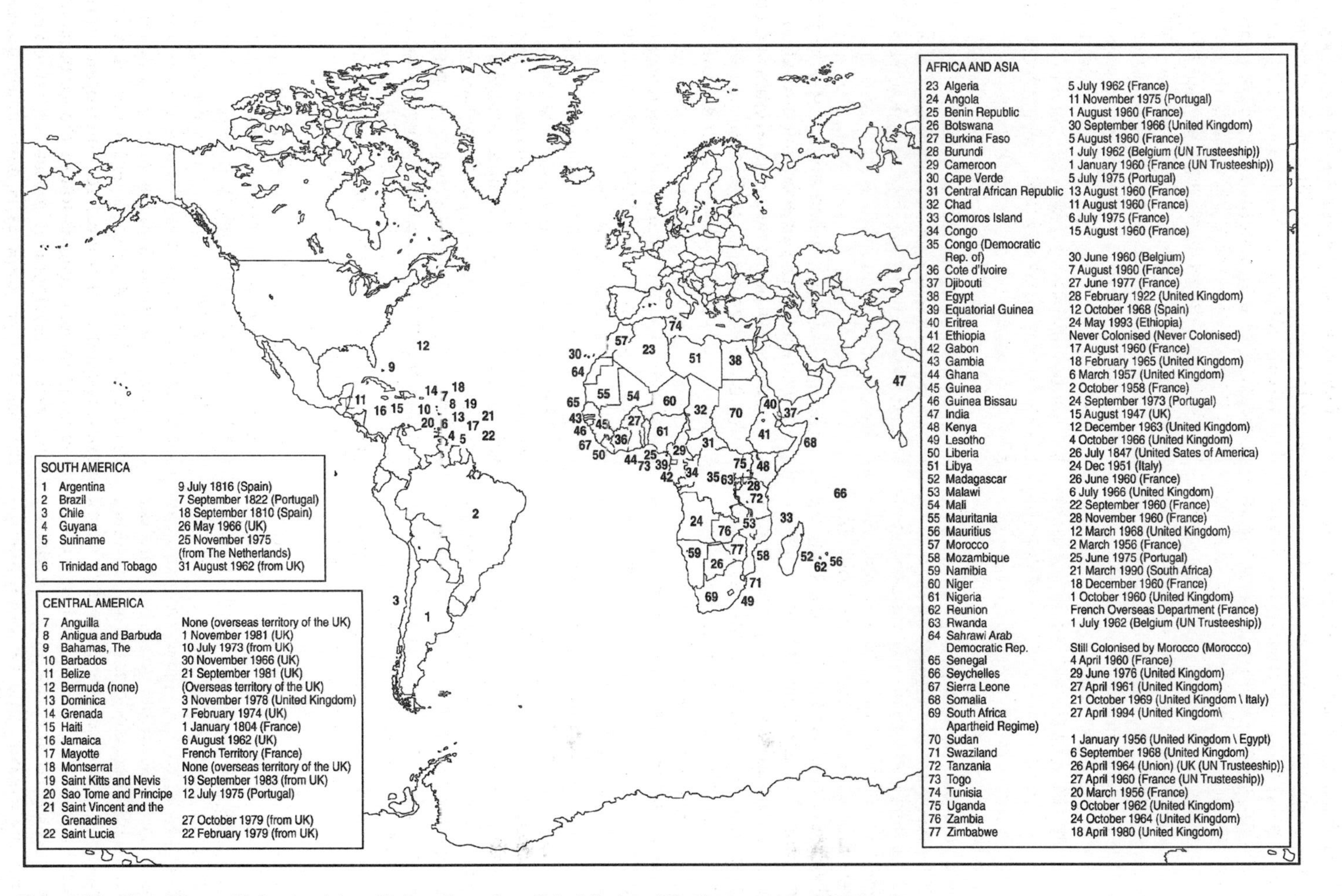

Figure 1.13 Map of the world showing dates of independence from Colonial rule and the European colonising country

are making ever less appropriate the conventional language used to interpret the geography of development in the world economy (Rich/Poor, North/South, First World/Third World, Developed/Developing).

For some commentators, what used to be called the 'Second World' of socialist revolutionary regimes has supposedly 'disappeared' from history (Dirlik, 2002), although Cuba, China, North Korea and Vietnam still retain a concern with 'socialist development' of various kinds. The more important point perhaps is that there is a sense in which the other two worlds are to an extent left seemingly face to face with one another, trying to renegotiate cultural relations in a new situation of 'globalisation'. Questions of geography are central to such renegotiations. One of the primary aims of this book is to provide readers with a sense of the changing ways in which geographers view development, particularly in the first three decades after the Second World War. It also aims to contribute to an understanding of the benefits of a geographical approach. In particular, questions are posed about whether since 1945 (or earlier) development has indeed failed in many respects, assigning 'developing countries' and the 'world's poor' to marginal positions within global economic and political structures while making 'prescriptions for advancement [which] were either half-hearted or completely misguided' (Kiely, 1999: 39). This book is thus divided into ten chapters, which, while each is designed to stand alone, can also be read in sequence as a number of different angles upon or approaches to the study of geography and development. This outline contains references to some of the more complex terms we will be using in the book and in this sense readers may find it useful to turn to the Glossary to grasp the meaning of such words from the very beginning. Here such terms are used more as summaries of quite complex ideas that this book will be exploring, but a key point is that it is *always* necessary to think about the context in which such terminologies are used. There are many overlaps as well as some discontinuities, 'gaps' or 'time-lags' between the chapters but the following is intended as a guide to what follows in the book, a map of the different directions taken:

- Chapter 2 begins by exploring the emergence of anti-capitalist protests in a number of places around the world and poses questions about the possible intellectual and political connections between 'postmodern' approaches to the study of development and the protests of the 'anti-globalisation movement'. The chapter also looks at the complex historical beginnings of development and how it is measured, focusing in particular on the case of foreign aid and its relations to the global war on terrorism.
- Chapter 3 moves a little further back in time, exploring the work of geographers in the 'Tropics', the 'Third World' or the 'South'. It raises questions about the ways in which geographers have written about and imagined 'modern' progress and development over time and in different spaces. The objective here is to examine the legacies in terms of definitions of knowledge (epistemologies) and ways of knowing geographical difference. Theoretically and methodologically many of these geographies were politically impoverished in their critiques and very empirical in their designs.
- Chapter 4 focuses on the question of theorisations of development. Geographers have been strongly influenced by development studies and the theoretical explanations it put forward – particularly theories of modernisation and dependency. In more recent times geographers have been at the forefront of some of the most critical work concerning the study of development. This chapter focuses on the legacies of development's invention in eighteenth- and nineteenth-century Europe and examines in further detail the strengths and weaknesses of a 'post-development' framework.
- Chapter 5 extends this concern with the geographical representation of difference by setting out to interrogate the mythology of the 'Third World' schema which rank-orders world regions according to specific sets of prescribed criterion. The relevance of 'geopolitics' (see Glossary for definition) to an understanding of development is explored here, as is the way in which traditions of 'Thirdworldism' produced certain kinds of political, cultural and economic struggles. A key question here concerns the extent to which the idea of three worlds represents diversity and difference in the South.
- Chapter 6 seeks to explore the interconnections between 'postcolonial' (see Glossary) theories, readings and interpretations and the study of development's complex geographies. It is possible that this framework can help to forge a more critical approach to development and its

subjects, spaces and places. The chapter focuses on the legacies of colonial representations and ways of thinking about distant others. This chapter also develops a focus on nationhood and debates about national belonging and citizenship after the formal end of colonialism, and explores new forms of cultural expression as ways of reading and interpreting development stories.

- Chapter 7 examines the important question of globalisation and development, focusing on the role of neoliberal frameworks in contemporary development thinking and their manifestation in particular localities and places. How is the 'borderless' world predicted by these debates likely to affect the South? A key question here revolves around the nature of 'free trade' and the impacts of globalisation on place-based identities and place-specific practices. Again the role of the state and national territory is central to this discussion.
- Chapter 8 looks at the postwar dissemination of development and explores the various ways in which language is used to communicate particular images and notions of progress. This chapter focuses in particular on conceptions of 'livelihoods' and 'partnerships' in development, again discussing the historical evolution of such ideas. A key objective is to interrogate some of the loaded and often quite superficial words of development and the ways in which they are popularised and disseminated.
- Chapter 9 examines resistances to neoliberal development and seeks to highlight how the practice and theory of development has been increasingly contested by social movements of various kinds and by growing coalitions of international solidarity movements which seek to connect struggles and particular places in opposition to certain kinds of narrations of 'development'. Examining a variety of case studies from India and South Africa to Brazil and Mexico, this chapter explores the possibility of transnational resistances and of people 'theorising back' from below in defence of their communities, places and ecologies.
- Chapter 10 concludes the book with a discussion of the millennium development goals of reducing world poverty by half by the year 2015, and offers a synthesis of the main themes explored, showing how they can be drawn together. This chapter also raises questions about working in development and researching development geographies, and looks at the shape of future agendas for development geographers.

2 Illuminating the Dark Side of Development

> The estimates of the number of poor people in the world also influence assessments of the seriousness of the problem of world poverty, the scale of resources that should be devoted to reducing it, and the regions to which these resources should be directed.
>
> (Reddy and Pogge, 2002: 1)

INTRODUCTION: AN INCOMPLETE AND CONTESTED PICTURE OF PROGRESS

In studying development one is often confronted with particular official definitions of 'sustainable development' or UN definitions of social and economic *progress*. Allen and Thomas (2001), for example, begin with the idea of a 'balance sheet' of development, where account is made of negative and positive change over time. They suggest that rather than viewing development as 'outright failure' we should construct a balance sheet of 'advances and setbacks' (Allen and Thomas, 2001: 6), but this would seem to imply a degree of confidence in how these advances are to be measured which is not shared by everyone. How do we know that the world is on the 'right track' in terms of poverty reduction strategy (Reddy and Pogge, 2002)? Assessments of the seriousness of the 'problem' of poverty reduction affect the scale of resources that are directed towards each region. The absence of clear, sound and consistent conceptions of poverty undermines the strength of many of these indicators, as does the reality that such estimates are dominated by certain institutions that see themselves as the sole producers of poverty numbers.

Despite the widely touted 'progress' and 'advances' that have been made, meeting basic needs such as water and sanitation provision remains far short of what we might have expected to have been achieved by the turn of the century. In this chapter we will examine how there are different senses of what constitutes an 'advance' or a 'setback' and why the statistical methods used to measure this can be called into question. A key pointer in this revolves around the relative importance accorded to statistics, numbers and figures. These types of information are often a very central component in constructing geographies of development, in determining which spaces of the world are 'rich' and 'poor'. In a massive variety of reports about development produced by institutions and organisations responsible for promoting images of progress (e.g. the UN, World Bank) this way of representing inequality is never far from view. According to Reddy and Pogge (2002: 1), the World Bank, for example, uses an ill-defined poverty line and misleading and inaccurate measures of purchasing power, giving the false impression of statistical precision. This in turn leads to 'an incorrect inference that it [global poverty] has declined' (Reddy and Pogge, 2002: 1). Arguably, because poverty is so widely underestimated and underemunerated, many of these indicators *undermine* the case for giving aid and assistance to poor countries rather than strengthening it. More and more people around the world are beginning to protest about this mistaken assumption that global poverty is (universally) declining.

In some ways the lack of an agreed set of indicators and measures or of common systems of data collection tells its own story of international development since 1945. According to the experts at the UN Statistics Division's thirtieth annual conference in March 1999: 'Attempts at defining what an indicator [of development] is have not as yet yielded a single definition text that has been widely applied' (UN Statistics Division, 1999: 4).

After more than half a century of international 'development' this must rank reasonably highly on the list of areas where the UN has failed to yet build international consensus. These sources of data can therefore be treated with the 'deepest suspicion' (Peet with Hartwick, 1999: 7). It is agreed, however, that statistics, 'facts', measurements and quantitative information yield evidence and perceptions of distant geographies, demonstrating 'progress', measuring social change and the dynamics of conditions and situations of development over time or across space. At the same time they are important because they convey messages about those geographies, about the level of modernity and its geographical diffusion, constructing and determining relationships between programmes and peoples. Through them, various *spatialisations of development* are produced which require further analysis. A lack of data, non-responses from countries or inadequate international systems for compiling these indicators mean that gaps and inaccuracies are plentiful, illustrating the partiality of the picture that is painted through them.

A look at the UNESCO 1998 Statistical Yearbook reveals that, out of 164 countries and territories covered, the only data available for some fifty-five 'developing countries' are either estimates or data referring to 1985 or in some cases even earlier. In some countries there are few if any established mechanisms for data collection required by the UN for the conference-linked indicators it seeks. If one looks at the comparable datasets available for countries such as Britain and the USA the statistical picture of these countries is also far from being complete and entirely accurate. In some ways aid and development communities are thus defined by a degree of 'statistical gimmickry' (Sogge, 2002: 29) where official data are inflated and misinterpreted in various ways. In addition to basic problems of accuracy there are also overlaps in the data collection procedures of international organisations, resulting in widespread duplication 'either in collection of data from countries or in the dissemination of the indicators' (UN Statistics Division, 1999: 7). Censuses, sample surveys and administrative records are all used to compile these figures, but after nearly sixty years of development there remains limited consensus on the key indicators and how data should be collected to measure them. The UN provides technical assistance to improve the capacity of poor countries to collect such data, but even the UN recognises that 'the sheer volume of development indicators and the lack of information on how similar indicators are related often make it difficult and confusing for analysts and decision-makers to use them' (UN Statistics Division, 1999: 11). These countries are burdened with multiple questionnaires and requests for data by the UN agencies. Some forty-two requests for data were sent to countries by the Food and Agriculture Organization of the UN in 1995 alone. Each of the major UN agencies has its own suite of indicators which are then mobilised to readily demonstrate the parts these agencies play in the story of global progress.

TABLE 2.1 INCONSISTENCIES AMONG INDICATORS IN UN PUBLICATIONS – ESTIMATED PERCENTAGES OF THE POPULATION WITH ACCESS TO SAFE WATER

Country	*Source A (1990–1996)*	*Source B (1995)*
Chile	95	Not available
China	67	90
Mali	66	36
Mozambique	63	32
Nepal	63	48
Senegal	63	50
Trinidad and Tobago	97	82
Uganda	56	34

***Source A*: United Nations Children's Fund (UNICEF), *State of the World's Children* (New York, Oxford University Press, 1998).**
***Source B*: World Health Organization (WHO), The World Health Report, 1996: *Fighting Disease Fostering Development* (Geneva, WHO, 1996); WHO expanded programme of immunization information system: WHO *et al.*, *Water Supply and Sanitation Sector Monitoring Report, 1996* (United Nations Statistics Division, Annex IX, 2000, 34).**

The major purpose of this chapter is to begin to raise questions about the predicted futures and expected outcomes of 'development' by exploring the past history of this incredibly contested idea. Just what is read into such statistics and how much do they actually have to do with the quality of life 'on the ground'? One of the reasons that statistics

BOX 2.1

Shaping the twenty-first century: global poverty eradication crusades and the international millennium development 'goals' (MDGs)

> The answer is needed urgently. While there has been more progress with poverty reduction in the past 50 years than in any comparable period in human history, poverty remains a *dire global problem.*
>
> (World Bank, 2000b: 2, emphasis added).

As we have seen, myths of global poverty present a technical problem which must be subject to rational decision and management, entrusted to the development professionals 'whose specialised knowledge allegedly qualifies them for the task' (Escobar, 1995: 52). The new agenda that is set forth in the millennium development goals (MDGs) adopted by donor communities recently 'in consultation' (OECD, 1996) could be seen in a similar way: as problems to be solved by donor communities for and on behalf of needy 'others'. The MDGs are estimated to cost US$50 billion, which would require a doubling of international aid just to meet the minimum estimates (Naschold, 2002). Costing and financing development goals such as this reveals only half of the story however; what about the wider structural commitments and engagements that are required to meet these objectives?

The MDGs are as follows:

- Reducing by half the proportion of people living in extreme poverty by 2015.
- Achieving universal primary education in all countries by 2015.
- Making progress towards equality of the sexes and the empowerment of women by eliminating disparities in primary and secondary education by 2005.
- Reducing the mortality rates for infants and children under the age of 5 by two-thirds and maternal mortality by three-quarters – both by 2015.
- Providing access through the primary healthcare system to reproductive health services for all women and girls of child-bearing age as soon as possible, and no later than 2015.
- Implementing national strategies for sustainable development in all countries by 2005 to ensure that losses of environmental resources are reversed both nationally and globally by 2015.

have assumed such significance is because they allow global institutions to illustrate that there has been forward movement over time and space. That there has been some sort of temporal and spatial improvement is central to the notions of progression and 'development'. Despite their many limitations, data about poverty do retain a degree of usefulness, however, in pointing to the 'gross inequalities emerging from processes of modernization' (Peet with Hartwick, 1999: 11). Thus where some global institutions have claimed that there has been a global progression of something or a reduction of poverty, data sources can also serve to highlight the failure of development strategies and ideologies over time and space.

In order to understand the importance of these ideas of progress and improvement it is also necessary here to explore their historical roots, to examine the extent to which they begin with the imperial encounter between West and non-West. This chapter looks at the specific question of historical geographies of foreign development aid and emergency relief, and their links to debates about international relations and foreign policy. In the first few years after the Second World War, many countries in Europe and North America began to turn their attention towards harnessing links with Third World countries by providing them with foreign aid and development assistance of various kinds. Here we can also begin to understand the political geography of postwar development, where development aid sometimes leads to a certain reworking of political and economic spaces in Third World countries. As we shall see, there is now a dominant 'borrow/invest/export/repay' model based increasingly on 'export successes' and becoming an 'emerging market' in the terms of international trade and investment corporations. This ideology is today also a 'pivotal issue in the debate over Third World debt' (Rowbotham, 2000: 70). Although advanced under the banner of financial 'aid' and 'assistance', market-based loans to the South produce a significant return for the World Bank, IMF and commercial banks involved. One of the main grounds for action in debt relief therefore has to be the recognition that loans were once advanced to Third World countries on the understanding that the money was repayable, which was in many ways a misrepresentation (Rowbotham, 2000). It is worth remembering that some of these institutions have lent money to all kinds of suspect and questionable political regimes in their first fifty years of existence, bedazzling recipients with logic:

> The World Bank in particular and the regional development banks have, since the 1960s, dazzled Third World leaders with irrefutably logical arguments whilst promising the development aid programmes. Many people were convinced by the ideology of credit-based development.
>
> (Gelinas, 1998)

This ideology of credit-based development needs to be radically overhauled. In order to do this we need to understand further its political, historical and geographical origins. Continuing relations of 'dependency' have been related directly to the 'debt crisis' which began in the 1970s and 1980s. As a consequence of heavy borrowing commitments many countries have become increasingly dependent on development assistance from international institutions. The severity of the debt crisis gave key financial institutions such as the IMF and the World Bank (see Box 2.2) renewed influence in the affairs of national governments (Potter et al, 1999). Between 1980 and 1991 southern countries suffered an estimated cumulative loss in total export earnings in real terms of some US$290 billion or an average annual loss of US$25 billion (Sogge, 2002: 36). A sharp deterioration in the terms of trade between North and South also made it very difficult for many countries to make debt repayments, further increasing dependence on the World Bank and IMF which then imposed Structural Adjustment Programmes (SAPs) on these countries. Debt is usually denominated in US dollars and thus 'the "demand for dollars" to repay debt actually helps maintain the value of the US dollar' (Rowbotham, 2000: 80). The 'neoliberal' perspective advanced by these institutions urges a withdrawal of state intervention in the national economy, a liberalisation of trade relations, the privatisation of government institutions and a devaluation of national currency.

SAPs have since been 'rebadged' as Poverty Reduction Strategy Papers (PRSPs), although the underlying principles remain essentially the same (see Chapter 8). They have often commanded a degree of submission to the ambitions of global capitalism and the expansion of transnational corporations, but it was through these programmes that many countries were encouraged to 'catch up' with the 'developed' and to deepen their victory over a colonial past by *modernising* and industrialising (Amin, 2001b). Everything that had been achieved over the centuries in popular struggles for freedom of various kinds (e.g. against colonialism) disappears from view in the face of the structural reform prescriptions advocated for the ailing economies of the Third World. In this sense it is necessary to understand that neoliberalism rose to prominence partly as a response to the wider crises of capitalist development in the 'postcolonial' world. Incredible as it may seem today, these institutions were once seen as progressive, deterring future conflict and building peace and stability for the majority of the world. Initially, however, these agencies had no control over individual governments' economic decisions; nor did their mandate include a licence to intervene in national policy decision-making around development issues. This chapter also explores the connection that can be made here between the recent protestations about the failures of global capitalist development and development institutions such as the World Bank (e.g. in Seattle) and the more polemical critiques of what have been called 'post-development' perspectives.

The World Bank and IMF agreed to reduce the debt burden of some of the poorest countries with the Heavily Indebted Poor Countries (HIPC) initiative which began in September 1996. This commenced when the IMF and World Bank finally decided to try to deal with the growing debt crisis, producing initially a list of forty-one countries that were both poor and paying more on servicing their debts than on the health and education of their citizens. Just as in colonial times, countries were required to demonstrate certain types of externally measured reforms, changes and levels of progress before a foreign audience that would assess their performance. After five years, only two countries (Bolivia and Uganda) saw their stock of total debt reduced, having reached what in World Bank jargon is arrogantly called 'completion point', or 'a sort of oasis for "thirsty" debtors' (Telatin, 2001: 1). Thus the Bank has decided where this 'oasis' should be located and what kinds of criteria should be used for deciding on a list of those who should be allowed to drink from it. Debt sustainability (a very

BOX 2.2

The Bretton Woods institutions – explaining the international architecture

The International Monetary Fund (IMF) is based in Washington under Managing Director Horst Kohler. The Fund was set up as a result of the Bretton Woods conference to maintain currency stability and develop world trade by establishing a multilateral system of payments between countries based on fixed exchange rates and complete convertibility from one currency to another. The IMF is not controlled by the people or the governments of low-income countries and is not even like the other UN agencies where every government has a vote (Wood and Welch, 1998). Instead the fund is run by a select band of high-income countries that fund its operations (the USA has an 18 per cent share of the votes and *de facto* control of the organisation). The World Bank (founded at the same conference) is now involved increasingly in macro-economic issues and has often been chaired by an American, while the President of the IMF is usually European. The latter is dominated by OECD countries and in particular by the USA, and is organised in a similar way to the Bank. The World Bank is also based in Washington under its Australian President James Wolfensohn. The International Bank for Reconstruction and Development (better known as the World Bank) was founded as a means of reviving war-damaged European economies, a mandate that was extended later to developing nations. The Bank is funded by dues from members and by money borrowed in international markets. It makes loans to member nations at rates below those of commercial banks to finance development 'infrastructure' projects (e.g. power plants, roads, hydro-dams) and to help countries 'adjust' their economies to globalisation. Like the IMF, the Bank was an early and enthusiastic supporter of the neoliberal agenda. The World Bank group also includes: the International Development Association (IDA) which makes 'soft' loans (no or very low interest) to the poorest nations; the International Finance Corporation (IFC) which tries to attract private-sector investment to Bank-approved projects; and the Multilateral Insurance Guarantee Agency (MIGA) which provides risk insurance to private investors in member countries. According to a Brazilian delegate at the Bretton Woods conference (where the IMF and World Bank were set up in 1944), these agencies were inspired by a desire to establish a central lending source to which governments would contribute such that 'happiness be distributed throughout the world' (IBT, 1998: 3).

In 1990 the Bank's then President, Barber Conable, remained confident of alleviating world poverty: 'we know a great deal about who the poor are, where they are and how they live. We understand what keeps them poor and what must be done to improve their lives.'

The 2000/2001 World Development Report (WDR) prepared by the Bank was entitled Attacking Poverty (World Bank, 2001b) and promised 'the most detailed-ever investigation of global poverty' (World Bank, 2001c). The result of more than two years of research, the WDR draws on a 'background study' which sought the 'personal accounts' of more than 60,000 men and women living in poverty in sixty countries around the world. This approach allows the Bank to claim that its vision of development is based upon 'the testimony of poor people themselves'. The 2002 WDR, *Building Institutions for Markets* (World Bank, 2002), focuses on weak domestic institutions 'which hurt poor people and hinder development.' Instead, the Bank argues, institutions suitable for (supporting) markets should be promoted so that all poor people can feel the benefits of the free market and free trade. This seems somewhat at odds with the Bank's historical insistence on deregulation of markets. The 2003 WDR, *Sustainable Development with a Dynamic Economy,* is currently in preparation. Out of an annual budget of US$15 billion the Bank reserves about US$12 million for project grants with adjustment lending coming to an estimated US$3 billion. In addition, the Bank annually concludes about 40,000 contracts in projects it finances in ninety-five countries worldwide (*Geobusiness*, 2001: 61). Projects must also rely on private investors and local governments for funds. Interestingly, 90 per cent of financing cases where the Bank intervenes are for an amount inferior to US$1 million, with contracts reviewed in Washington. Latin America receives US$4.1 billion. The debts of the world's poorest countries have been reduced (although there is still some considerable way to go). The Bank has contacts and business dealings with some 300 businesses across Europe, representing 47 per cent of the World Bank's contracts. Consultants in borrowing countries cream off another US$1.2 billion a year (*Geobusiness*, October 2001). While the agency listens to the 'voices of the poor', consultants are paid handsomely for their work and perspectives on development projects.

value-laden concept itself) was defined as the capacity of each country to repay debts using export resources received. The Bank's President even claimed in 1998 that the initiative was always 'good news for the world's poor' (IBT, 1998: 12). If debt is merely reduced and conditionality imposed, indebted countries will still be obliged to struggle to compete for export revenues and to extricate themselves from a complex web of debt. A similar criticism has often been directed at the provision of overseas development assistance or aid by the world's richest economies.

The first section of this chapter looks at the growing sense of dissatisfaction with development ideas and outcomes, discussing the recent wave of protests and demonstrations against capitalism and globalisation and exploring the potential overlap with 'post-development' critiques and theorisations. The following section examines the ways in which contemporary development needs to be 'decolonised' given the important legacies of imperial assumptions and mindsets in framing global development thinking today. The next section then moves on to explore further the political and psychological rationale for 'helping' the poor and donating aid and assistance. In many ways examining the case for aid can tell us a great deal about the history of development theory and practice, and provides a useful opening on to wider discussions of North–South relations. The final section then introduces the concept of 'geopolitics' in relation to these debates about foreign aid and colonial concepts of helping the poor in order to save the 'civilised world'.

THE GLOBAL 'PROTEST CARAVAN'

> It is a social decision to classify people [as poor] in such a way. Since it has little to do with who those people are, or even what they do, the expectation that 'the poor will always be with us' can hardly be faulted. . . . Such people can then be declared a 'problem' which the rest of society must and should cope with.
>
> (Bauman, 1999: 20)

When the World Trade Organisation (WTO) met in Seattle in November 1999 (for its Third Ministerial Conference) pictures of the violently subdued protest riots were shown across the world, with the story running in many national newspaper headlines for weeks (Rowbotham, 2000). There had been large militant protests before against the WTO and its sister organisations the World Bank and IMF, but these were different in that they attracted major amounts of media attention. The protests also showed that the US economy (often held up as something that poor countries should try to mimic and copy) was not without its own internal contradictions. Activists disrupted the WTO meeting in Seattle, which, coupled with rebellion from a variety of Southern countries, meant that the WTO failed to land a new round of trade liberalisation, celebrated as a historic defeat for corporate rule and the global trade regime. A huge number of international solidarity actions took place throughout the protests which were particularly significant in terms of the connections being made by activists from around the globe. A well-organised and seemingly transnational movement (including many NGOs) had thus started to try to place important questions about who controls global regulatory institutions and for whom firmly on the global agenda. As we shall later in the book, a rejection of 'corporate liberalism' linked these street-based demos to some of the radical protests of the late 1960s (Watts, 2001).

At precisely the same time that the Seattle demos unfolded, the millennium countdown was closing in on the Jubilee 2000 campaign which had sought, since 1994, to address escalating levels of global indebtedness. By using high-profile public figures such as U2 lead-singer Bono, the campaign managed to gather some 25 million signatures for their petition for debt cancellation (*The Economist*, 1 June 2002). While the problem of what has been called 'Third World debt' remains completely unresolved (and institutions such as the WTO still need urgent reform), these events at the close of the twentieth century, the 'American century', heightened consciousness about the 'future of development' and its (im)possibility in a highly uneven twenty-first-century world. Since Seattle, the 'global protest caravan' (BWP, 2001a), which includes a variety of movements under the banner of anticapitalism, has stopped off to protest and disrupt a number of high-profile IMF, G8, OECD and World Bank conferences and meetings. After Washington, New York, Montreal, Melbourne, Genoa, London, Prague and other clashes with protestors, these organisations now consider meeting on remote islands (e.g. in Japan) that cannot be reached easily. Many host nations have surrounded themselves with immense and excessive policing and protection so that whole spaces of these cities are not accessible to people seeking to voice their concerns, which are often portrayed by the media as those of random 'anarchists' looking for a fight. More than 200 people were injured in clashes between demonstrators who tried to enter the G8 summit HQ in 2001 and a further 280 were arrested (BBC, 2001). At the G8 2002 Summit in Western Canada some US$300 million was spent on securing the site of the meetings, which was more than the amount agreed for Africa at that particular gathering. Many media agencies have often talked of the 'senseless' protestors in their representations of these demonstrations:

> If your only source of information was the slanderous coverage of the 'free press', you'd think the protestors were nothing but a bunch of violence-prone, middle-class 'anarchist' Birkenstock-wearing, neo-hippies with nothing better to do than destroy the property of good, clean, hard-working small business owners.
>
> (Gonsalves, 2002: 1)

There are a variety of perspectives on the protests however, and it is important to look beyond the problematic coverage of some media agencies. The globalisation of the neoliberal project has therefore also been connected to the beginnings of a (partial) globalisation of resistance, and important networks are beginning to link places and localities as sites of protest around the world (see Chapter 9). Neoliberalism also takes many different forms however, but what we are seeing are new forms of solidarity and consciousness emerging where people are marginalised and excluded on a global basis (Hardt and Negri, 2000). Neoliberalism has proven to be so enduring and pervasive partly because as a system of diffused power it is quite literally everywhere and, as the protestors are beginning to learn, 'the underlying power sources of neoliberalism remain substantially intact' (Peck and Tickell, 2002: 400). The core institutions of neoliberalism (the WTO, World Bank and IMF) are complex and changing institutions however, and we need to move beyond the simplistic images of these organisations as simple essences, monolithic and all-powerful. The notion of 'entanglements of power' (Sharp et al, 2000) encourages us to think about the manner in which domination and resistance in development are interconnected, involving a complex interweaving of the 'power over' with the 'power to'. In focusing on neoliberalism, it is thus necessary to highlight the close connections between the Banks' 'power over' development thinking and the emergence of powerful resistances to this (see Chapter 9). Although many discussions of this power may appear sometimes reductionist and 'over the top' or too cut and dried, clearly the enthusiasm of the early 1960s for development 'has crumbled and faded away' (Rist, 1997) as this fragile bubble has quickly been burst (Escobar, 1995). Too many over-ambitious development projects have ended in failure, and in any case why do we need the concepts and strategies of yesteryear if our aim is to radically rethink and reconstruct development geographies?

One of the most important themes this book explores is the idea that development is in crisis, that there is an absence of alternatives to development and that existing theorisations have been found wanting in some way. Rowbotham (2000) articulates the growing view that at the end of the twentieth century international development is failing and in a kind of crisis. The views of people that are the individual human subjects of development are very important to any study of its meaning and interpretation, yet all too often people have been left out of consultations and meaningful participation in their own development. In this way, the post-development approach sets out to deconstruct the development discourses of the postwar period, to critique 'development' as a system of knowledge produced by the 'First World' about the underdeveloped 'Third World'. It begins by questioning terms such as 'progress', 'development' and 'improvement', and why and how progress has been seen as universally beneficial by asking who determines what 'beneficial' means (Peet with Hartwick, 1999). This involves a new kind of methodology in that everything that had previously been taken for granted or seen as untouchable is called into question as long-standing certainties are destabilised. Thus development is seen here not just as an instrument of economic control and management, but also as a knowledge 'discipline' which marginalises peoples and cultures and precludes other ways of seeing and doing development. For Hoogevelt (1997: 255), this is rather like 'stripping the walls before putting fresh paint on'. Rather than seeking development alternatives, this approach calls for a rejection of the entire paradigm, freeing up the imagination for other ways of thinking about development. What these approaches would seem to share is a common rejection of existing theories and practices and a common concern to develop new and more effective forms of resistance.

Attention is focused in particular in much of the literature on the tendency of post-development schools to 'romanticise' the lives of the poor and their preoccupation with language over the material 'realities' of global development today. Potter et al (1999) describe the anti-development approach as little more than a 'Utopia for New Age travellers', for example. For some commentators then the global protest caravan has been mistaken for something radical and useful when actually what started it was a bunch of hippy New Age travellers with utopian dreams. Other commentators see the 'alternatives' put forward by post-development thinkers as having a 'high New Age-like content clad in Third World clothes' (Schuurman, 2001: 6). Thus some critics are cautious of 'favouring' the views of oppressed people and the 'New Age romanticism' this might lead to (Peet with Hartwick, 1999: 198). Other critics of this approach, such as Corbridge (1999: 73), have argued that these writings downplay or ignore the 'very real successes' that can be associated with development since 1950, in terms of better health or education, for example. Similarly some authors point to the simplistic posture of the 'post-development' writers that see development as a 'hoax' that was:

> never designed to deal with humanitarian and environmental problems, but simply a way of allowing the industrialized North, particularly the USA, to continue its dominance of the rest of the world in order to maintain its own standards of living.
>
> (Allen and Thomas, 2001: 19)

Thus, in this context, what should we make of Jubilee South's claim that the HIPC and PRSP strategies of debt relief initiatives formulated by the International Financial Institutions (IFIs) are in fact nothing more than a 'cruel hoax' (quoted in Bond, 2002b: 3)? These institutions and their 'relief strategies' are examined throughout this book in an attempt to interrogate further the assumption that the world is on the 'right track' in terms of poverty reduction strategy. Kiely (1999: 36) argues that post-development critiques focus exclusively on the 'dark side of development' and are vague about alternatives while romanticising local cultures. There is an over-simplification here in that these writings seek to explore not just the 'dark side' of development and its failures but rather the 'other side of the story' about the development project or 'the view from below, the views of women, the view from the South' (Munck, 1999: 200). This involves more of a concern with 'history from below' (Dirlik, 2002) revealing stories and histories that had been suppressed in earlier accounts and challenging the claims of some development 'experts'.

None the less, the 'postmodern' turn in the social sciences during the 1990s was based on a recognition of the shortcomings of modernism, on the ways in which modernist theories were focused narrowly on the state or were themselves part 'of the technologies of colonial and imperialist governance' (Bhaba, 1994: 195). To suggest that development is an illusion and has been illusory is not to suggest that complicated and cunning plans were devised to mislead the world's poor so that the USA could continue to have its high living standards. This is a massive simplification of a complex and emerging set of positions around the idea of seeking alternatives to development. These are not just a series of points about the derivation of terms or language or simply arguments for different 'labels' but rather raise important and fundamental questions about the power and knowledge that make international development, about the assemblage of knowledge concerning 'poor' peoples and 'poverty' and about the construction of development projects that seek to intervene. They are also of interest because (among other approaches) they represent a challenge to hegemonic institutions and Eurocentric or even 'Americo-centric' approaches (Kiely, 1999) which go beyond the nation-state. Rahnema, writing in The *Post-Development Reader*, argues that the approach can give rise to new forms of international solidarity: 'the end of development should not be seen as an end to the search for new possibilities of change . . . for genuine processes of regeneration . . . to give birth to new forms of solidarity' (Rahnema, 1997b: 391).

It is important to remember then that it is possible that what is *intended* by development theories and strategies is often confused with the question of 'what is development' in terms of its outcomes (Cowen and Shenton, 1996: viii). The ambiguity of the term thus sometimes leads to a degree of confusion and a failure to find common ground for debate. What *is* becoming clear is that development theories remain intrinsically important (despite their multiple ambiguities and confusions), 'so embedded in our thinking that they have a life of their own, a life quite divorced . . . from development practice' (Apter, 1987: 9). In this sense it is necessary to distinguish here between the post-development concern with development as a discourse and the more political and polemical approaches that these writings seem to suggest and imply. As we shall see later in the book, resistances to 'global capitalism' and neoliberalism share a common rejection of existing theories and practices of development with the post/anti-development school. It may be argued that discrepancies between policy and outcome have often been so disastrously immense therefore that another *round of development thinking is called for*, one which transcends disciplines but recognises that it is not possible to opt suddenly for discarding the languages of development (without first understanding further how they work) since it is already such an important part of our thought about global social and geographical change. No one such single language (theoretically) should be able to monopolise so complex a subject.

As Cowen and Shenton (1996: 4) point out, development comes to be defined by a 'multiplicity of "developers" who are entrusted with the task of development'. Despite being fraught with a (Faustian) level of ambiguity, development is still seen as something which happens simultaneously to individuals, communities, nations and regions, as something which happens cumulatively and builds step by step in an organic and natural fashion. In many ways then, development is associated with unfurling and growth, which suggests that the term has directionality, continuity, cumulativeness and irreversibility (Rist, 1997). In this way the idea of development was caught up very closely with economic and political changes in Europe in the eighteenth and nineteenth centuries, amidst the 'rough and tumble' of European industrialism. *Developmentalism* is a term that refers to the view of Third World spaces and their inhabitants as essentialised, homogenised entities. This was also very closely associated with an unconditional belief in the concept of progress and the 'makeability' of society, which was in turn a product of the dominance of evolutionary thinking in the Enlightenment era of seventeenth- and eighteenth-century Europe (see Chapter 4). Thus the relationship between capitalism and development has always been a central one (Cowen and Shenton, 1996). In some ways the use of organic metaphors naturalises this history, obscuring much of the complexity involved in 'positive changes'. This book examines the geographical nature of these historical imbrications and interrelations of capitalism and the Third World, seeking to reject conventional and mainstream assumptions about capitalist development and to develop and explore the realm of alternative geographies.

As we have seen, there are different senses in which the term 'development' might be used here, to see both an intellectual project which is continuing and a material process which is ongoing (Apter, 1987; Thomas, 2000). In some ways 'Development' may be seen as a vision, a description or measure of the state of being of a desirable society. In other ways we are talking about a historical process of social change in which societies are seen to be transformed over long periods. What we need to shift our attention towards here is the idea of development as consisting of deliberate efforts aimed at improvement on the part of various agencies, including governments, all kinds of organisations and social movements. Again this involves a distinction between what Hart (2001: 650) calls 'big D' Development and 'little d' development, between the history of interventions in the Third World and the processes of capitalist development (and the development of capitalism) as contradictory and uneven. The post-development perspective illustrates how 'developing societies' are not organic wholes produced around structures and laws (as has often been assumed) but rather they may be seen as fluid entities, stretched in various ways and no longer bounded, discrete and localised. How have the wishes and worldviews of some groups and societies become universalised or 'imperialised' (Escobar, 2002: 194) and how might it be possible to think about alternative social orders? In this way, we need to take a step back in time to consider colonial visions of progress and the beginnings of an imperial development mindset.

DECOLONISING THE MINDSET OF DEVELOPMENT

We will be exploring the post-development approach in more detail in Chapter 4, but one common implication that does emerge from these critiques is the importance of *thinking historically* as well as geographically about development. One of the main themes of this book is the continuing importance of colonialism in development thinking today, in seeking to understand how the values of colonialism have spread widely and seeped into many cultures and identities in a variety of complex ways. As we shall see in Chapter 3, this has also had a very diverse range of impacts on geography as a discipline and its heritage as a western colonial science. Colonial domination varied enormously, as did the multiple geographies and histories it created. For some authors 'nothing has really changed' in that when colonialism 'left off', development 'took over' the management of non-western countries, but this is a somewhat simplistic image to convey. None the less, development ideas and practices remain in need in of *decolonisation*.

In many ways development is primarily 'forward-looking', imagining a better world, and does not always examine issues of historical and geographical context (Crush, 1995). The historical process by which a 'gap' emerged and has been sustained between 'North' and 'South' has been interpreted in a variety of ways but a key question has been to what extent did European expansion and colonialism 'underdevelop' (Frank, 1969) large areas of the world? Between 1800 and 1878, European rule, including former colonies in North and South America, increased from 35 per cent to 67 per cent of the Earth's land surface, adding another 18 per cent between 1875 and 1914, the period of 'formal colonialism' (Hoogevelt, 1997: 18). In the last three decades of the nineteenth century, European states thus added 10 million square miles of territory and 150 million people to their areas of control or 'one fifth of the earth's land surface and one tenth of its people' (Peet with Hartwick, 1999: 105). In this period the periphery was inserted and brought into an expanding network of economic exchanges with the core of the world system. Colonial rule also led to forms of developmentalism being put into practice, homogenising other spaces and their inhabitants as essentialised entities. A new sense of responsibility for distant human suffering first emerged during this time as the societies of Europe and North America became entwined within global networks of exchange and exploitation in the late eighteenth and early nineteenth centuries (Haskell, 1985a, 1985b). Thus the origins of a humanitarian concern to come to the aid of 'distant others' lay partly in response to the practices of slavery in the transatlantic world and to the expansion of colonial settlement in the 'age of empire':

Figure 2.1 A monument to the Portuguese imperial discoveries in Lisbon, Portugal
Source: The author

> Not only did colonisation carry a metropolitan sense of responsibility into new Asian, North American, African and Australasian terrains, it also prompted humanitarians to formulate new antidotes, new 'cures' for the ills of the world
> (Lester, 2002: 278)

These new antidotes, cures and remedies were to have an enduring significance for the shaping of twentieth-century global development theory and practice, which also often carried an implicit 'metropolitan sense of responsibility'. Colonial developmentalism was also associated with an unconditional belief in the concept of progress and the 'makability' of society, being heavily conditioned by the dominance of the evolutionary thinking that was popular in Europe at the time. Imperialism was viewed as a cultural and economic necessity where colonies were regarded as the national 'property' of the metropolitan countries and thus needed to be 'developed' using the latest methods and ideas. With this came a missionary zeal to 'civilise' and modernise the colonised and their ways of life. An important contention here then is that colonialism 'conditioned' the meanings and practices of development in a number of important ways. It is important, however, to avoid the argument that because colonisation was 'bad' decolonisation must automatically be 'good', just as we need to avoid seeing development as 'bad' and post-development as 'good' (Mercer *et al.*, 2003).

Under President Truman, 'underdevelopment' became the incomplete and 'embryonic' form of development and the gap was seen as bridgeable only through an acceleration of growth (Rist, 1997). What happened after 1945, however, was that development acquired a transitive meaning and was reconceived as an action performed by one agent upon another (Rist, 1997). As we shall see, there was an important sense in which, after 1945, 'development' was viewed as a process that was entrusted to particular kinds of agents who performed the acts and power of development upon distant others. Globally, development would have its 'trustees', guiding 'civilised' nations that had the 'capacity' and the knowledge or expertise to organise land, labour and capital in the South on behalf of others. This is a crucial idea then that quite a paternal and parental style of relationship was established through the imperial encounter between coloniser and colonised in ways which continue to have a bearing on the definition of North–South partnerships in the postcolonial world. In addition, what is also relevant here is that many postcolonial states continue to maintain important political, cultural and economic ties with their former colonial rulers.

Decolonisation is thus simultaneously an ideological, material and spatial process. Just as complicated as colonisation (Pieterse and Parekh, 1995), we will be examining this concept throughout the book, exploring the possibility that it can help us to reconstruct new forms of alternatives to development. Colonialism put in place important political and economic relations but the cultural legacies of colonialism bequeathed deep social and cultural divisions in many societies. In the process of decolonisation 'development' became an overarching objective for many nationalist movements and the independent states they tried to form. Although experiments with development were tried in many colonies, the idea of development was invested with the hopes and dreams of many newly emerging states which wanted to address these inequalities and divisions in their societies (Rahnema, 1997b). An important issue here concerns the extent to which colonial state machineries were reworked and transformed after independence. The colonial state had rested on force for its legitimacy, a legitimacy which was thus highly superficial. Colonial states also played a role in creating political and economic communities, defining the rules of the game and the boundaries of community while creating power structures to dominate them. The colonial state was also the dominant economic actor, creating a currency, levying taxes, introducing crops, developing markets, and controlling labour and production. Above all, colonial state administrations sought the integration of the colonial economy into the wider economies of Empire, to make linkages with the metropole and to establish flows of peoples and resources. After the formal end of colonialism, new states had to formulate alternative methods of garnering legitimacy for their authority (i.e. other than the use of force preferred by the colonists).

As the European capitalist system expanded and became ever more global in its reach, the structures of economic, social and political life that existed in colonies (before colonialism) were often radically remade. In India, after independence in 1947, Jawaharal Nehru dominated the country's politics, attempting to direct national development through a reinvented postcolonial state. India under Nehru (1947–1964) provides 'a clear example of the interaction of the colonial legacy and the Cold War' (Berger, 2001: 225) in that Indian politicians such as Nehru were trying to build states on the foundations of colonial institutions but found that these trajectories were being shaped by wider international political struggles during the Cold War. Under Nehru's tenure, the state in India stabilised and became a developmental agency that aspired to penetrate all areas of Indian society (Corbridge and Harriss, 2000; Ghosh, 2001). Nehru believed that only a national state, centrally responsible for directing economic development, could safeguard the postcolonial progress and independence of India. As we have seen, global inequalities have widened since 1945 and this has complicated many postcolonial development strategies. Several recent shifts in the global economy and the emergence of what has been termed 'neocolonialism' have undermined attempts to meet promises made in the early postcolonial era (see Chapters 7 and 8). Particularly in the past three decades, a number of issues have served to sustain the linkages between the 'core' and the 'periphery', usually to the detriment of the latter. In a way imperial rule created the very idea of an imperial centre and a colonised periphery and established a whole variety of important binaries, divisions and constructed boundaries between civilised and non-civilised, between the West and the Rest.

After the Second World War vast areas of Western Europe had been destroyed and cities lay in ruins with millions of displaced people. As a consequence, the USA launched its first major foreign aid programme – the Economic Recovery Program proposed by Secretary of State George Marshall. This proposed US$13.3 billion of spending over four years in quite 'recipient-friendly terms' (Sogge, 2002: 1). After the war Southeast Asia had also become an area of immense social and political upheaval within the British Empire. The region had been badly damaged by war (particularly in terms of transport and communications) and by 1948 India, Ceylon, Pakistan and Burma had all become independent of Britain. The British government was concerned that, unlike the case in its colonial territories, it could not direct the development paths of postcolonial countries and that because of this, these paths to development might lead these independent countries to align themselves with the USA or USSR and thus not with Britain. The Commonwealth was an effective tool in this process (the Commonwealth had existed since the Second World War and was created partly for the purpose of securing that alignment through development finance).

More of the aid programmes undertaken by Britain after the Second World War involved a greater commitment of funds to independent former colonies, organised through the Commonwealth. Thus while 'aid' was avowedly about *giving* to the newly independent countries, it was also very much about *taking* from them. A language was constructed around the need to fill 'gaps' and 'shortfalls' in the kind of finance and capital available to developing countries but also around the nature of their governance and social organisation. Declining competitiveness in British industry led successive British governments to relate the provision of aid to the industrial needs of Britain. A Ministry of Overseas Development was established in 1964 with this aid–trade provision as a key principle of its approach. The purchase of UK goods and services was thus to become a key objective of the UK's distribution of overseas development assistance (ODA) for many years. All this meant that the initiative for allocating aid came not from the ODA but rather from an industrial or commercial concern which believed it needed assistance to win an order (as was the case with the infamous Pergau dam project in Malaysia where a dam was added as a sweetener to a British arms deal). As the world economy slipped into recession in the 1970s this linkage became more explicit, not just in Britain but elsewhere.

In the USA, the Marshall Plan had sketched out the principles of 'helping' poor countries to 'recover' and restructure their economies and societies. Presenting itself as the antithesis of empire, the USA 'lavished billions on corrupt regimes that ignored poverty' during the Cold War (Engardio, 2001: 58). None the less, many observers have argued that the USA was taking on the mantle of imperialism by interfering in the political affairs of countries around the world after 1945. In 1961 the US Congress passed the Foreign Assistance Act (FAA) which distinguished different forms of aid (from military aid) and emphasised wide-ranging economic and social assistance efforts in the 'underdeveloped world' that President Truman had identified years previously. Under President J. F. Kennedy, the United States Agency for International Development (USAID) was created to administer the programmes suggested by the FAA. Kennedy outlined how 'collapse' in developing countries would be 'disastrous to our national security' and 'harmful to our prosperity' (quoted in USAID, 2002b: 2). Kennedy justified 'helping the poor' as being morally right in terms of 'saving the few who are rich'. For its part, USAID adopted notions of countries becoming more 'modern', through a series of stages, in ways which drew directly upon the development theories of the early 1960s. An 'Alliance for Progress' was formed, for example, with certain Latin American countries in a hemisphere-wide commitment of funds aimed at setting up new development planning mechanisms. Even this language of helping through generous aid expresses power relations however, and so by deconstructing these languages and statements we can reveal the knowledge bases, power sources and motives that lie behind them (Peet with Hartwick, 1999).

A key emphasis after 1947 was countering the spread of communism, particularly the influence of China, which intensified after the war with Vietnam and lasted until the mid-1970s. In the USA the rhetoric of anticommunism and superpower rivalry was critical in harnessing support for aid in the country.

THE POLITICAL AND PSYCHOLOGICAL SATISFACTION OF 'HELPING' THE POOR

Critics of foreign aid have argued that aid has been less effective than private investments and commercial loans in stimulating long-term economic growth. Other critics point to the dubious Cold War record of foreign aid in subsidising autocratic regimes and inflaming regional conflicts. Thus, many critics of foreign aid have sought to highlight the 'strategic interests' at work in its distribution as well as the inequality and unpredictability of aid provision. The Cold War provided an important rationale for foreign aid in this sense in that it was often based around ideological sympathies and connections between western and non-western countries. Both the United States and the USSR used aid to further their geopolitical positions and ideologies across the global South.

Strangely, except for a brief period during the mid-1970s, anti-poverty measures have not been an important focal point of foreign aid, while aid has led to many reversals as well as to advances. Seen as a (simultaneous) remedy to problems of growth, governance, poverty and inequality, it has become (not unlike the idea of development) overburdened with expectations (Sogge, 2002) and an over-ambitious enterprise (Rist, 1997). In Canada, for example, the evidence suggests that much of the bilateral aid programme run through the Canadian International Development Agency (CIDA) has been associated only rarely with poverty alleviation strategies, while CIDA itself acknowledges that more than 70 cents of every dollar of aid benefits Canada, creating 36,000 jobs in aid and NGO sectors (De Silva, 2002).

Since 1970 the strategies for successive UN 'Development Decades' have recognised that economically advanced states such as the USA have a

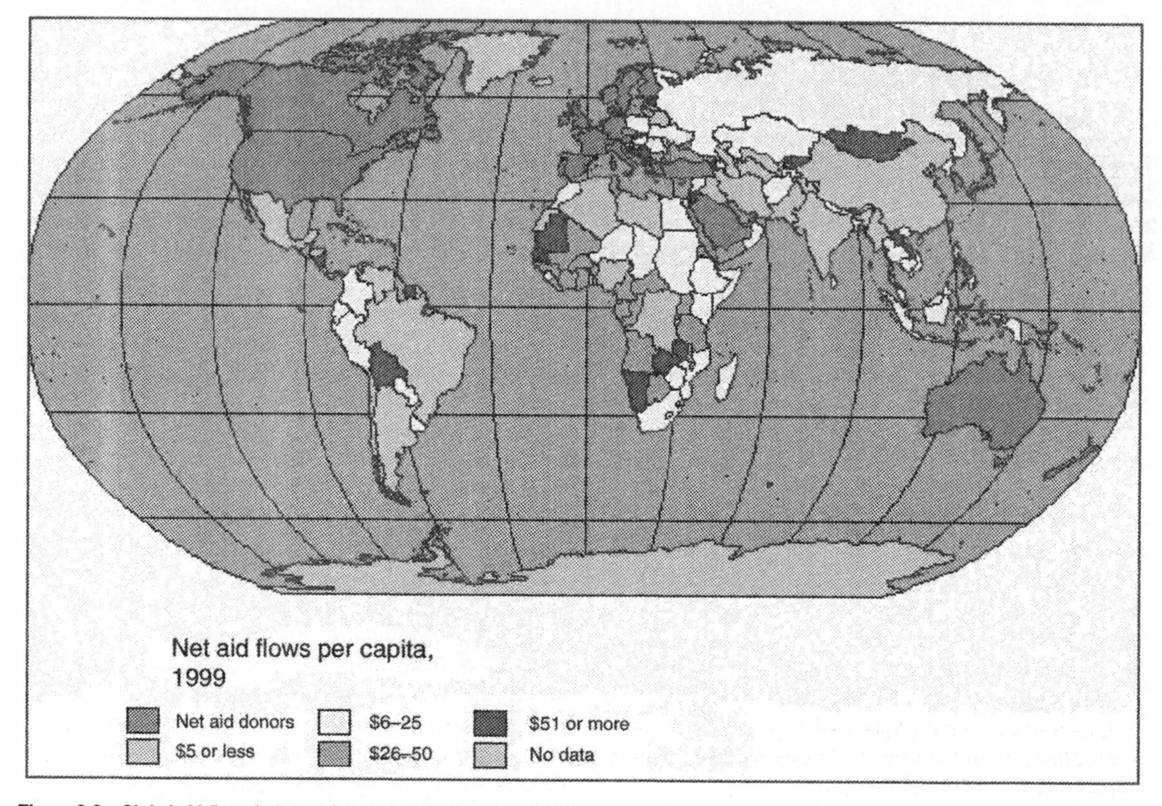

Figure 2.2 Global aid flows between donors and recipients in 1999
Source: World Bank (1999)

BOX 2.3

Geopolitics and the dissemination of Plan Colombia

Plan Colombia is a US-authored and promoted policy formulated at the beginning of the twenty-first century which was 'directed toward militarily eliminating the guerrilla forces in Colombia and repressing the rural communities that support them' (Petras, 2001: 1). The two major guerrilla movements in the country, the Revolutionary Armed Forces of Colombia (FARC) and the National Liberation Army (ELN), exert political and economic influence in areas of Colombian territory which house important centres of drug production. In recent years the USA has increasingly acknowledged that it is waging a war against this guerrilla insurgency. The issue is larger than simply the flow of narcotics to US consumers that US policy-makers seek to stem, but rather relates to the importance of the geopolitical context of this particular plan for Colombian development. US geopoliticians are concerned with a number of key issues involving Colombia that could 'adversely affect US imperial power in the region and beyond' (Petras, 2001: 1).

Over the past twenty-five years the USA has increasingly stepped up its 'war' against production of cocaine (and more recently opium poppies) in the Andes, particularly since the beginnings of Plan Colombia. Consequently there has been an increase in the production of coca (the shrub from which cocaine is extracted) and poppies (the source of heroin) in neighbouring Peru while the drug industry has simply shifted across the border into Peru and then also Bolivia. There are plenty of economic incentives for drug production in Peru and Colombia; the farm-gate price for coca leaves has risen to an all-time high of around US$3.50 per kilo, against just 40 cents at the low point in 1995 (*The Economist*, 16 February 2002a: 55). Coca is therefore very competitive against other crops at $1 per kilo, whereas the price of coffee (the obvious coca substitute for Andean farmers) has slipped to below the cost of production in recent years. American officials recognise that Plan Colombia will affect neighbouring countries, and as a result there was a tripling of anti-drug-related aid flows to Peru in 2002 to US$156 million, including US$30 million to upgrade police helicopters for opium areas and more than US$80 million for what is described as 'alternative development'. Bolivia is also involved in the trade, bringing in a 'dignity plan' under former President Hugo Banzer.

The government of Colombia is thus caught between immense pressures from the USA to adopt a zero-tolerance attitude, while growers continue to see the crop as a kind of economic magnet, such that more clashes between the two seem inevitable (*The Economist*, 16 February 2002). Officially, the USA provided some US$1.3 billion in aid to Colombia in 2000, with US$300 million more given in 2001, most of which must be spent on anti-drug operations, although some has been used in the counter-insurgency campaigns against the Revolutionary Armed Forces of Colombia which have been in conflict with government forces for four decades. As a result of US aid the army now has 50,000 salaried soldiers, a modern helicopter fleet and better intelligence (*The Economist*, 19 January 2002b: 47). Recent work on 'geopolitics' also suggests the need to focus on the ways in which states and governments construct and attempt to popularise their authority.

In May 2002, the Bush administration redefined the objectives of the Colombia Plan through supplemental appropriations submitted to the US Senate (CIP, 2002). The changes were intended to reflect the new priorities of Bush's global 'war on terrorism' which had not been written into the objectives of the Plan. The revisions enlarged the mission of US aid to Colombia to include a unified campaign against drug trafficking and activities of 'terrorist organisations' such as FARC and the ELN. In March 2002, the Bush administration also applied for additional funds for 'supplemental' aid to Colombia, part of a much larger request (for some US$27.1 billion) used in funding the war on terrorism. The aid was intended not only for Colombia but also for much of the surrounding Andean region. The request for aid funding to Colombia in 2003 seeks to provide some US$374 million in military aid and a substantially lower contribution of US$164 million for social and economic aid. Between 1999 and 2003, between 69.5 per cent and 75.1 per cent of aid to Colombia was for military and police assistance rather than for 'development' objectives, such as dealing with the many hundreds of thousands of Colombians internally displaced by this long-term conflict. Of the total of US$860.3 million provided in aid to Colombia for 2000 and 2001, only a pitiful US$68.5 million (8 per cent) was given to what the US Security services referred to as 'alternative development'. Will the drugs trade ever disappear when there are so many users in the USA willing to pay good money for this crop? This funding is intended to establish

BOX 2.3 (contd.)

alternative development and crop substitution programmes and also includes a US$10 million contribution which aims to aid rural peasants displaced as a result of the 'push into Southern Colombia' and its disputed territories (CIP, 2002). Bolivia received more than Colombia for such 'alternative' development proposals (US$85 million) as part of the wider regional strategy of Plan Colombia. Thus it becomes clear, if one considers the changing geographies of US aid recipients in Latin America, that these provisions are clearly motivated by strategic geopolitical motivations in the 'civilised world'.

responsibility to contribute to the financing of development and have specified that each 'developed' country should provide *at least* 0.7 per cent of its annual GNP as official development assistance. While still the world's biggest aid donor, the USA has slashed overseas development assistance since 1990 by some 34 per cent and foreign aid accounts today for just 0.1 per cent of GDP – the lowest level of any industrialised nation (Engardio, 2001). There are also plans to limit foreign aid further in order to minimise the US budget deficit (Zuza, 2002). In 1997 OECD donors gave the smallest share of their GNPs in aid since comparable statistics were first collected in the 1950s – less than a quarter of one per cent (World Bank, 2000a). Meanwhile again only a pitiful 4.7 per cent of major aid donors actually reached the UN target of 0.7 per cent of GNP in 2002 (Morrissey, 2002).

There are two major kinds of development aid. *Multilateral aid* is funded through contributions from wealthier countries and administered by agencies, such as the United Nations Development Programme or the World Bank. *Bilateral aid* and assistance is administered directly by donor governments (such as CIDA or USAID). Aid means lots of different things to different people but down through the decades we can discern a number of common features. According to Sogge (2002: 23) aid represents:

- A financial services industry, promoting loans on easy terms, and quietly insuring creditors against bad debts.
- A technical services industry, improving know-how and infrastructure.
- A 'feel-good' and image industry that can relieve guilt and subtly pander to the satisfactions of parental/paternal authority.
- A political toolshed stocked with carrots and sticks to train and discipline clients.
- A knowledge and ideology industry, setting policy agendas and shaping norms and aspirations.

In many ways some of the above can be useful in our understanding of the idea of development more generally. Development might thus be viewed as a knowledge industry, or as an ideological, financial and technical services industry which generates 'feel-good' images and understandings with the help of a toolshed stocked with the means of administering discipline and training to developing countries. Total aid currently amounts to about US$56 billion from twenty-one major OECD countries and some others (Maxwell, 2002). Two-thirds of this is government-to-government or bilateral aid and the remainder is multilateral aid disbursed by agencies such as the World Bank group, the UN bodies and the EU. Responding to humanitarian crises is becoming a more important component of all aid, more than doubling as a share of overseas development aid between 1980 and 1999 (Macrae, 2002).

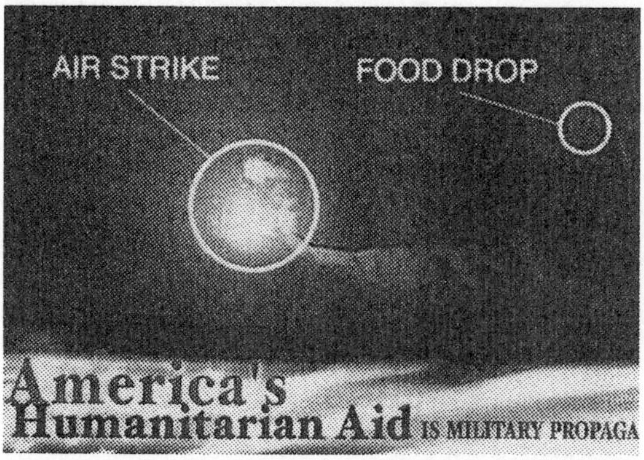

Figure 2.3 Aid or propaganda?
Source: © Protest Graphics

The number of states giving aid has continued to grow despite intermittent periods of major reductions in spending (e.g. during a recession). Between 1973 and 1992 a number of oil-producing countries (Saudi Arabia and the Gulf states) emerged as quite substantial donors, making grants and loans to countries in Asia and Africa with large Islamic populations. Ireland, Portugal and Spain are all now donors even though they themselves are recipients of EU funding and assistance. South Korea also became a donor in 1996 when it joined the OECD. Japan emerged as the leading source of development aid in 1994, overtaking the USA as the major contributor. In 2001 USAID claimed, however, that it was the world's largest provider of ODA at around US$11 billion (USAID, 2002a). The geography of foreign aid flows and the geographical spread of their destinations is far from even however. Recipient countries form a mixed group where 'former or current neo-colonial relationships strongly determine who gets what from whom' (Sogge, 2002: 28). The US assistance centres on a sphere of influence in the Middle East and Latin America, while France, Britain and Portugal focus mainly on their ex-colonies. Australia dominates aid flows in Papua New Guinea and Fiji, while historical ties link German aid to Turkey (Sogge, 2002). Initially Canadian overseas development assistance went mostly to Asia (and was allied to the war against communism there in the 1950s and 1960s) but has come to include Africa and particularly the Caribbean countries in close proximity to Canadian borders (De Silva, 2002). A large volume of bilateral aid from Canada since the early 1960s has been directed towards boosting the Canadian economy and creating jobs for Canadians (De Silva, 2002). In 2001 the OECD signed an agreement designed to untie bilateral aid to poorer countries, but it remains a purely voluntary agreement that few countries will seek to implement fully. There are also important new South–South flows of aid, pioneered by Cuba, China, Vietnam and Taiwan (among others).

As net reductions in aid were recorded in the 1990s, private capital flows from rich to poor states increased rapidly. Official private flows expanded from just over US$250 billion in 1996 to around US$300 billion by the turn of the century. Indeed, official assistance has begun to account for less than a quarter of all finance available to 'developing countries' (World Bank, 1999). It is worth remembering that when we talk about the 'debt crisis' much of this indebtedness comes from private capital flows to countries of the South, of which the USA alone generated some US$36 billion annually between 1997 and 2000 (USAID, 2002a). Private aid flows, according to critics of aid cutbacks, cannot have the same 'integrity of objectives' associated with public development aid and assistance. Private capital flows (it is argued) are driven by commercial motivations and are thus less strongly connected to human needs and the need for sustainable development of the environment. Most North–South private capital flows in recent years went to the Newly Industrialising Countries (NICs) of Southeast Asia rather than to Africa.

Examining European Union aid to Africa we can see that public opinion on aid and development is not some unchanging, undifferentiated mass, but dynamic and complex. What is clear, however, is that 'development was not a particularly important issue for Europeans in the 1990s' (Olsen, 2001: 664). Public opinion has also been shaped by a succession of aid scandals, together with tales of waste, failed projects and plush lifestyles for aid workers, all of which have undermined public confidence (Sogge, 2002). More than a quarter of the population of EU member states surveyed at the turn of the century did not even know if the EU actually gave aid or not (Olsen, 2001). The EU's development policies are potentially of great significance however, with EU countries controlling some 51 per cent of foreign direct investment (FDI) outflows and 56 per cent of all overseas development assistance (Van Riessen, 1999). EU states also represent the largest block of members with voting rights in the WTO, the IMF and the World Bank. Coordination and mobilisation of EU actions in these realms could thus create a powerful base of support for 'development' in the South. Brussels politics and policy-making is changing all the time however, while coherence and consistency have often been lacking. Lister (1997) looks at how the EU sought to foster cohesion between and among other EU members through a 'low policy instrument' such as aid which is seen as somewhat less controversial than other issues. For this reason the development policy of the community is itself 'a cornerstone of European integration', more about Europe and Europeans than a concern for its intended recipients. At an EU summit in Seville in June 2002 many EU states even tried to attach a condition to their aid donations which aimed to hit the pockets of those countries not deemed to be doing enough to stop the flow of asylum-seekers to European countries, although proposals were later watered down (BBC, 2002d).

One crucial motivating factor in giving development aid was that a 'certain psychological satisfaction' is gained by Europeans from helping the poorest states in the world (Ravenhill, 1985: 35). Humanitarian assistance from the EU also seems to depend in part on the media, which can build mass public interest in distant geographies, because of the strong emotions involved in representing human suffering on a massive scale (Ignatieff, 1998: 291). Humanitarian emergencies partly become 'common knowledge' through a kind of 'CNN effect', from being screened by news channels such as CNN (Shaw, 1996; Robinson, 1999). Similarly in a narrowly focused situation such as humanitarian emergencies, the media play a 'decisive role in informing the public and stimulating action' (Rosenblatt, 1996: 139). This is a complex issue, but it is clear that certain kinds of 'crisis' are overlooked in a world that can manage only one problem situation at a time (Livingston, 1996). The role of aid workers in conjuring the imagery of emergency and humanitarian crisis is important here:

> We cannot have misery without aid workers. They conjure away the horror by suggesting that help is at hand. . . . Television coverage of humanitarian assistance allows the West the illusion that it is doing something.
>
> (Ignatieff, 1998: 298)

Thus there are multiple images of 'Third World poverty' and humanitarian crisis which tend to paint a picture of negativity, of failure, of something lacking. According to the Voluntary Services Overseas (VSO) some 80 per cent of the British public associates the developing world with 'doom-laden images of famine, disaster and Western aid' (VSO, 2001: 3). The VSO survey found that nearly two decades on from the high-profile Ethiopian famine (and the Live Aid relief concert organised to provide for famine victims) these images were uppermost in the popular British imagination of Africa and 'maintain a powerful grip on the British psyche' (VSO, 2001: 3). Aid and development are by no means the same thing, but the way their interrelationship is imagined is important in that it tells us a great deal about how people envisage the 'problems' of other spaces and worlds of difference. One of the surest signs that development is indeed in crisis is the fact that for many development practitioners the concept has increasingly been quite narrowly focused on short-term, humanitarian and emergency relief. Many African countries (Rwanda, Somalia, Liberia, Sierra Leone) are represented by the media through the myth of Africa as the 'Heart of Darkness' or as the 'hopeless continent' (see Figure 2.8). As Smillie (1996: 28) argues, 'on the subject of public opinion and development assistance: the rationale for aid in the public mind was and remains emergency relief' (cf. Jarosz, 1992; Robinson, 1999). Arguably this is a kind of throwback to the age of Empire. What is especially interesting about aid in this respect is that it represents 'another important instrument to project power beyond national borders', just as colonialism had done previously (Sogge, 2002: 13). Thus while aid is also very much about foreign policy and the strategic interests of donors, it is also partly about projections of economic, cultural and political power beyond the boundaries of certain countries which establish important chains and connections, linking donors and recipients in interesting and important ways.

Figure 2.4 Unloading EU-donated food aid in Huambo, Angola
Source: Alex Hofford

Many of the millennium development goals set forth by the international donor community in the 1996 OECD report *Shaping the Twenty-first Century* (OECD, 1996) are unlikely to be met because aid agencies arguably suffer from similar 'disbursement-driven dynamics' as multilateral donors (see Box 2.2). Kofi Annan's Global Fund for HIV/AIDS, TB and Malaria, for example, had a target of raising US$10 billion a year between 1997 and 2002, but only US$1.5 billion has been raised as the first five years neared its end (Morrissey, 2002). Is this annual US$50 billion increase in aid flows such a

Figure 2.5 Are you now or have you ever been a terrorist/communist?
Source: © Protest Graphics

vast sum of money (especially when one considers how much is spent each year by the same governments on arms purchases)?

When the aid industry arrived in Russia in the early 1990s and after the end of state socialism, the commanding heights of the international development industry (the World Bank, the IMF, USAID) quickly came to spread the doctrine of neoliberalism or the virtues of the free market. Not for the first time there was little recipient control or public accountability. According to then World Bank senior economist Lawrence Summers, these agencies came to '[s]pread the truth [that] the laws of economics are like the the laws of engineering, one set of laws works everywhere' (quoted in Sogge, 2002: 2). Interestingly, when directed towards the countries of Africa, Asia or Latin America, this aid is called overseas development assistance, but when it is directed towards Eastern Europe or the ex-Soviet Union and other 'countries in transition' it is given the much less condescending label of official aid (OA) (Sogge, 2002). This book seeks to develop an exploration and interrogation of the psychology of 'helping' the poor and tries to understand how this operates in particular spaces and places or can be considered the product of colonial and imperial histories of exploitation and underdevelopment in the South.

THE (GEO)POLITICS OF AID

The word 'geopolitics' relates to a branch of political geography that deals with international relations, international conflicts and foreign policies. The term refers partly to the way in which space is important in understanding the constitution of international relations. As a school of thought it was shaped originally by its emergence as a discipline in Britain, France, the USA, Germany and Russia. Today the term is used to focus on political meanings and representations and to explore the contestation of the political world in a variety of spaces and places (see Chapters 5 and 9). In particular, certain trends and patterns in international relations are important to the way in which we understand development. A transition from one geopolitical world order to another was very much at the core of development's postwar invention as an arena of state practice. Thus with the end of colonialism one set of political practices and representations begins to break down and replacements were constructed. In this way international development always involves a geopolitical imagination and carries within it implicit assumptions about the state of the world political order and the role of these political practices and representations. This book argues that it is in many ways impossible to understand the contemporary making of development theory and practice *without* reference to geopolitics and the geopolitical imagination of non-western societies. This is not to say that development is simply the continuation of politics by another means, since we cannot dismiss aid as simply part of some past and therefore 'outdated sideshow in the repertoire of geopolitics' (Sogge, 2002: 10). A more nuanced understanding is required here.

In 2001 the World Bank President and the IMF Managing Director issued a joint statement which cancelled the 2001 annual meetings of their respective Boards of Governors and related committees in light of the terrorist attacks in Washington, DC and New York City (World Bank, 2001d). At the time of the attacks, and for many months after, World Bank and IMF leaders made numerous points about the links between poverty and terrorism, suggesting that this was to become the key issue of the twenty-first century. The ensuing global war on terrorism led by the 'free' or 'civilised' world began in Afghanistan, where, even before 9/11 and the beginning of this 'just war', the UN estimated that millions were being sustained (barely) by international food aid.

When the allied attacks against the retrograde Taliban regime began, US air drops of food aid were condemned scathingly by aid agencies as 'barely concealed propaganda tools' (Chomsky,

Figure 2.6 Investing in global war
Source: © Protest Graphics

2001), causing more harm than benefit in some cases. These appeals coincided with World Food Day and seemed to illustrate the variety of ways in which aid and development (as ideals) are manipulated in many supposedly 'humanitarian causes'. Why have repeated attempts by the UN to bring peace to Afghanistan failed over the past three decades? The Soviet invasion of Afghanistan in December 1979 and the US and Arab arming of the Mujahideen resistance brought 'total war' to Afghanistan, leading to the criminalisation of the country's war economy and the militarisation of Afghan society. Pakistan and Iran were also drawn into the various periods of conflict through refugee movements which spilt over into their countries. A succession of grandly titled UN bodies have come and gone (the latest was the UN Special Mission to Afghanistan, UNSMA), while the UN's overall approach has changed little despite the high turnover of staff. Interventions by the UN were thus often based upon a poor understanding of the historical processes which created and sustained the Afghan conflict (which does not begin and end with the Taliban).

Cluster bombs and 'daisy-cutters' seemed to be guaranteed to reach their intended destination, whereas much less could be said about desperately needed food and medical supplies which quickly became bargaining points. If there are 'smart bombs' why is there no 'smart aid'? (Engardio, 2001). President Bush has pledged to help rebuild the economies of post-Taliban Afghanistan and Pakistan, yet the USA has been providing 'aid' and foreign assistance to these countries for years, so we may well ask why have these efforts not *already* been successful in alleviating poverty and rebuilding communities? Both Russia and the USA are at least partly responsible for the (historical) creation of this 'humanitarian disaster' since during the Cold War they both intervened in Afghanistan's political affairs, particularly during the 1980s and 1990s. Anti-Americanism in 2001 was therefore not based on a random hatred of western modernity or technology envy that came from nowhere but rather:

> [o]n a narrative of concrete interventions, specific depredations and in the cases of the Iraqi people's suffering under US-imposed sanctions and US support for the 34-year old Israeli occupation of Palestinian territories.
>
> (Said, 2001)

When the United Nations Development Programme (UNDP), the World Bank and the Asian Development Bank came together during the Afghanistan crisis it was to discuss the terms of reconstruction and development in Afghanistan after the war, focusing on options for private investment in infrastructure which were actively encouraged (as if there were no alternatives) (Chomsky, 2001). The UNDP estimated that the reconstruction of Afghanistan would cost around US$7 billion to 12 billion for the first five years alone (World Bank, 2001b). Only loans and not grants are awarded by these bodies and private companies tend to be seen as most capable of organising production and service delivery.

In discussing the role of the global war on terrorism and how it shapes contemporary development thinking and practice we need also to raise

Figure 2.7 Who said that foreign policy was hard for world leaders?
Source: © Patrick Chappatte, in *International Herald Tribune*

questions about the role of Saudi Arabia, Iran, Iraq, Turkey, Sudan and Israel. In relation to debates about foreign aid and development more generally we can explore the experiences of Egypt, the Emirates, Oman, Tunisia and Yemen (among others). These countries are particularly relevant to this objective of rethinking development geographies in a number of ways. As Aarts (1999) has shown, since 1945 the Arab world has passed through several phases of development ranging from cold war to consensus. Middle-Eastern states are also seen by many commentators to lack necessary degrees of 'order', hierarchy and societal organisation present in its alter-ego, 'the West'. The CIA report *Global Trends 2015*, for example, paints a negative picture of the Middle East, pointing out that 'most regimes are change-resistant' (CIA, 2000: 11). Just as in the age of Empire, distant others are seen in an unchanging, timeless stasis, as outside of history and permanently resistant to change. The states of the Middle East, however, must also be understood as the historical and geographical products of the modern state system and owe much to it. It has long been problematic, however, to see the Middle East as a single political or socio-economic whole, to collapse contradictory trends between its countries, cultures and places. As Halliday (2000: 213) has argued, one of the great distortions to beset the region is that its politics and history are often explained by timeless cultural features: 'a Middle Eastern essence', 'rules of the game', or an 'Islamic mindset'. The Middle East still barely figures in discussions of foreign direct investments (FDI) flows to the Third World, for example (Halliday, 2000: 214), despite a fivefold increase in FDI inflows during the 1990s:

> On the map of globalization, except as a source of oil and investment funds, the Middle East hardly figures ... Trade and Investment have most certainly not acted as solvents in relations between Israel and Egypt, or the Arab world more generally.
>
> (Halliday, 2000: 214–215)

This is a key theme here: To what extent does development studies or development geography have its own margins, where certain countries 'slip off the map' of research and discussion about development? The Middle East thus is often seen as an 'exceptional' case 'eternally out of step with history' and immune to the trends affecting other parts of the world (Aarts, 1999: 911). Aid flows to the Middle East have declined but they have always been based on the perceived interest of the donor states, which could thus turn the supply on and off as required. As elsewhere in the South, gross inequalities of wealth, power and access to rights – aka imperialism – persist (Halliday, 2000: 219). This suggests the need to reformulate anti-imperial critiques of development around a sense of the indigenous roots of aggression and dictatorship, not simply seeking to blame them all on an externalised process of (US) imperialism seen as somehow totally foreign to Middle-Eastern and other cultures.

One estimate has it that a world total of US$1,000 billion has been spent on aid since the 1960s (Easterley, quoted in Gonsalves, 2002: 1). Although debates about assessing the effectiveness of aid have been well rehearsed elsewhere, they have not always been properly historicised or always seen as fundamentally related to geopolitical transitions and transformations. These debates are important because they frequently promised growth and poverty reduction and therein often masked deeper 'strategic' geopolitical motivations from the very start. Few aspects of world politics in the late twentieth century were as controversial as foreign aid. Foreign aid represents a transfer of resources that would not have taken place as a result of market forces and includes grants and loans provided by governments and the IFIs (international financial institutions).

Some critics have even seen aid as a kind of political narcotic, fostering addictive behaviour among states that receive it and thus come to depend on it. States are thought to exhibit the symptoms of dependence – a short-run 'fix' or benefit from aid, but external support sometimes does 'lasting damage to the country' (Goldsmith, 2001: 412). In feeding this 'addiction' the aid donors have supposedly weakened the resolve of states to act on behalf of their own citizens. How then to kick the habit and reduce dependence? The practice of sharing wealth with impoverished peoples has emerged as a norm among many western countries and nearly every state in the world has participated as a donor or as a recipient of foreign aid since the Second World War. Amid the growth and institutionalisation of aid flows, the practice of giving aid has also been subject to intense criticism and debate. Some development theorists, for example, have dismissed foreign aid as an instrument of capitalist control for maintaining dependency and the exploitation of the 'periphery' by the 'core' (see Chapters 3 and 6). President Nixon's injunction to 'remember that the main purpose of American aid is not to help other nations but to help ourselves' is a reminder of some of the considerations and motivations that underpin the actions of many western states. Despite the controversy around foreign aid, few donor countries have held widespread public debates about its necessity and purposes. We can critique the case for aid (morally and politically) as a way of beginning to explore the ways in which development is defined, thinking about how the effectiveness of aid in facilitating development has been assessed and how the relations between aid and trade have been historically conceived.

TABLE 2.2 ESTIMATED US SECURITY ASSISTANCE TO THE WESTERN HEMISPHERE, BY COUNTRY, 1997–2000 (GRANT MILITARY AND POLICE ASSISTANCE) (US$)

Country	*1997*	*1998*	*1999*	*2000*
Antigua/Barbuda	336,000	508,000	519,000	202,000
Argentina	1,615,418	6,062,635	7,783,733	1,420,000
The Bahamas	1,104,000	899,000	1,455,000	2, 437,000
Barbados	354,000	260,000	478,000	348,000
Belize	432,000	654,000	644,000	395,000
Bolivia	22,600,000	48,852,000	35,648,878	64,267,000
Brazil	3,460,000	5,831,000	2,615,000	5,200,000
Chile	544,000	17,548,050	993,000	878,671,000
Colombia	86,000,000	110,232,000	305,777,788	798,290,000
Costa Rica	333,000	451,000	701,078	2,905,000
Dominica	233,000	147,000	195,000	111,000
Dominican Republic	1,169,600	2,058,000	2,015,961	1,554,000
Ecuador	2,757,250	5,270,000	12,243,068	24,428,000
El Salvador	621,000	783,000	550,999	3,982,185
Grenada	149,000	216,000	249,000	274,000
Guatemala	2,158,000	2,847,000	3,115,000	3,250,000
Guyana	178,000	281,000	316,000	321,000
Haiti	500,000	940,000	550,000	1,143,000
Honduras	719,000	2,923,404	803,000	872,000
Jamaica	1,489,000	2,601,650	2,481,000	1,969,000
Mexico	87,044,000	26,238,000	21,244,000	16,618,000
Nicaragua	201,000	74,000	200,000	204,000
Panama	2,534,000	2,591,000	4,597,570	5,609,000
Paraguay	1,229,000	736,000	508,000	388,000
Peru	33,969,000	38,299,000	65,561,200	57,628,000
St Kitts and Nevis	158,000	245,000	218,000	77,000
St Lucia	142,000	279,000	252,000	168,000
St Vincent and Grenadines	193,000	217,000	179,000	77,000
Suriname	149,000	82,000	100,000	155,000
Trindidad and Tobago	568,000	2,625,394	598,000	893,000
Uruguay	353,000	1,198,000	1,596,000	330,602
Venezuela	6,081,000	7,182,000	4,011,000	6,475,000

***Source*: adapted from Center for International Policy (2000, 12), http://www.ciponline.org/ The figures given are estimates and cannot be regarded as exact figures as they do not include arms sales or training exercises and military deployments. Some of the total amounts in some cases include contributions which are for more than just military and police aid which can be variously defined by the US state depending on the strategic security objectives it has in each country.**

CONCLUSIONS: GEOPOLITICS AND THE 'APOTHECARY' OF DEVELOPMENT REMEDIES

At the beginning of the twenty-first century development geographers need to examine some of the striking contradictions in the nature of international institutions of governance, to think critically about the ways in which international development agencies invent, construct and name 'poor' countries, places and peoples. We have seen that the picture of progress held by such agencies is often an incomplete and very partial one due to the fragility of their data collection procedures and their distance from events 'on the ground'. This book does not set out to tell yet another story of uniquely predatory northern actors victimising uniquely defenceless southern actors since 'predation is worldwide' (Sogge, 2002: 36). Instead it is to the complex connections and collusions between North and South that we should turn our attention. Post-development perspectives, as we shall see in Chapter 4, hold out the possibility of radically rethinking the power, place and spatiality of development in the contemporary world. In this sense there may be lessons to be learned from the coalitions that are being formed across cultural and political differences in and through a global protest caravan that is explicitly linking struggles at various spatial scales. It is also necessary to listen to the *other side of the development story*, the views of women, the views of people in the South, the views of people who are excluded socially, and the views from 'below'. Arguably, in this way there remains a deepening contradiction within a modern development paradigm driven by 'techno-scientific and economistic variables' (Schmitz, 1995: 55–56). This dominant paradigm rests on grossly undemocratic and inequitable global political economies that are increasingly being disputed and contested.

Questions must also be raised about the *geopolitics of development* and about the changing international geopolitical frameworks in which development is continually remade. Development is imagined geopolitically and has been conditioned by some important geopolitical discourses, particularly during the Cold War. During the Cold War, the USA developed something of a 'special relationship' with Israel, for example, based around the perception of shared strategic interests between both states (Riech, 1996; Jones and Murphy, 2002). It has been estimated that Israel received some US$65 billion in economic and military aid from the USA alone between 1948 and 1996 (Bar-Siman-Tov, 1998). The fact that the Middle East and North Africa have just over 68 per cent of all world oil reserves has been crucial in the establishment of this relationship between the USA and Israel. A range of 'soft' variables such as Israel's democratic status, the idea of Israel as 'western' and the influence of the Jewish lobby in the USA have also been important here.

What is particularly interesting about USAID today is that it still draws heavily on the principles of the FAA, despite numerous attempts at overhauling and rewriting the act, while its basic philosophy remains much as it did just over forty years ago when the organisation was created. Arguably few if any presidents after J. F. Kennedy have shared this humanist concern with other regions. According to the Development Assistance Committee of the OECD, in the post-Cold War world aid donors have also often been motivated by 'national concerns' and are sometimes now 'more concerned to show the national flag on their development projects than to join collective sector improvement efforts in which donor identities are merged' (OECD Development Assistance Committee, quoted in World Bank, 1999: 13).

In thinking critically about geopolitics and development, this book argues that it is possible to re-examine debates about statehood, civil society and citizenship, to open up the ideological spaces(s) of development thinking. As Radcliffe (1999: 84) has argued, '[r]ethinking development means reconsidering the categories we use in development geography, and unpacking the power relations that shape them'. This question of the power of development and the unpacking of the power relations that are inscribed in development practices is a crucial one. In some ways polymorphous notions of the nature of power and power relations in development in different spaces and places are also needed here. How have these relations evolved historically and what have been the legacies of the past given that development was preceded by colonialism in a number of cases? A project of decolonising the mindset of development thinking is thus required because regions such as the Middle East are not alone in being seen as eternally out of step with progress and modernity and there are important historical and geographical reasons for this.

According to the 2002 World Development Report *Building Institutions for Markets*, the World

Bank's primary concern remains that state institutions should serve and support trade, investment and markets. People will benefit from all that participation in the global capitalist market can offer, while 60,000 'voices of the poor' are assembled to underwrite this claim with a degree of popular legitimacy. *Rethinking Development Geographies* seeks to contest this vision of institutional reform and is interested in a whole variety of other voices, including those of the many hundreds of thousands of anti-capitalist or 'anti-globalisation' demonstrators who have besieged the Bank's 'public' meetings in recent years. A short thirty-one-minute video produced by the Bank in 2001 entitled 'Voices of the Poor' has little to say about the *international* causes of impoverishment, about the specific political or economic structures that exist *between* nations and how they maintain relations of neocolonialism in particular places.

Connections are made in the film which seek to link the poor in different regions of the world, but poverty within the core of the world economy, in some of its so-called 'developed' regions, is not really addressed. Even though the film recognises that the meaning of poverty varies according to each country's culture and history, various parts of the 'developing world' merge together into a kind of distorted collage of underdevelopment and impoverishment which universalises experiences of deprivation, the very thing the film sets out *not* to do. Interestingly, few of the 'voices of the poor' the Bank consulted are actually named in the film as the subjects or interviewees are identified simply by their citizenship in a 'poor country'. Throughout the film the Bank emerges as the triumphant slayer of the dragon of impoverishment through its help and assistance to the weak and the needy, and its conclusions suggest that we can feel good about the work they are doing in the name of progress. Watching this particular film one could easily leave with the (misguided) impression that the world is very definitely 'on the right track' in terms of a poverty reduction strategy.

Opposition to the Bank's discourse of development is dubbed 'anti-globalisation', yet this is misleading as most of the activists are not opposed to growing international economic and cultural interconnections among peoples, but are opposed to the way those connections are being structured 'benefiting large firms, while creating hardship and instability for many, many people' (MacEwan, 2002: 4). A large number of these demonstrators have not set out to oppose international connections and flows of ideas between people *per se* but rather to harness some of these global relations and interconnections for the purpose of mobilisation against neoliberalism. Policies such as those adopted for Argentina by the IMF (see Chapters 7 and 8) typify the problem. Argentina's economic collapse and continued recession in the national economy show that IMF policies not only fail to bolster economic development but also often lead to social and political disintegration (MacEwan, 2002). In rethinking development geographies we need to continue to pressure institutions such as these which have a profound impact on the political and economic trajectories of what they call low- and medium-income countries. The key point here is that neoliberalism systematically produces these kinds of political and economic crises and thus faces 'permanent revolt and inevitable explosions' (Amin, 2001b: 20).

As millions of investors have painfully learned from the evaporation of their investments in the 'emerging markets' of Asia during the financial crisis of 1997, globalisation is a 'two-way street' (De Soto, 2001). Western Europeans now vote for

Figure 2.8 The hopeless continent
Source: *The Economist*

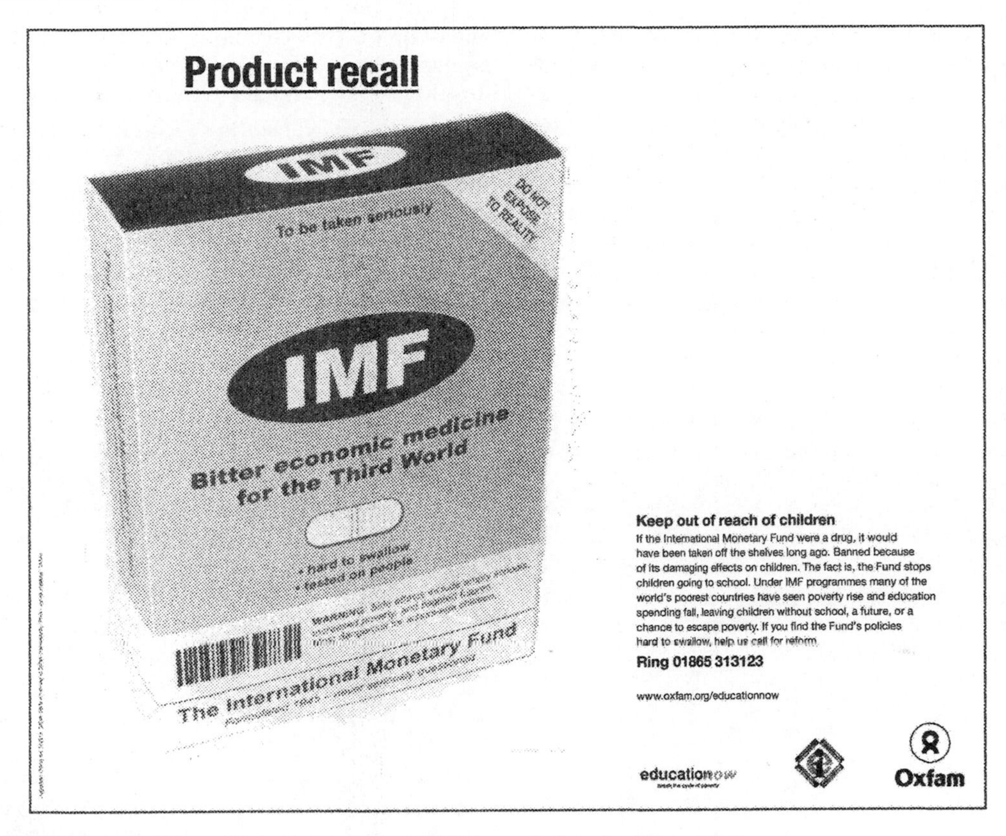

Figure 2.9 Not to be tested on children: Oxfam argues that IMF prescriptions should be recalled
Source: *Oxfam*

politicians who promise them a 'third way', but Peruvian writer Hernando De Soto argues that the demonstrations against globalisation and capitalism have only forced American and European leaders to trot out the same old wearisome lectures:

> Stabilize your currencies, hang tough, ignore the food riots, and wait patiently for foreign investors to return. Foreign investment is, of course, a very good thing – the more of it the better. Stable currencies are also good, as are free trade and transparent banking practices and the privatisation of state-owned industries, and every other remedy in the Western pharmacopoeia.
>
> (De Soto, 2001: 45)

In Latin America, De Soto argues that these policies have failed repeatedly to build successful capitalist structures and that rather than questioning the adequacy of this *pharmacopoeia of remedies*, it needs to be recognised that policies and prescriptions 'fall so far short as to be almost pointless' (De Soto, 2001: 45). The word 'pharmacopoeia' refers to the idea of an officially published text or book which contains a list of medicinal remedies and drugs and prescribed directions for use. This would suggest, however, that the Bank and the Fund are quite open and democratic institutions with a list of remedies that can be negotiated. Perhaps it is better to refer to an apothecary of development remedies, to a kind of storehouse and chemist which claims to be licensed to dispense medicines and drugs. Under these institutions neoliberalism was seen as 'rational' practice; it was orthodoxy within a short period and has since become the ideology of a small army of economic 'experts', who claim that their remedies and the 'truths' that underscore them are the only way forward.

This allows Third World peoples to be blamed for their lack of entrepreneurial spirit or 'market orientation'. If these countries fail to close the gap between themselves and the richer western capitalist economies, it is not because capitalism does not work effectively in these contexts but because the culture of the people there is seen to prevent them from doing so 'or their IQ's are too low on the Bell curve' (De Soto, 2001: 45). The poor are seen to be trapped in the 'obsolete ways' of the past, to be part of dysfunctional cultures or are simply 'pitiful beggars'. In turn this comes back to the vision of development outlined in the imperial encounter, where non-western peoples were seen as trapped in the outdated co-ordinates of the past, as noble beggars with dysfunctional societies. The poor already have capital however, and the value of their savings can be much higher than is often assumed, but the problem is that this capital is seen to be held in 'defective forms' (with unrecognised legal rights to land and houses or unrecorded businesses and a variety of assets that are not easily recorded). Thus people lack the capacity to *represent their capital*, the titles to their homes or land, the formality of recognition for their business enterprises, 'crops not deeds, businesses but not statuses of incorporation' (De Soto, 2001: 47).

How can we begin to dismantle and think beyond the western idea of development remedies? One fundamental starting point is to accept that the global apothecaries of development medicines, potions, remedies and drugs developed a 'licence' to dispense their prescriptions as a direct result of histories of western imperialism and Cold War geopolitics.

BOX 2.4

Useful chapter-related websites

http://www.brettonwoods.org/
The Bretton Woods organisation organises campaigns about the activities of the Bretton Woods organisations, the World Bank and the IMF.

http://www.50years.org/
The '50 years is enough' campaign focuses on the first fifty years of global development and the failures of global development institutions.

http://www.globalizethis.org/
Globalise This! campaign site.

http://www.foreignaffairs.org/articles/
Associated with the publication of the journal *Foreign Affairs*.

http://www.imf.org/
The International Monetary Fund's website.

http://www.Southcentre.org/publications/
The South Centre.

http://www.g7.utoronto.ca/
University of Toronto site on the G8.

http://www.worldbank.org/poverty/voices/
The World Bank website and the section dealing with its 'voices of the poor' survey.

http://www.jubilee2000uk.org/
The Jubilee 2000 campaign concerning the global debt crisis.

http://www.saprin.org/voices_from_the_south.htm
Voices from the South.

http://www.challengeglobalization.org/
Challenge globalization site.

http://www.onetheline.org/
On the Line – critique of the World Bank.

http://www.usaid.org/
Official USAID site.

http://www.globalissues.org/
Discusses a range of global issues and contains useful material on geopolitics.

http://www.abolishthebank.org/
Contains background information on the World (WEF) and news of anti-capitalist and anti-WEF protests with some useful links. This site talks of making 'resistance visible'.

http://www.studentsforglobaljustice.org/
Based around the Students for Global Justice organisation which links student resistance to anti-war, anti-capitalist and global justice campaigns.

http://www.socialanalysis.org/
On how not to count the poor.

3
Geographers and 'the Tropics'

INTRODUCTION: ORIENTALISM AND TROPICAL GEOGRAPHIES

> The identification of the Northern temperate regions as the normal, and the tropics as altogether other – climatically, geographically and morally – became part of an enduring imaginative geography, which continues to shape the production and consumption of knowledge in the twenty-first-century world.
>
> (Driver and Yeoh, 2000: 1)

> Geography, like many other academic disciplines, has been seriously contaminated in recent years by elements derived from other essentially marginal disciplines . . . [resulting in] various types of mathematical mystification.
>
> (Llwyd, 1968: 58)

Our knowledge of the world and its geographies is constructed in a variety of ways, for example through experience, learning, memory and imagination. This chapter seeks to examine the question of what 'experiences' and whose 'learning' has been brought to bear in understanding the world and its geographies. How then have geographers constructed 'the Tropics' as a particular space for investigating progress and development in non-western regions and how have geographers viewed 'the Tropics' as a world of particular forms and relations? As we have seen, western colonial projects, despite being far from homogeneous, were based partly on an imagination of the world which legitimised and supported the power of the West to dominate 'others'. By representing other societies as 'backward' and 'irrational', the West and Europe in particular emerges as 'mature', rational and objective. Here the notion of 'the Tropics' was invented alongside a range of other labels to view certain societies as 'other' and different from the self of Europe or the 'West'. This difference was often seen as something which could be interpreted and represented by rational, even-handed social scientists in the West in a supposedly 'purely objective' manner.

In this chapter we will be examining the ways in which geographies of development have contributed to the formation of a particular kind of Orientalism (Said, 1978) or an imperial vision of particular places and subjects (that displaces other voices) which had important material consequences. In an important and influential text on Orientalism, Edward Said (1978) suggested that *Orientalism* and Orientalists analysed a large corpus of representations of the East produced by academics, novelists

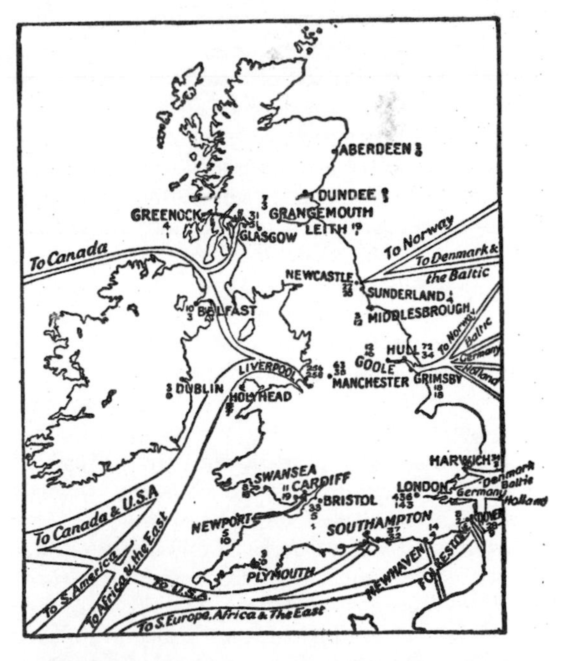

Figure 3.1 Map showing the principal ports of the British Isles and their global connections
Source: Stamp (1934)

and others situated in the 'West' during the eighteenth and nineteenth centuries. Said argues that these representations comprised a discourse which he defines as 'a tradition . . . whose material presence or weight, not the originality of a given author, is really responsible for the texts produced out of it' (Said, 1978: 94). An Orientalist discourse then transcends the individual or the institution and represents a kind of archive of images and statements. In an important way this provided a common language for knowing a particular region and its peoples. 'The Orient' was thus established in the European geographical imagination as an image, an idea, a personality, an experience (Lester, 1999). Said's approach to Orientalism had provided an exceptional account of the working of the unequal relationship between the western colonial powers and the peoples of the 'East' during the nineteenth and twentieth centuries. *Orientalism* looks at how this manifested itself in the production of knowledge about this particular part of the world, thereby reinforcing a sense of *difference* between the Orient and the rest of the world based around a sense of distances and inequalities (Bilgin and Morton, 2002).

Europeans sought knowledge about the Orient and through their Orientalist discourse they produced what Said (1978: 3) calls a 'Western style for dominating', inventing the authority and legitimacy to subjugate and oppress non-western peoples. This chapter argues that 'the Tropics' were also endowed with a similar condition of 'otherness' which was ascribed to tropical peoples and sought to view the tropical as a twin to the temperate: the opposite of all that was civilised, modest and enlightened (Driver and Yeoh, 2000). In his later work *Culture and Imperialism* (Said, 1993), Said also makes a passing reference to an 'Africanist discourse' which had emerged around the study of Africa, based upon Orientalist attitudes and practices towards the continent. What is particularly important about this reference to an emerging Africanist discourse is that it soon became (particularly after 1945) a 'systematic language for dealing with and studying Africa' (Paolini, 1997: 83). What Said was concerned with was the Eurocentric discourses which enframe regions such as Africa and other parts of the 'Third World' and the way they 'contain' and 'normalise' the region through a certain kind of hegemony. Through such discourses, acceptable ways of speaking about and representing Africa are established which were very influential, and impacted on the way the region was viewed and understood.

Associated with these Africanist and Orientalist discourses was the idea of 'the Tropical', a term which is still routinely attached to a whole range of terms, everything from 'rainforests to rainstorms, from resorts to beaches, from urbanisation to soils' (Livingstone, 2000: 95). As we shall see in this chapter, the notion of a 'tropical geography' then had a somewhat complex history through the twentieth century and became more formalised and more widely recognised after the Second World War. The French-speaking geographer Pierre Gourou was one of the most famous exponents of these kinds of geographies (Gourou, 1947, 1960). From 1953 new possibilities began to emerge for the publication of such geographies with the creation of the *Malayan Journal of Tropical Geography*. A distinct field of geographical enquiry had thus begun to emerge, supported by new conferences, journals and funding possibilities, which would have an important impact on the nature and terms of geographers' contributions to the study of 'development'. In a way the very notion of 'the Tropics' was crucial to the very creation of the discipline of geography, imbuing it with a sense of itself as a 'sternly practical science' (Livingstone, 1992). Geographers themselves, however, also invented areas and regions of study that would concentrate their 'expertise' – frequently in the service of colonial rule (Driver, 1992; Watts, 1993a, 1993b; Smith and Godlewska, 1994). What began to emerge was a new spatialised domain of intellectual enquiry and imagined constructions of otherness around which crystallised a distinctly 'modern' set of truths, assumptions and hierarchies.

Our interest in this chapter is in tracing aspects of how 'colonial' and 'tropical' geography as practised after the Second World War became 'development geography' in the 1960s and 1970s. This is a complex story, related as it is to the varied course of radical geography (Power and Sidaway, 2003). This chapter shows that from the vantage points of 'tropical geography' or later 'development geography', important paradigmatic shifts, such as the rise of quantitative positivist geography, are deeply inscribed within other global dramas, notably decolonisation and socialist revolution. The story here focuses primarily on the British case but this necessarily takes us into other national traditions and into the complex spatiality of what have been variously called the 'tropical/developing/third/postcolonial' worlds. As Driver and Yeoh (2000: 2) have noted, 'to problematise the notion of "the tropical" is necessarily to open an historical enquiry into the term and its enrolment in a variety of scientific, aesthetic and political projects'.

This chapter seeks to explore the invention of a particular geographical domain by geographers, 'the Tropics', and to examine how this defined and shaped early geographical contributions to the theory and practice of postwar development. Important questions need to be asked about this 'enrolment' of geography and geographers in the political projects of intervention in other worlds. How did the *geography* of the idea of tropicality contribute to the invention of 'development' and how was this dependent on interactions with indigenous peoples and places? What is important here is the question of how geographical practices and knowledge provided a set of lenses through which 'the Tropics' were known. Although there are a whole variety of geographical traditions of writing about development, we will primarily examine the work of geographers in Britain and the United States.

Despite the many opportunities presented by colonial possessions and emerging debates about colonial development, few geographers had worked overseas by 1939 (Farmer, 1983). Geographers were thus 'a bit slow off the mark' (Butlin, 2000: 10) and were quite few in number at this time. Many French and American geographers made important contributions to these debates and thus it is necessary to discuss a range of national 'traditions' of geographical enquiry about other worlds and to ask questions about non-western geographers working in the South in contemporary contexts (Cline-Cole, 1999). The extent to which geography as a discipline has itself been decolonised is important here. Inventions and constructions of 'the Tropics' have an important bearing on the contemporary writing of development geographies and on the definition of geographers as 'experts' in 'tropical affairs'. The imagination of tropical worlds and the process of inventing 'others' that inhabit these spaces requires a kind of deconstructive engagement whereby this process is opened up and problematised. What emerged after 1945 was a spatialised domain of intellectual enquiry, research and speculation alongside constructions of tropical difference underscored by notions of certain kinds of 'truths'.

In the nineteenth century the tropical had been enfolded within racial imaginaries, especially in debates about the prospects for white settlement, 'degeneration', evolution and 'races' which were rendered 'scientific' by the cloak of Darwinism. Geography participated in this via studies of climate and land-use, rendered through theories of 'determinism' and 'possibilism' (Livingstone, 1992). After 1945 however, these approaches were fading as 'development' and 'environment' began to inhabit the discourse of 'the Tropics', defining arenas of the discussible for such zones. As a discipline, geography retained a relatively weak position within British universities at the beginning of the twentieth century and this often extended to the newly established colonial universities (Forbes, 1984). Despite limited support for overseas research during the recession years of the 1930s, some geographers had developed an interest in non-western geographies, such as Sir Lawrence Dudley Stamp (1898–1966) who went to Burma (then part of the Indian Empire) as an oil geologist between 1921 and 1923 and was appointed Professor of Geology and Geography at the University of Rangoon in 1923, at the age of 25. Stamp returned to the University of London in 1926 but went on to become one of the most well-known British geographers writing about 'the Tropics' and was knighted in 1965. Stamp's regional and economic geographies of Asia were widely read and tended to run, chapter by chapter, through the main countries and regions or were incredibly descriptive in content (looking at geology, minerals, soils, vegetation, fishing, industry, communications, population and foreign trade). The concept of the region is central throughout, although there is no explicit reference to any conceptual material on regions. The material presented is inventory-like, and processes and relations between the imperial power and the colonised are reduced to a summary of trade links of limited significance (Forbes, 1984).

Prior to 1945 it was geographers in a number of other European countries, however, who first began writing tropical geographies. In France much research focused on Southeast Asia through the important works of Charles Robequain and Pierre Gourou (Gourou, 1931, 1940; Robequain, 1931, 1958). The Germans also had an imperial interest in Southeast Asia, leading to research on Thailand (Credner, 1935) and the Philippines (Kolb, 1942), as did Dutch geographers who worked for the Netherlands Indies Topographical Service and wrote about demographic congestion in Java, for example (Van Valkenberg, 1925).

In each case the connections between geography, geographers and empire was always a close one (even if this was not always acknowledged), as the maintenance of European empires created a number of new opportunities for geographical research. Most of this work was conceived of as part of the quest for the 'Enlightenment' (see Chapter 4) and more scientific knowledge of colonial territories with rich potential for further colonial progress and development. Particular geographical ideas, to do with land-use and agrarian change, population growth and mobility and environmental conservation, run through much of the early thematic approaches of these kinds of geographies. At the end of the Second World War many geographers could easily and immediately see the case for what Gilbert and Steel (1945) called 'colonial geography' (geography in the service of colonialism) and were consequently often uncritical of the impact of colonial rule on indigenous societies. Some geographers, such as Robequain (1958), even dedicated large sections of their texts to 'colonial achievement' and (in general) viewed colonialism as something of a material success.

Pretensions of modernising African and Asian regions in particular structured many of the early postwar assumptions about modernity and progress away from conditions of underdevelopment. This way of viewing non-western regions began to break down and degenerate after 1945 with the widespread desire to overcome European notions and idioms of imperialism such as 'the white man's burden', the 'civilising mission' and the idea of 'civilisation'. In this way social scientists in various western countries began to adopt much less avowedly procolonial notions of 'modernity' and 'development' (Luke, 1991) and to engage with more radical streams of social theory. In addition, these geographers (perhaps more so than many of their contemporaries) were influenced directly by debates and publications in other associated disciplines (such as economics, sociology, anthropology, and the emerging fields of development studies and international relations).

The Second World War was a key turning point in that it had led to much wider, unexpected and in some cases involuntary foreign travel by geographers. A number of British geographers such as C. A. Fisher (who was posted to Singapore) and B. H. Farmer (who served in India, Ceylon and Singapore) found themselves in the service of the military (often the Royal Engineers or sometimes the Inter-Services Topographical Department) and even in some cases became prisoners of war:

> By an accident of war it happened that my first experience of South-east Asia, which may perhaps be described as a case of love at first sight, came in the course of military service in Malaya in 1941–2. Not even the ensuing three-and-a-half years of prison camp life succeeded in appeasing my appetite, which had thus been whetted.
>
> (Fisher, 1964: vii)

Other geographers worked in England for the Naval Intelligence Division (NID) which produced a series of geographical handbooks intended to provide commanding officers in the Navy with information on countries they might be called upon to serve in. Farmer (1983: 73) points out that these handbooks 'were very useful indeed to the first generation of post-war British geographers struggling to write lectures in their demob suits and to prepare themselves for fieldwork overseas'. Indeed, they were incredibly influential and possibly also served to inspire geographers to enter the brave new world of reconstruction in the colonies and the newly independent states of India and Pakistan in particular. The Geographical Section of the NID was formed in 1915 and from then on the purpose of these handbooks was to provide a discussion of the naval, military and political problems of countries such as the Belgian Congo, Turkey and China. The handbooks were used in government posts and embassies throughout the world and were also used widely by the League of Nations (the UN's predecessor). The geography, ethnology, government/administration and resources of each country were thus detailed in a convenient and digestible form (Godfrey, 1944). The series of handbooks produced during the Second World War thereby drew on the work of geographers from universities and military sub-centres and were aimed at military personnel who might be called upon to serve in these locations.

One handbook (NID, 1944a), concerning the Belgian Congo, constructed a time-scale which begins in 1482 and the arrival of Europeans rather than with any of the country's pre-colonial history. It pinpoints 'typical Arabs of the region' in one photograph and provides a simplistic map of 'tribal distribution' in another. The handbook concerning French West Africa which was also written in 1944 (NID, 1944b) offers a simple illustration of a thousand years of West African history condensed into one page, which again has little to say about anything that happened there before 1500. A picture of the 'natives' lining up for a census in Togo is included, as is a picture illustrating progress in the form of improvements to log-hauling, both 'old' and 'new' style. The machinery here will produce a much newer and more modern form of transportation that will supersede the traditions of the 'natives'. A further volume, concerned with French Equatorial Guinea (NID, 1942), includes a map of exploration and occupation which only really begins with the arrival of Portuguese caravels in the fifteenth century. The huts of 'tribal people' are also detailed through a series of pictures and drawings. Another envisions a 'track through the forest', depicting the modern vehicle traversing the landscape. Out of a total of fifty-eight volumes prepared, some sixteen dealt with parts of the European empires and colonies (Butlin, 2000).

Charles Fisher, W. B. Fisher, B. L. Johnson, O. H. K. Spate, J. T. Coppock, A. T. R. Learmonth and R. O. Buchanan were all in active service in India, Africa and the Middle East, and some of the initial work on what would be called 'the developing world' was conducted by this generation of geographers who had held various roles in the military services (Butlin, 2000). Spate and Learmonth in particular both worked on India and Pakistan, Farmer on Ceylon, Johnson on India and Southeast Asia and Fisher on Southeast Asia, all of whom have published in the *Geographical Journal*. Oskar Hermann Khristian Spate (1911–2000) one of the founders of the discipline in Australia, worked at the University of Rangoon and published papers on Burma before being conscripted into the Burma Volunteer Force to work on anti-aircraft artillery (Ward, 2001). Oskar Spate later worked for the Inter-Services Topographical Department, studying India and learning Portuguese from a Goanese teacher. In his later career Spate advised the Punjab Boundary Commission at the time of the partition of India, before writing a 'General and Regional Geography' of India and Pakistan (which was actually banned in India until 1985). From Australia he worked on studies of Papua New Guinea and Melanesia in addition to helping to establish the University of the South Pacific. Spate also advised the Fijian government about the country's economic problems and prospects in 1959. By the time of his retirement in 1976 he had written a dozen books, on India, Australia, Fiji and the wider Pacific in addition to work on exploration and Portuguese poetry (Ward, 2001). The divergent trajectories of postwar geographers such as Spate were soon to produce a range of important approaches and conceptions of Truman's bold new programme for eradicating backwardness.

The first part of this chapter is concerned with the beginnings and early trajectory of this subdiscipline and looks at the movements and writings of Keith Buchanan (a pioneering radical geographer trained at the School of Geography at the University of Birmingham, England, who later worked in Nigeria, Singapore and New Zealand). Examining Buchanan's work, we can formulate a means of exploring the contested and uneven evolution of development geography. The next section of the chapter explores the radicalisation of geography in the 1960s and the emerging engagements with radical streams of social theory. A further section focuses in particular on the emergence of a 'geography of anti-development' in the early 1970s and examines the possibility of decolonising geographies. The conclusions return to the theme of the status of development geography today in the context of key questions of cultural difference and Eurocentrism.

GEOGRAPHIES OF COLONIAL MODERNITY

> What has been of special interest for me has been the extension of post-colonial concerns to the problems of geography.
>
> (Said, 2000: 350)

> Roughly 1955–1975 – from the Bandung conference to the call by the non-aligned movement and group of 77 for a new international economic order – this was a period of extraordinary global change and political realignment. In it, the only recently constituted 'Third World' became the site of intense debates regarding options for 'development'.
>
> (Scott, 1999: 221)

For the next twenty-five years after the end of the Second World War numerous bulky, regional geographies of the non-western world were published, stimulated in large part by the naval handbooks (under the general editorship of H. C. Darby). Spate's (1954) volume on India and Pakistan is a good example of this way of writing about 'the Tropics' and was part of a series of books written by British geographers working in the new universities of Southeast Asia (Dobby, 1950; Fisher,

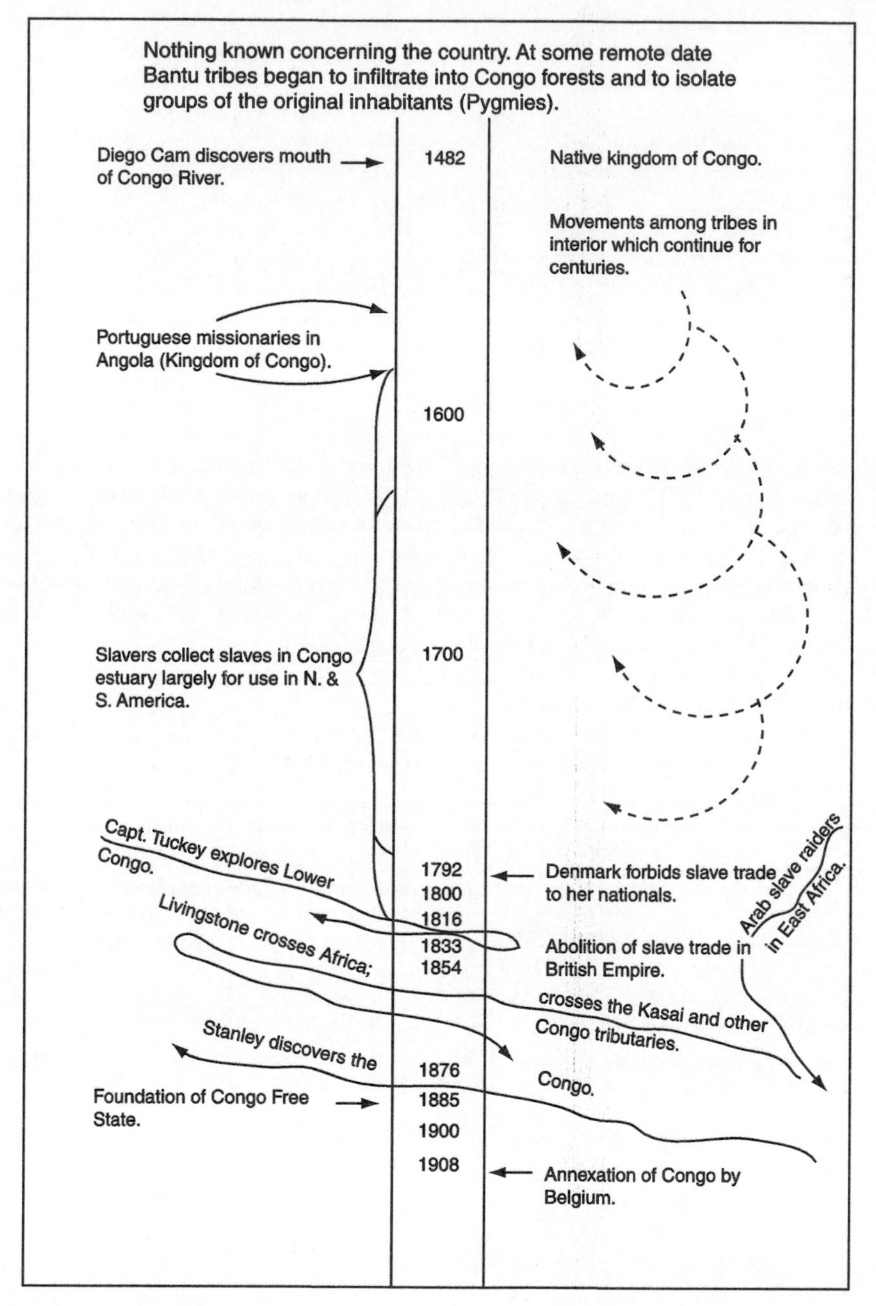

Figure 3.2 A European view of an African 'time-scale'
Source: Naval Intelligence Division (1944a)

1964). Although more information was available for postwar 'regional geography', there was limited methodological or theoretical progress (methodology was largely intuitive and few attempts were made to elaborate a theoretical context). As such a great deal of this work was a 'dreary' descriptive nature (Forbes, 1984) or characterised by a 'deadening stylization and factual aridity' (Farmer, 1973: 10). It is important to remember here that the regions chosen by these authors were not pre-given

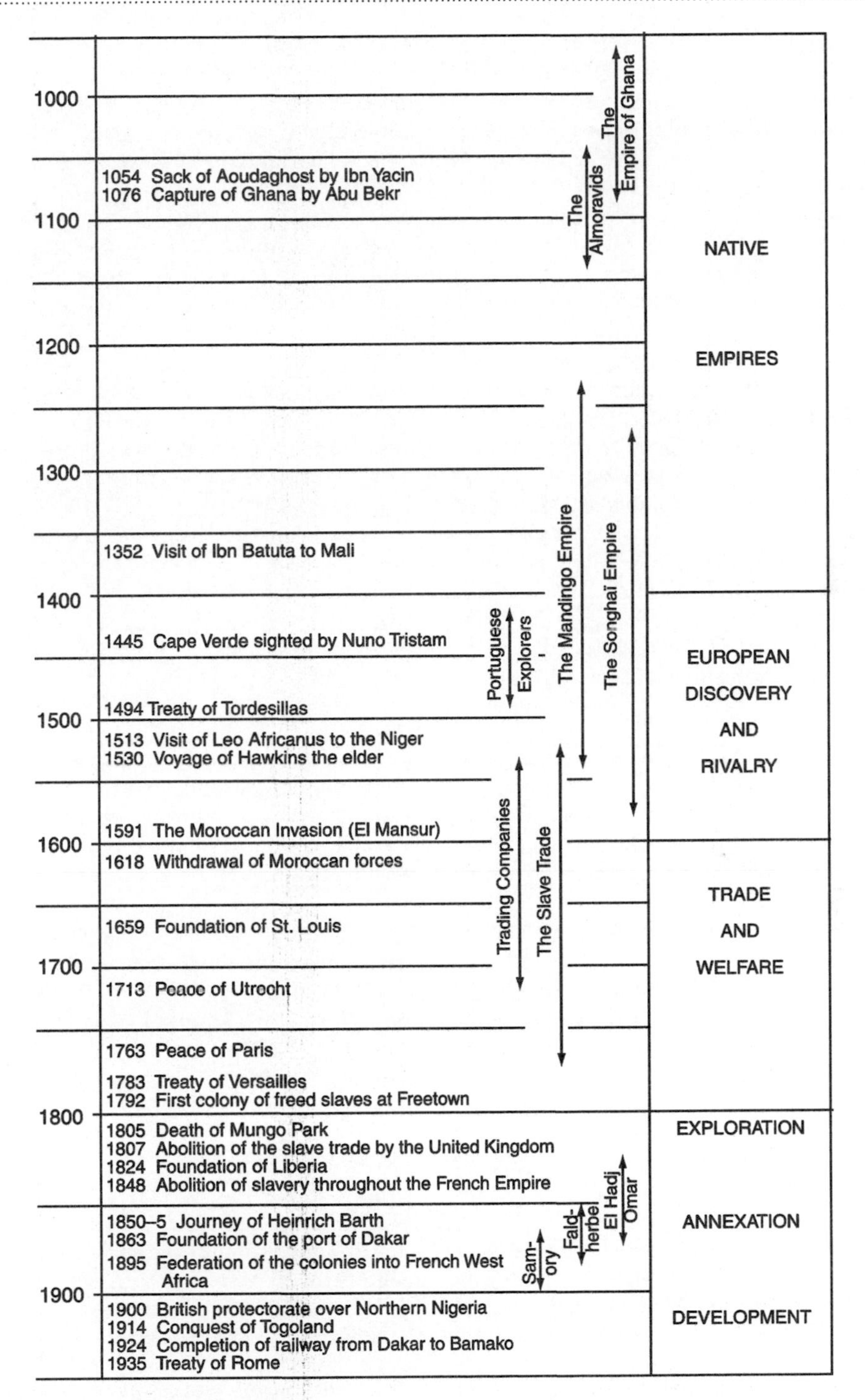

Figure 3.3 A thousand years of West African history
Source: Naval Intelligence Division (1944)

but were shaped by each geographer in pursuit of knowledge ('scientific' rather than indigenous), and thus there were no formal rules for recognising, defining, delimiting or describing the region or its broader position within the imagined 'Tropics'. Some of these inherent contradictions meant that by the end of the 1960s (and in the context of a move to more quantitative techniques) this kind of regional geography had become increasingly unpopular, leading some of its leading advocates to mount a stringent defence of its honour.

Debates revolved around whether regional synthesis was a valid academic study given the tendency to divide the world into unique and seemingly unrelated chunks and not to seek to identify more general laws governing development in all regions. These kinds of simplified images of difference, which assumed away much of the geographic variation within the world, became less popular partly because (despite being very normative) they offered no immediate ethical and socially responsible solutions to social ills (Golledge, 2002). For some geographers what was lacking was a 'dedicated specialization by area':

> By 'dedicated specialization by area' I mean the willingness to devote a lifetime, or the best part of a lifetime, to a specific area, so that one becomes thoroughly familiar with its physical geography and natural resources, its civilizations and culture and its history.
>
> (Farmer, 1973: 11)

Thus the implication was that work conducted before the early 1970s was characterised by a lack of specialisation and depth, or as Farmer (1973: 11) put it, by an 'unsatisfying unreality'. The curiosity aroused by war travels and experiences, coupled with the provision of new funding opportunities for graduate students at British universities to conduct research overseas thus began slowly to increase the interest of geographers in other, distant worlds. In addition, the emerging notions of underdevelopment and underdeveloped areas, in conjunction with the creation of new national and international agencies such as UNESCO and the FAO, also provided a stimulus to geographical research (see Chapter 4). New universities in the colonies such as the University College of East Africa, University College of the Gold Coast and the Universities of Ceylon, Colombo and Malaya in Asia were soon staffed in whole or in part by British geographers. S. J. Baker, for example, worked at the University College in Makerere and W. B. Morgan worked at the University of Ibadan in Nigeria. This body of knowledge thus represented a particular branch of Orientalism and its archives, bringing together a disparate group of academics, colonial functionaries, military men, businessmen, missionaries and adventurers whose primary objective was to reconnoitre the ground they were to occupy and to 'penetrate the consciousness of the peoples to better ensure their subjection by the European powers' (Abdel-Malek, 1977: 296). The postwar revival of Orientalism, area studies and regional geography could be ascribed to similar motives, although by this time the objective was the management of 'Third World peoples' through the invented idea of 'development' (Escobar, 1992). The idea of 'geography militant' (Driver, 2001) is important here: many of these geographies were written by former military personnel and in some cases former prisoners of war, and they were inspired by war experiences and geographical handbooks produced by military institutions. Each expedition and exploration of 'the Tropics' or its constituent parts also often represented a kind of militarised crusade dedicated to progress and modernity, the realisation of 'tropical potential' and to the enlightenment of darkened tropical spaces.

In France and Germany the end of the Second World War also created a new wave of geographical research on 'the Tropics', among which Pierre Gourou's *Les Pays Tropicaux* (Gourou, 1947) was particularly influential and was later published in English (Gourou, 1953). Gourou offered a powerful framework for geographical writing, for geographing 'the Tropics', which impacted on the postwar work of a number of European geographers:

> This book, among other things, presented a framework into which returning warriors could place their experience of the humid tropical environment and of the societies that wrest a living from it; a springboard with which some of them could launch themselves into research in the Tropics. The book strongly influenced my own work in Ceylon in 1951 (Farmer, 1957); though it perhaps gave me too pessimistic a view of tropical potential and the notion that the tropics constitute an intelligible field of study, of which I am now not so sure.
>
> (Farmer, 1983: 75)

This question of the 'springboards' used by geographers and the points of entry where they chose to make their contributions and interventions is important here. We also need to look at the extent to which 'the Tropics' and 'development' had begun to emerge as 'intelligible fields of study'. Published three years earlier, the final (fourth) English-language edition of Gourou's (1953) influential *The Tropical World* had scarcely a word to say on colonialism, its contestation or its legacies. Indeed, Gourou's text in many ways tells us as much about a collective northern-centred worldview as it does about the 'tropical world' it seeks to depict (Arnold, 2000: 6). Given the impact of earlier editions of Gourou's book in fostering 'tropical geography', it is worth pausing here to consider this final edition, published at a time when tropical geography was (as we shall see below) already past its prime. There are many pages on tropical diseases, soils, plantations and population densities, on the importance of scientific knowledge for tropical 'development', and occasional references to the potential for white settlement and the relations between lesser and greater civilisations.

Echoing the ideas of nineteenth-century philosophers, Gourou sees the West as the epitome of 'civilisation', with India and China in secondary roles and the rest more or less outside history – the subjects of 'tropical geography'. Such ideologies were soon being contested through the rise of a number of national liberation and anti-imperialist movements in a variety of locations. An important backdrop to this was the widening struggle in the remaining European colonies of Africa (e.g. Rhodesia, Guinea Bissau and Angola), and the widening conflict and deepening American intervention in Indochina. In addition, the rise of Third World nationalisms which were associated with revolutionary pressures across much of Asia, Africa and the Americas was also significant. Later on in Gourou's book, the rubber plantations of Malaya are praised for their positive environmental and commercial impacts. Next to the type of graphic portrayal of population density that peppers the text (and which became recurrent features of tropical geographies), readers learn that 'as most of the land was underutilized, the establishment of plantations gave rise to no territorial problems' (Gourou, 1966: 173). No reference is made to the fact that these were already in the independent state of Malaysia, or to the fact that after Malaysian independence in 1957:

> London found, after the spring of 1948, the longest and fiercest resistance it ever faced in the history of its modern empire – from a Malayan communist party which grew out of the Malayan People's Anti-Japanese Army.
>
> (Anderson, 1998: 7)

It is useful here to consider briefly some of the words and geographical references which were made in some of the titles and naming of research in 'the Tropics' (see Box 3.1). This partly reveals the predominant focus on the regions of 'the Tropics' (Southeast Asia or West Africa) as the primary focus of analysis. Authors often offered a regional, systematic, 'short' or tropical geography. Often the 'focus' included whole continents in a single study (usually Africa) or authors claimed to know whole worlds (e.g. the 'Tropical' or 'Southeast Asian' world). The geopolitics of naming transnational regions was, however, usually led by military strategies, as in Southeast Asia and the Middle East (Lewis and Wigen, 1997; Philpott, 2000; Sidaway, 2002). Concerns ranged from a region's 'fundamental characteristics' or its physical geography to later work in the 1960s on economic development and the spatial dynamics of modernisation.

In the USA, geographers such as William H. Haas, who worked at North-Western University and conducted a Rural Land Classification Project in Puerto Rico in 1952 (Haas, 1952), drew their inspiration from the work of writers such as Malcolm J. Proudfoot, Preston E. James, L. Dudley Stamp and others. Tropical studies of this kind also adopted a sort of 'inventory mapping' approach (Proudfoot, 1952). Malcolm Jarvis Proudfoot (1907–1955) worked for the US War Production Board between 1943 and 1945 and also served with the allied expeditionary forces in Europe, and later worked as a consultant to the Caribbean Commission in Trinidad (then the British West Indies) on 'problems of surplus population, employment opportunity and migration' (Jones, 1957: 4). Manfred Schaffer was another US geographer (based at North-Western University) who was influenced by these same types of approach. Writing in 1965, in a piece entitled 'The competitive position of the Port of Durban' (Schaffer, 1965), his contribution was based on fieldwork in Johannesburg and Lourenço Marques (1959–1961). In the Acknowledgements, a whole range of influential US geographers such as William A. Hance (at Columbia) and Edward J. Taaffe (Ohio) are credited for their guidance. Again

BOX 3.1

Key contributions to 'tropical geography': a selection of authors and titles 1934 to 1973

Barbour, K. M. (1961) *The Republic of Sudan: A Regional Geography.*
Benneh, G. (1972) 'Systems of agriculture in Tropical Africa'.
Boateng, E. A. (1966) *A Geography of Ghana.*
Buchanan, K. M. (1967) *The Southeast Asian World: An Introductory Essay.*
Buchanan, K. M. and Pugh, J. C. (1955) *Land and People in Nigeria: The Human Geography of Nigeria and its Environmental Background.*
De Blij, H. (1964) *A Geography of Sub-Saharan Africa.*
Dobby, E. H. G. (1950) *South-East Asia.*
Fisher, C. A. (1964) *South-East Asia.*
Fitzgerald, W. (1967) *Africa: A Social, Economic and Political Geography of its Major Regions.*
Forde, C. D. (1934) *Habitat, Economy and Society: A Geographical Interpretation to Ethnology.*
Gould, P. (1970) 'Tanzania 1920–1963: the spatial impress of the modernization process'.
Gourou, P. (1953) *The Tropical World.*
Grove, A. T. (1970) *Africa South of the Sahara.*
Hance, W. A. (1964) *The Geography of Modern Africa.*
Harrison-Church, R. (1971) *Africa and the Islands.*
Morgan, W. T. (1973) *East Africa.*
Mountjoy, A. B. and Embleton, C. (1965) *Africa: A Geographical Study.*
Pollock, N. C. (1968) *Africa: A Systematic Regional Geography.*
Prothero, R. M. (ed.) (1969) *A Geography of Africa: Regional Essays on Fundamental Characteristics, Issues and Problems.*
Pugh, J. C. and Perry A. E. (1960) *A Short Geography of West Africa.*
Riddell, B. (1970) *The Spatial Dynamics of Modernization in Sierra Leone.*
Robequain, C. (1944) *The Economic Development of French Indo-China.*
Spate, O. H. K. (1954) *India and Pakistan.*
Stamp, L. D. (1953) *Africa: A Study in Tropical Development.*
Steel, R. W. and Prothero, R. M. (Eds.) (1964) *Geographers and the Tropics: Liverpool Essays.*
Varley, W. J. and White, H. P. (1958) *The Geography of Ghana.*
Wellington, J. H. (1955) *Southern Africa: A Geographical Study, Volume One, Physical Geography.*

in a rather apolitical manner this study looks at the efficiency of the Port of Durban and at its 'rich hinterlands', and explores its 'traffic-generating abilities' (Schaffer, 1965).

One important exception to these kinds of approach came from Buchanan and Pugh's *Land and People in Nigeria* (Buchanan and Pugh, 1955). The Foreword to the book, written by L. D. Stamp, affirms the importance of knowledge to enlightenment and progress and, rather oddly for Stamp, asserts that 'no country can be treated as a blank sheet of paper' (a feature of many of Stamp's own tropical geographies) (Stamp, in Buchanan and Pugh, 1955: v). However, the book argued for what could be achieved by 'harmonious inter-racial cooperation' rather than colonial subjugation of one group by another. Further, more so than many of its predecessors, it does explore some of the variations in human responses and makes some interesting connections between the authors' British and African experiences. For Stamp, the combination of these different worlds meant that the authors were well placed to study factors 'governing development':

> For their task the authors are particularly well qualified . . . [Buchanan] studied in minute detail the agricultural geography of Worcester for the Land Utilisation Survey of Britain. It was natural that when he moved to South Africa he should initiate similar studies, and when appointed Head of the Department of Geography at Ibadan, he turned his attention to almost untouched fields of study in Nigeria.
>
> (Stamp, in Buchanan and Pugh, 1955: v)

Although the work is still intended for use by officials in British colonial service, Buchanan and Pugh (1955) appear much less directly motivated than many of their contemporaries (e.g. Stamp) by this desire to venture into 'untouched fields of study', often drawing upon local sources of scholarship and regularly trying to tease out the various problems and aspirations of different human communities. Wellington's *Southern Africa: A Geographical Study* (also published in 1955) is dedicated not to the people of the region about whom he wrote but 'to the glory of God and to my fellow workers for the progress of Africa'. Similarly, Stamp (1948) affectionately dedicated his main (encyclopedic) book on Asia to his wife 'in memory of bullock-cart days and irrawaddy nights'.

Keith Buchanan's work provides a rare and striking contrast to much of the research that was conducted in the 1960s (see Box 3.2). His death in 1998 (at the age of 78) produced several obituaries and retrospectives on his wide *oeuvre*, including a consideration of his satirical critique (written under the name of an 'exiled celt') of apolitical regional geographies which was published in 1968 (Watters, 1998; Wise and Johnston, 1999; Johnston et al, 2000; Moran, 2000). After graduating from the University of Birmingham in 1940 Buchanan moved to Africa, first to the University of Natal (Durban) (1948) and then to Ibadan in Nigeria (1948–1951), before returning briefly to the UK (LSE 1951–1953) and then taking up a post at Victoria University in Wellington, New Zealand, where he spent the rest of his academic career (Buchanan retired early in 1975 and went to live in Wales for a time – where he continued to write journalism and political commentary – before returning to New Zealand). Subsequently, Buchanan's years in Africa coincided with the beginnings of anti-colonial nationalism. A brief encounter with apartheid South Africa sharpened his distaste for colonialism and his understandings of the logic of racism. Buchanan's later experience in Nigeria is also significant, for at Ibadan he encountered an intellectual milieu where nationalist ideas were increasingly influential.

Although there was not as yet an African faculty in the department, geography too registered something of this shifting orientation – of which Buchanan and Pugh's (1955) *Land and People in Nigeria: The Human Geography of Nigeria and its Environmental Background* is an early manifestation. And while the full radical expression would only come later, Buchanan's extensive travels in late colonial Nigeria helped to crystallise his later critiques of colonial claims. In the words of Akin Mabogunje (2002: 1–2), then an undergraduate student at Ibadan:

> Buchanan . . . saw the country as . . . being underdeveloped by colonialism and he was concerned about how to transform the situation . . . his lectures always stressed the importance of the human agent in transforming the environment and therefore of the need to enhance the quality of this agent through education. We always felt as students that he was already way left of centre in his general unconventional comportment such as jumping through the window into the lecture room when he felt he was late. Even though his lectures were not stridently radical in the sense of being Marxist or socialist, they were not patently pro-colonial.

Perhaps, too, Buchanan's earlier training and experience at Birmingham of geographical fieldwork and local contextual study generated scepticism towards the kinds of ('scientific') knowledge that he would later dismiss as spurious objectivism in the service of imperial and neocolonial power. The continued attachment to fieldwork in geography suggested that there were growing 'doubts about the universality of experience upon which positivist accounts of the other are ordered' (Philpott, 2000: 31). A commitment to fieldwork, in and on the margins of tropical geography, is one way in which the field maintained links to wider geographical method and debates. Jim Blaut (1953: 37) extended this concern in the first issue of *The Malayan Journal of Tropical Geography* where he developed a 'micro-geography' approach (then quite popular in the USA) which he saw as particularly applicable in situations where geographers cannot obtain adequate background data on the economic, cultural or environmental characteristics of a region.

After he left Ibadan, Keith Buchanan soon became a prolific author. In the early 1950s, amidst publications derived from his work in Birmingham, such as agricultural geographies of the English Midlands (1950a), are early critical papers on the status of 'coloureds' and 'Indians' in South Africa (1950b), on 'internal' colonialism in Nigeria (1953), followed, in the 1960s, by a steady flow of papers, reviews and essays on China, Southeast Asia, revolution, development and environment, amidst occasional works elaborating the framework of internal colonialism with regard to the status of Britain's 'Celtic Fringe' (Ireland, Wales, Scotland and Cornwall). Buchanan was clearly influenced by these revolutionary transformations and upheavals, and he saw them as directly relevant to the way in which he understood and sought to theorise the 'majority world'. Later, in Singapore, where Buchanan

BOX 3.2

Geographies of modernisation and dependency

Debates about 'modernisation' and 'dependency' also had a major impact on geographers (see Chapter 4). Critiquing a paper by De Sousa and Porter published in *Antipode*, David Slater (1976) lambastes the authors (also geographers) for drawing upon modernisation theory which consequently produced an analysis that suffered from 'a debilitating vagueness in the presentation of ideas' (Slater, 1976: 89). The authors were taken to task for their vagueness about colonialism and colonial social relations in particular, producing an ideology of geography that excluded an analysis of class. They are challenged in particular for their suggestion that quantitative development geographies, working in conjunction with government organisations, would blaze a trail for committed humanist researchers:

> [S]ubordinating geographical enquiry to the needs of the state presupposes that the state in capitalist society is somehow part of the solution to the 'development problem' whereas in fact it is an essential part of the problem itself.
>
> (Slater, 1976: 92)

Brookfield (1975) is also criticised by Slater for failing to 'successfully break away from orthodox geographical formulations' (Slater, 1976: 92). His key contention is that evidence had shown increasingly that 'poverty was created and institutionalised through external capitalist penetration' (Slater, 1976: 89). Slater argues that it was necessary to also ask which social class controls power and to question the reality of the relations between states and international capital. For their part De Sousa and Porter (1976) responded that Slater was viewing development 'solely in terms of socialism' and were critical of Slater's adoption of Marxist theory, outlining the dangers of drawing too heavily on Marx's writings about Europe written a century before. In particular they cautioned that '[a] straight transfer of Marxian class analysis to the African situation is likely to obscure more than it reveals' (De Sousa and Porter, 1976: 96).

McGee (1974) also criticised some of these geographies of modernisation which had ascribed quite negative interpretations to the meanings of 'tradition' which he argued needed to be rescued by the formation of a 'geography of anti-development'. Although this is not really elaborated on much by McGee in his paper, it is clear that he rejected the idea of a 'neat unilinear change from traditional to modern systems, even within such a dynamic growing economic complex as Hong Kong' (McGee, 1974: 41). Gilbert (1971), Connell (1971), Logan (1972), Slater (1973), Blaut (1973) and Buchanan (1972) had thus all begun to contest the hegemony of modernisation discourse as an approach to the 'geography of development,' representing a new generation of geographers that was concerned 'with the spatial manifestations of Western impacts on indigenous society' (Logan, 1972: 176). McGee contested the primacy of the 'American–European' experience in modernisation discourses and the assumption that 'modernization variables such as capital, technology and institutions, will operate in the same manner in LDCs' (McGee, 1974: 31). The focus instead shifted towards a concern with the relations between developed and developing regions and on economic structures and relations between them. The models of 'tradition' and 'modernity' advanced by the geographers of modernisation presupposed a consistent body of norms and values and quite static and geographically homogeneous traditional societies which were criticised from the early 1970s onwards. Far from the perceived clash and conflict between traditional and modern systems (leading to the disappearance of the latter) the 'new geographers', as they are referred to by McGee (1974), began to recognise that it was also important to look at the 'emergence of syncretic forms of behaviours and organization' (McGee, 1974: 33). The conclusions drawn about the causes of poverty and underdevelopment were often portrayed in quite blunt and uncompromising terms however.

worked as an external examiner of the Department of Geography at what is now the National University of Singapore (1967–1970), he clashed directly with a department that retained a substantial foreign (colonial) staffing and a commitment to tropical geography conceived in the tradition of regional analysis (Power and Sidaway, 2003, forthcoming). Buchanan's work, however, in particular the shift from a traditional regional geography to more systematic treatments of development and underdevelopment (Buchanan, 1967a), made a significant impact on students reading for geography degrees in the department, even if few contemporaries embraced the more radical aspects of his work.

In some of his teaching and research (and particularly through his experiences of fieldwork) Buchanan sought to disrupt and displace the familiar tropes of development as western modernity, looking beyond the imagined geographies outlined by many of his contemporaries to alternative readings and interpretations. In a similar vein, Blaut (1953), who had also worked in Singapore, shared this experience of having been profoundly influenced by his research in the field. Both Buchanan and Blaut found in fieldwork a way of disrupting the conventional narrations of development they had encountered, transgressing some of the spatialised boundaries that this puts in place and the imagined geographies of communism. A few years previously, Buchanan had published three papers in *New Left Review* (*NLR*) (1963a, 1963b, 1966) shortly after contributing to the journal *Monthly Review* (Buchanan, 1963c). Before the 1990s (when interventions and reviews of the work of David Harvey, Mike Davies and Doreen Massey were published) no other geographer had ever been published in the *New Left Review*, a journal that soon became an influential academic/cultural marker of the New Left in the UK. Buchanan's articles are strikingly at odds with the then dominant tones of tropical geography and the vanguard of geographies of modernisation (Power and Sidaway, 2003, forthcoming).

In many ways it is possible that Buchanan's experiences in China were a key turning point in his understanding and conceptualisation of 'development'. In the same year that Buchanan responded to some of the criticisms of his work in *NLR* his book *The Chinese People and the Chinese Earth* was published (Buchanan, 1966), which focused in some detail on the culture and histories of Chinese people to a greater extent than had been the norm with many traditional regional geographies of Asia. The following year Buchanan published *Out of Asia: Essays on Asian Themes* (1967a), which also seemed to further extend his wish for a more imaginative study of regions and their peoples and cultures. The year 1968 was an important one for China, Vietnam and also the USSR (about which Buchanan knew a great deal according to David Hooson's obituary in 1998), while a 'new geography' was also emerging with a concern for quantitative knowledge. For Buchanan, these statistics and quantitative methods were of less importance than his 'radical, humanist concern for the people and places he wrote about' (Hooson, 2000: 630) and the revolutionary insurgencies across the Third World (see Chapter 5) whose contours he sought to map and understand. According to Mansell Prothero and Andrew Learmonth, Buchanan's work on South Africa and Nigeria was not the only form of geographical research on political geographies in the 1950s and 1960s; Michael Barbour's work on international boundaries in Africa and his work on Sudan, Barry Floyd's work on land apportionment in Southern Rhodesia and Victor Prescott's work on boundaries in Nigeria were other (if often much less critical) examples. None the less, Butlin (2000: 13) warns of the danger of creating a 'false counterfactual historiography, both in terms of the amount of work done and the opportunities and the context'. Thus, despite the dreary and descriptive nature of many early 'tropical geographies', geographers such as Buchanan were beginning to explore other, more radical, approaches.

Figure 3.4 Colonial-style Singaporean building
Source: Lily Kong

FROM TROPICAL TO DEVELOPMENT GEOGRAPHY (VIA IMPERIALISM)

> It may be asked why British geographers should devote themselves to overseas areas. Cannot India, it may be argued, be left to the Indians?
>
> (Farmer, 1973: 13)

Published at the end of the decade, Prothero's (1969a) edited collection on the *Geography of Africa* was a good example of the sometimes superficial interpretation of 'backwardness'. Less critical than Keith Buchanan's work, it consists of ten 'regional essays' and begins from the assumption that 'our former ignorance of Africa has been replaced by a great deal of enlightenment during recent decades' (Prothero, 1969a: xiii). All the authors in this collection had taught and worked for considerable periods in Africa and the text reflects their concentration in British Imperial Africa. In the first chapter of the book 'Understanding Africa' (written by Prothero himself) the continent is firmly located in the 'Third World' which is seen to derive unity and coherence from comparable 'problems'. 'Underdevelopment' characterises much of Africa, and the continent is seen to face five major thematic questions; environmental, historical, social, political and economic. The influence of physical factors predominates in a way that frequently verges on environmental determinism. Climates rather than capitalism impoverish the people of 'the Tropics' in these accounts. Historical factors are seen to consist of large-scale mineral discoveries, historical research on deserts and the historical particularity of Egypt and Ethiopia. Neither in the first section nor in subsequent sections on political and economic change is imperial rule and colonial domination elaborated upon as a major influence on African geographies. The reasons for this are complex and multifaceted, however. Geographers, according to Butlin (2000:12), 'were overtaken by events'. New technologies and new agricultural techniques were seen as the way forward for African rural economies. Where colonialism *is* seen as relevant the focus is on white settler states such as Rhodesia and the needs of white colonists rather than on African stories of colonial rule.

In an interesting review of publications produced through the Institute of British Geographers (IBG) between 1946 and 1960, R. W. Steel (himself an Africanist geographer of some distinction) looked at the geography of areas covered by the Institute of British Geographers publications and discerned an uneven spatial distribution of papers (Steel, 1961). Steel identified a strong concentration on the British Isles and concluded that the IBG journal *Transactions* had included only ten contributions on Africa during this period (which he argued reflected university expansion there), a further two for Latin America, three for South Asia and two for Southeast Asia, and none for China, Australasia other parts of East Asia. Steel's review then highlighted the incredible geographical unevenness of this kind of work in the overseas territories, locating Africa as the predominant focus of attention. What seems particularly interesting here is the persistent lacunae formed by the lack of interest among geographers in the Middle East. In many other disciplines the Gulf figured prominently in debates about world wars, capitalist development, imperialism and resource economics (Vitalis, 2002) but hardly at all in geography.

Another important dimension of the history of tropical geographies is the impact their construction had on the British universities where their authors held teaching and research positions. Steel and Prothero's *Geographers and 'the Tropics': Liverpool Essays* (1964) highlights the importance of particular institutions and localities in the making of this kind of geographical tradition. The book opens with an essay by R. W. Steel on Liverpool and 'the Tropics' which focuses on Liverpool as a decisive link in British 'tropical trade'. Different universities prioritised the study of particular regions at different times, but for Liverpool the 'tropical world' was deeply inscribed in the city's cultural landscapes, in its economy and across its history.

This importance of particular places to the making of early geographies of development was further highlighted when, under the chairmanship of Sir William Hayter, the British University Grants Committee became concerned with the state of regional studies in British universities from the early 1960s and in particular with Oriental, Slavonic, East European and African studies. The Committee's report of 1961 argues that while enough was being done in terms of language and classical literature, 'modern area studies' (including geography) had been neglected and Britain was in danger of falling behind the USA, France, Germany and the USSR in terms of academic studies of the 'Third World' (University Grants Committee, 1961). The Hayter Committee advocated that a moderate amount of finance be made available for this kind of research and that centres be created within British universities specifically dedicated to this purpose and adopting an interdisciplinary framework. Further support to geographers interested in researching 'the Tropics' came with the establishment of a geography department within the School of Oriental and African Studies in London in 1966 under Charles Fisher (who had worked in Southeast Asia). Fisher's obituary notice in the *Geographical Journal* for July 1982 emphasises his 'anti-parochial vision' and highlights the significance of his 'first hand experience of the problems of de-colonization in the tropical world' (1982: 297).

The Hayter Committee drew heavily on their review of area studies centres in the United States which had been set up partly to meet wartime needs for intelligence reports (Farmer, 1983; Forbes, 1984). In America, many geographers worked for the Office of Strategic Services (OSS) and provided intelligence analysis during the Second World War, working with political scientists, historians and economists such as Walt Whitman Rostow (CIA, 2001). The OSS's predecessor, the Office of the Coordinator of Information, had also recruited Americans who had travelled abroad or studied world affairs, many of them geographers. Several geographers went on to work for the successor to the OSS, the Central Intelligence Agency (CIA). As Peet (1977b: 10) has argued, the 1950s in the USA was a key period for spatial theory and scientific methods, as geographers 'responded to the call for spatial decision makers like old soldiers returning to a long lost army'.

The creation of these area studies centres was intimately bound up with Cold War geopolitics (Pletsch, 1981; Luke, 1991) and impacted upon a wide range of social sciences. Geographers soon became active in many such centres and also found roles in newly established schools of development studies. Many others remained convinced, however, that geographies of 'the Tropics' were of little interest to their contemporaries (who supposedly preferred Somerset to Somalia) and bemoaned the continued absence of sufficient scholarship in this area (Woodridge, 1950; Steel, 1961; Farmer, 1973). Some critics had more fundamental problems with the very conception of these centres, for example H. Brookfield, who argued that area studies, as then practised (particularly in the USA), was 'colonially conceived' (Brookfield, 1973b: 1) and that the 'appropriate centres of Asian studies are in Asia, or Pacific Studies in the Pacific'. Until the 1940s the study of foreign cultures in the USA remained an intellectual enterprise pursued largely by 'amateur enthusiasts' (McCaughey, 1984; Bilgin and Morton, 2002).

A lack of area experts to inform policy was noticed during the 1940s as 'the War and the Armed Forces reawakened students to study geography and to learn about the "funny people" of the world' (Wallerstein, 1997: 199–200). In the 1950s six area studies associations were formed in the USA as social scientists took up the mantle laid down by President Truman's injunction to reach out to the 'backward' regions of the world in his Point IV Program (Bilgin and Morton, 2002). Thus specialist research into 'backward' regions began to increase in volume as more disciplinary generalists entered the debate, 'initiating the three worlds schema and a focus on states within the crude classification of "Third World" studies' (see Chapter 5). During this early period of the Cold War then, development and area studies programmes became increasingly initiated by US state security discourses which began a certain kind of relationship between scholarship and policy-making centred around debates on 'development' and 'modernisation'. These attempts could thus be viewed as designed to produce knowledge which would enable the maintenance of political control over societies that threatened the institutional capacities of certain states (Bilgin and Morton, 2002). A new insistence on the importance of strong 'capabilities' for postcolonial states thus soon began to be legitimised.

Not surprisingly, much of the geographic scholarship of the 1950s and 1960s was framed by some variant of modernisation theory, or the presumption that processes of modernity were shaping indigenous institutions and practices (see Chapter 4). Many geographers interested in Africa sought to model modernisation surfaces and attempted to map patterns of modernity by charting the diffusion of indices of modernity (such as schools or mailboxes) through the settlement pattern (Soja, 1968; Riddell, 1970). This work at best raised only limited questions about the legacies of colonial transport systems or the character of African urbanism. Geographers (some local, some expatriate) also worked on sub-national studies, sometimes adopting ethnographic approaches based around careful local empiricist research. One important focus was marketing (Hodder, 1965; Scott, 1972), another the question of urbanisation and human mobility (Prothero, 1957; Mabogunje, 1962, 1968). For both British and French geographers land-use was perhaps the most important object of study (Sautter and Pelissier, 1962; Hunter, 1967; Morgan, 1969; Benneh, 1972). The more influential work focused increasingly on 'modernisation', seen less as a tropical phenomenon and more as part of a *universal* process of 'development' (Gould, 1960, 1970; Soja, 1968; Riddell, 1970; Berry, 1973). Peter Gould's study of transportation patterns in Ghana (1960) is a good example of this approach. Written while Gould was at North-Western University in 1960, the study is dedicated to his 'Geography

Master' at Nautical College in Berkshire. Also influenced by the work of US geographer E. J. Taaffe, it explores shipments, commodity flows, road and rail networks, passenger and traffic flows and imports/exports for Ghana. This work starts historically with the arrival of Portuguese traders and theoretically with the modernisation approaches to economic development where transport is singled out as being one of the most important keys to the development process. Geography would, therefore, help to 'unlock' such potential.

Despite the civil war in Nigeria in 1966, despite the emergence of political corruption (very relevant to the pursuit of tropical potential) and despite the manipulation of ethnicity by leaders such as Amin in Uganda, Bokassa in Central Africa and Nguema in Equatorial Guinea, geographers had little or nothing to say about these themes. Many of the contributions to what was becoming known as 'development geography' lacked any major theoretical undercurrents (Forbes, 1984) and often concentrated on broad, traditionally defined questions. During this time however, the number of geographers based in the South began to increase, in countries such as Nigeria, partly as a consequence of the establishment of new universities in what was becoming the British Commonwealth. These geographers had an increasingly important impact on the formation of national development policy (particularly in the area of planning) from the late 1960s and early 1970s onwards (Cline-Cole, 1999). As the numbers of geographers from non-western regions began to increase, so the theoretical impoverishment of development geography was soon brought into stark relief:

> There was no substantial, explicit discussion of theory until the early 1970s . . . but, by then, there had been a notable shift in the nature of development studies and, simultaneously, in development geography.
>
> (Forbes, 1984: 58)

Major shifts in development thinking were thus beginning to influence researchers in a number of disciplines and the way they viewed the hallowed 'development process'. In the 1960s this began to change the nature of the discipline as geographers were 'propelled into a heightened state of social awareness' by political and economic events around the world (Peet, 1977b: 10). Dependency theorists such as A. G. Frank had begun to offer popular structuralist explanations of the causes of 'underdevelopment', and the idea of development as 'economic growth' had been subjected to radical critique by theorists who sought to locate the production of dependent relations within the nature of the capitalist world economy.

Although we will later consider geographers' readings of and contributions to dependency theories, we must leave aside the wider evolution and trajectory of dependency theory here. Brazilian geographers, for example, provided one pathway through which dependency ideas and geography were articulated – but the points of contact and circulation were complex. Development promised 'trickle-down' effects which never materialised and hence attention began to shift to the material sphere and the provision of basic needs. Development programmes, however, continued to fail to induce substantive improvements in the well-being of the world's poorest countries, which were gradually seen less as anomalies of modernity or 'special' problem cases and more as part of a global system which actively produces inequality.

FROM 'PIOUS EUROCENTRISM' TO GEOGRAPHIES OF 'ANTI-DEVELOPMENT'

These shifts in development thinking during the 1960s and 1970s had a profound impact on geographers, particularly after the establishment in 1969 of the radical journal *Antipode*, which soon began to publish important articles about geography and uneven development (Slater, 1973, 1976, 1977; McGee, 1974; Peet, 1977c). This journal was founded, upon principles of radical geography, by a group of faculty and students at Clark University in Worcester, Massachusetts in 1969 (Peet, 1977a). Ben Wisner, editor of the first issue, argued that value questions needed to be asked within geography and that geographers should focus on institutions and 'their rates and qualities of change' (Wisner, cited in Peet, 1977b: 11). The discipline's traditional concerns with descriptive and apolitical methods were thus increasingly being seen as 'eclectic irrelevancies', quirks of interest which might include, for example, a study of the 'Goldfish Industry of Martinsville, West Virginia'! (Peet, 1977b: 11). The importance of 'relevancy' was thus being increasingly highlighted by 'action-oriented geographers' in the words of Richard Peet, editor of *Antipode* in the mid-1970s.

Many of these geographers met in the United States from 1967 onwards at the annual geography conferences, but the discipline of geography in the USA was slow to respond to their radical discussions, dominated as it was until the early 1970s by 'the quantifiers' and a kind of ephemeral spatial science. The work of Buchanan, for example, only really became known in the USA from about 1972 onwards (Peet, 1977b), but his work and that of other geographers such as Jim Blaut (1970, 1973) began to illustrate a new awareness of the ethnocentrism of 'western development' (issuing from a 'white North' as Buchanan put it). In addition, David Harvey's writings about social justice (1972) raised associated issues about cities and the unevenness of capitalist development. Suddenly, disenfranchised groups were offered geographical 'expertise' by a community of geographers seeking to become less engaged with states and other monopolisers of power.

Some geographers studied Wales (Buchanan, 1977a) and the Republic of Ireland (Regan and Walsh, 1977), illustrating how cases of poverty and underdevelopment were not exclusive to the 'Third World'. Radical geographers also raised questions about warfare in North Vietnam (Lacoste, 1973) and the geopolitics of imperialism (Abdel-Malek, 1977). Other accounts (Cannon, 1975) were concerned with exploring the geographies of dependence. One of these articles, by Terry McGee, was subtitled 'Towards a geography of anti-development' (McGee, 1974), which the author recognises must have seemed 'positively reactionary' at the time. The phrase 'anti-development' only really appears again in the final sentence of the paper. McGee explains how geographers interested in development and modernisation needed to rethink their typically narrow and negative approach to 'tradition' in non-western countries:

> This then is an appeal for geographers who work in LDCs to return to the grassroots; to reassert the 'tradition' and core of the discipline as a field subject, not simply a data manipulation subject; in this fashion I believe we can contribute.
>
> (McGee, 1974: 42)

Geographical research in the years leading up to McGee's article is depicted rightly as based around a notion of the discipline as some kind of 'data manipulation' subject, adrift from the 'grassroots' concerns of LDCs (the language used here was very typical of that used in a number of development debates at this time). McGee also talks of the importance of understanding development 'in each ecological milieu' (McGee, 1974: 42). Instead this work called for a broader 'geography of anti-development', arguing that most of the geographers of the western world 'have been educated in the liberal ethos that development is a necessity' (McGee, 1974: 30) as if they had all somehow been collectively duped. In his article McGee criticised the modernisation approach for its reassertion of the primacy of the 'American-European experience' (McGee, 1974: 31) and argued that it was important to remember that other regions had different 'temporal, demographic, economic and social dimensions to that of the American-European experience' (McGee, 1974: 31). Gould's (1970) uncritical list of modernisation variables presented in his study of Tanzania included, for example, such things as the number of high court circuits, churches, the size of the prison population and the number of criminal cases. This approach was criticised by Brookfield (1973b) and McGee (1974) for failing to recognise that these 'variables' were conditioned in part by the legacies of colonialism and because there 'was no real questioning of whether these so-called "modernization variables" are functioning in the same manner in LDCs as they are supposed to do in the West' (McGee, 1974: 32). Quite interestingly, McGee refers to these geographers as the 'distance-decay geographers' that:

> Are like pre-historians plotting the distribution of some artefact whose function they assume from analogy with a similar artefact found in another culture.
>
> (McGee, 1974: 32)

Thus some geographers were looking to reconstruct the history and geography of non-western areas from the artefacts of their own past, from the experience of their home countries and western regions. McGee tries to argue that through his work in Hong Kong and the informal sector it had become clear that there was 'no neat unilinear change from traditional to modern systems' (McGee, 1974: 41). The hybrid forms of tradition/modernity in 'the Tropics', he argued, had not been adequately theorised by geographers and deserved further attention. While McGee was pointing to other directions that would emerge more fully only in the 1990s amidst debates about 'postcolonialisms', 'post-development' and (alternative) 'modernities' (which we explore later in the book), the critique of geographies of modernisation proved effective. By the 1980s, renditions of 'development geography' could not credibly ignore power relations, dependency and imperialist legacies. At the same time 'modernity' and 'modernisation' were increasingly being seen as multiple, hybrid and heterogeneous. Radical geographers had pointed to resistance, disjuncture and difference amidst the enduring power relations (Buchanan was not reticent to call these 'imperialist') of domination and dependency.

The philosophical and theoretical foundations for this radicalism were found in the work of Marx, which radical geographers began to study seriously in 1972 (Peet, 1977b: 21). As outlined in the work of Karl Marx, capitalism was commonly understood as a system of economic production which thrives on the inequality arising from the continual need to develop production forces (labour power, raw materials, transport and machinery) and to make commodities which can be sold at a profit. This system is shown in Marx's work to be inherently growth-oriented and prone to crisis. Many geographers influenced by Marxist thought have thus illustrated how capitalism attempts to surmount problems by expanding and shifting production into new territories and forming new international divisions of labour. Indeed, the very expansion of European empires into Africa, Asia and the Americas from the fifteenth century onwards is seen as driven by this desire to expand capitalism, to extend production into new territories, to open up new markets for capitalist goods and to find new global sources of profit. The development of central regions was therefore seen as predicated on the underdevelopment of peripheries in all capitalist societies.

The idea of a 'map of development discourse' (see Table 3.1) developed by Michael Watts (Watts, 1993b) is useful here in furthering our understanding of the emergence of Marxist debates in development geography during the 1960s and early 1970s. Marxist development writings suggested that conditions in what were then often termed 'LDCs' could not be understood without an examination of their roots and basis in a global system of exploitation and unequal exchange. Although maintaining a notion of a 'Third World periphery', they focused

TABLE 3.1 A MAP OF DEVELOPMENT DISCOURSE

NORMATIVE ASPECT OF DEVELOPMENT THEORY			
PERIODISATION	*The State*[1] (Market failures, regulations for growth with equity, institutional and political capacity)	*Civil Society*[2] (Association life, households, communities, lobbies, non-governmental organisations, non-state economic and cultural production)	*The Market*[3] (State failures, separability of equity and efficiency, harmony)
Phase 1 1760–1890 First Industrial Revolution	*Relative backwardness and catching up* Protectionism, forced savings (Meiji reforms, Witte's Russia, List and Bismark in Germany)	Marx proto-socialists and European populists Artisanal production, small-scale co-operatives, collective control (Sismondi, Owen Proudhon, Fourie, Herzen)	*Classical Political Economy* *Laissez-faire* division of labour, comparative advantage (Smith, Ricardo, Malthus)
Phase 2 1890–1945 Classical imperialism	*Soviet Socialism* Nationalisation, central planning, collectivisation and primitive socialist accumulation (Preobrazhensky, Stalin) *Keynesianism* State role in crisis regulation through fiscal and monetary policy (Keynes, Gerschenkron)	Gramsci, Arendt, neo-Marxist theory Autonomy of civil society solidarity, pluralistic rights *Neo-populism* Russian populists, Narodniks, East European Green uprising, Gandhism (Chayano, Gandhi, Chernyskevski)	*Neoclassical Economies* Harmony and just returns, general equilibrium models (Marshall, Austrian School, Schumpeter, Pigou)

THE 'DEVELOPMENT' REVOLUTION

Phase 3 1945–1980 Growth to crisis	Third World socialism and radical dependency Maoism, Ho Chi Minh, Che, Debray, Indian Marxism/ Nehru, Komai (delinking, basic needs, central planning)	*Development economics and growth with equity* Import substitution industrialisation protection, bug push, linkages, (Lewis, Myrdal, Mahalanobis, Economic Commission for Latin America), redistributive strategies, basic needs	*Development populism, agrarianism, small is beautiful* Community development informal sector appropriate technology, African socialism (Nyerere, Lipton, Mellor)	Weberian modernisation Rationality, calculability, modernity, institutional capacity (Hirchmann, Elias, Pye, Ong, Buttel)	Neoclassical economic development Pluralist state theory, agriculture and innovation, aid and trade (Bauer, Schultz, Ruttan, Myint, Jorgensen)	Modernisation theory Human capital, formation, stages of growth, diffusion, savings, need achievement (Rostow, Chenery, Rodgers)
Phase 4 1980–1990 Crisis, stabilisation, adjustment	*New political economy* Developmental state: neo-Weberian state capacity, relative autonomy, embeddedness (Wade, Chalmers Johnson, Evans, Amsden)	New growth theory: endogenous government behaviour, collective action theory, multiple equilibria (Becker, Bates, Krugman)	New institutional economics Transaction cost approaches, imperfect and asymmetrical information (Stiglitz, Bardhan, Nugent, de Janvry, Williamson)	The public sphere Local knowledge/ peasant science, new social movements, non-governmental organisations/ private voluntary organisations, feminisms, post-Marxism (Escobar, Shiva, Hettne, Otte, Laclau, Kothari, Habermas)	Neoliberal counter revolution Price distortions, rent-seeking, market strategies, trade theory (Timmer, Krueger, Berg Report, Lal, Little Balassa, Bauer)	

Notes: **[1] The State is understood as a set of institutions which act as a system of political domination/regulation with specific effects on class and class struggle (see Jessop and Brown, 1990: 28).**

[2] Civil society is understood in the Gramscian sense as a non-state sphere of organisation – 'the ensemble of organisms commonly called private' – where hegemony and consent are organised, possessing the potential for rational self-regulation and freedom (Gramsci, 1971).

[3] The market is understood as a nexus between buyers and sellers (but an institutional nexus that has to be made, that is to say, an auction in which buyers and sellers bid against one another or as a broker-organised market).

Source: **adapted from Watts (1993b).**

on the supply of raw materials to the metropolitan countries, pointing out that core and periphery were 'locked together' by the need for accumulation and to appropriate surplus value from the periphery (and all the contradictions this involves). The periodisation formulated in this map begins in 1760 with industrial revolutions in Europe but discerns a 'growth to crisis' period running from 1945 to 1980 during which different streams of radical dependency and 'Third World' socialisms impacted upon the way in which social scientists viewed (capitalist) development. Although the intellectual origins of development have a long and complex genealogy or intellectual history (Rist, 1997), the cartography of development discourse constructed here highlights the fruition of 'development' as an idea after 1945 and outlines some of the complex contextuality of its invention and institutionalisation, through economic crises and depressions, through anti-colonial struggles across the 'Third World' and as a response to the revival of neoclassical economics: 'the naked emperor of the social sciences' (Keen, 2001). Human geographic scholarship was thus always a part of these dynamic contexts:

> Human geography was part and parcel of these transformations which were both in a sense rooted in the world system: foreign aid in the choppy waters of Cold war geopolitics, and theory in the growing sensitivity to the demands imposed by the world market and by dependent locations in the world system of transnational capitalism.
>
> (Watts, 1993a: 180)

Africa was central to some of the quite ferocious debates which began to emerge within geography about the political economy of underdevelopment (Cooper, 1980), led in part by the work of Dakar-based African scholar Samir Amin. Work by Wisner (1976, 1977) and O'Keefe and Wisner (1977) grounded the impact of recent droughts not in tropical climate variations but in class relations, looking at famine as socially and politically produced. Other work on dependent industrialisation, peasant differentiation, state intervention and African/global accumulation thus began to emerge alongside something resembling a rewriting of colonial history. Watts (1983), for example (working in Northern Nigeria), traced the history of famine and food security by drawing upon Marxist debates about agricultural modes of production and markets. A kind of new 'political ecology' approach had thus emerged (Blaikie, 1985), combining work on natural hazards with theories of political economy. Francophone geography was also important here with new work undertaken during the 1970s on the dynamics of rural space in Sahelian communities and on the problems of urban growth and health. In theoretical terms the 1980s and 1990s witnessed a retreat from structural Marxism and dependency theory to a concern for gender, social history, and community-level and household dynamics (often linked to agrarian issues). Human geography also emerged more prominently during this decade in a number of parts of the non-western world such as with the 'new wave' of South African geography (Crush, 1986) which raised important questions about economic and political restructurings in Southern Africa and often involved off-campus participation in political struggles throughout the region. The neglect of the Middle East continued however, which was strange and difficult to explain given the importance of OPEC members (e.g. Saudi Arabia) in the emerging world economy of the 1970s when oil price rises had given rise to debates about multinational corporations, the 'relative decline' of America, and (particularly on the Left) debates about 'dependency' and the 'internationalisation' of capital (Vitalis, 2002).

Selected British geographers continued to debate the perceived parochialism of their contemporaries into the 1980s, however (Farmer, 1983; Johnston, 1983; Potter and Binns, 1988; Potter and Unwin, 1988). Examining the 1987 Register of Research Interests for the Developing Areas Research Group of the IBG, Potter and Unwin (1988) found that relatively few British geographers were working on development topics and that planning, industrialisation and physical-environmental issues had been neglected. Interestingly the geographical coverage of this research appears to have changed only very slightly from that of the early 1960s (the leading countries for research projects in 1987 DARG registers were Nigeria (6), Kenya (5), India (3) and South Africa (3), although Brazil (5) and China (3) had been chosen more frequently as project locations.

One geographer, David Stoddart (1996), wrote a letter to the Editor of the *Geographical Journal* in response to a review of British geography published between 1992 and 1996. The only locality identified specifically during this period (other than Europe) was Antarctica, leading Stoddart to ask the question: 'Has British geography forgotten the world exists?' (Stoddart, 1996: 355). Thrift and Walling (2000) and Potter (2001b) have picked up these issues in their work, looking at the virtual demise of many area studies departments in British geography departments. Thrift and Walling (2000: 106) write: 'much is written in theory concerning the necessity to appreciate difference, but this is too rarely articulated in practice.' These authors also talk of a kind of 'pious Eurocentrism' that characterises much of the discipline, similar to what Potter (1993, 2001b) calls 'Little-Englander' geography. Potter (2001a) also notes that many geographers are not cited in development-oriented social science journals (Potter, 2001a: 189). None the less, the insightful work of geographers such as Richard Peet, Mike Watts and Jonathan Crush is being read and cited in other social science disciplines and there are signs that geographers are beginning to become more influential in debates about development theory and practice. The critical approaches to studying development geography that emerged in the 1980s and 1990s are explored in further detail in later chapters. Potter (2001a) makes some concluding points which are particularly interesting, referring to a conference he attended recently where one delegate had argued 'vociferously' that geographers concerned with development issues could not possibly be compared with the luminaries of other disciplines in terms of their overall impact on the study of development:

> It seems that some geographers think dichotomously about the contributions made by those who work in Africa, Asia, Latin America and the Caribbean. This seems oddly misplaced in a world system dominated by globalization and transnationalism, together with their highly uneven geographical outcomes.
>
> (Potter, 2001a: 189)

In a further article published later the same year Potter (2001b) explains how those who specialise in what are known as 'peripheral' areas are often seen to belong to some distant relative of the discipline, to area specialities which mark their contributions out from other parts of the discipline. For example, Alan Gilbert (1987) has observed that had he done the bulk of his work in the UK, he would be described as an 'urban geographer', but because he has carried out most of his urban research in South America he is seen as a 'Latin-Americanist'. Potter's critique suggests that it is necessary to ask why so many texts on urban and social geography neglect the 'majority world'. Why examine only those areas of advanced capitalism? (Sidaway, 1994). Potter (2001b) argues that 'the world's "have nots" also receive a disproportionately low share of the attention of geographers' (Potter, 2001b: 423) and asks why it is that China and India (with a combined population of over two billion people) receive relatively little attention by geographers in comparison to the attention focused on European and North American countries. Mansell Prothero is also quoted as sharing Potter's sense of dismay with contemporary attitudes, arguing that 'they are much more endemic and have characterised British geography over all the time that I have been involved, from 1947 onwards' (quoted in Potter, 2001b: 424). Thus many themes of interest to contemporary development geographers (e.g. globalisation, neoliberalism, postcolonialism) should 'also be seen as fertile grounds for sub-disciplinary interaction' (Potter, 2001b: 424). Referring to the important work *Geographies of Development* (Potter et al, 1999) Potter writes:

> Thus, Geographies of Development, specifically took this remit and stressed that the study of inner city areas, lagging regions and Eastern Europe are all part of development geography . . . the text uses as examples the case of aboriginals in Australia, and mass tourism and Southern Europe via the case of Spain.
>
> (Potter, 2001b: 425)

In a similar vein Potter (2002) argued in a recent editorial of the new journal *Progress in Development Studies* that the journal was keen to publish materials that deal with issues of development and change in all areas of the globe 'including eastern European states, the post-Soviet arena and, indeed, development issues in so-called "developed societies" ' (Potter, 2002). This work suggests that geographers interested in development must move to 'encompass issues and policies of development wherever they occur' (Potter, 2001b: 3) and also mentions the need for ' "more talk about development" in Geography' (Potter, 2001b: 425). In a related sense, Robinson (2002) writes of the persistence of a split in geography and other disciplines between accounts of cities in countries which have been labelled 'Third World' and those of the 'West', indicating how these sharp divisions in scholarship can reinforce the notion of these two categories being entirely distinct. Focusing in particular on the vast and expanding 'world cities' literature, Robinson looks at the continuing importance of categories in thinking about urban areas which 'privilege the West as the source of economic dynamism and globalisation' (Robinson, 2002: 539). By contrast, the 'developmentalist' approach to cities sees all poor cities as infrastructurally poor and economically stagnant. Robinson's work calls for a more cosmopolitan approach focused on what can be learned in each context, with US and African examples informing each other rather than being kept separate:

> One of the consequences of these overlapping dualisms, is that understandings of city-ness have

> come to rest on the (usually unstated) experiences of a relatively small group of (mostly western) cities, and cities outside the West are assessed in terms of this pre-given standard of (world) cityness, or urban economic dynamism.
>
> (Robinson, 2002: 539)

Thus substantial areas of the globe are confined to structural irrelevance in much contemporary thought about cities and urban life. Robinson thus calls for 'views from off the map'. This has particular relevance to our discussion of the changing map of development geography and suggests the need to focus on wider issues about so-called 'alternative modernities' (Ganokar, 2001) and on articulations of 'race', development and postcolonialism (Peake and Kobayashi, 2002). In this way it may be possible to extend the lines of enquiry anticipated by Buchanan, McGee and other radical geographers. Feminist accounts of tropical and development geography have also yet to be written, but important feminist work on the wider history of the discipline, such as that of Domosh (1991), Rose (1993) and McEwan (1998), are relevant and instructive here. The male domination of development geography comes through quite clearly in some of the masculinist assumptions made about 'development' and its subjects. Even the brief and partial account offered here indicates the consequences and merits of beginning to take into account the margins in histories of twentieth-century geography.

CONCLUSIONS: GEO-WRITINGS AND THE TROPICAL WORLDS OF DEVELOPMENT

> Rather than an objective, detached intellectual endeavour, international relations scholarly discourse on North–South relations becomes imbued through and through with the imperial representations that have preceded it.
>
> (Doty, 1996: 161)

> 'The Tropics' . . . on a metaphorical level this phrase partially substitutes for *the Third World* – that is to say, 'developing' regions, largely composed of former colonies. While not always geographically accurate (any more than the terms *the Non West* or *the South*), it serves as a reminder that people and machines operate in different natural and cultural environments, even as it recalls a telling complex of metaphors inherited from empire.
>
> (Redfield, 2000: 21, emphasis in original)

For a few short years development geography was infused with the sort of spatial algebra and 'mathematical mystification' then driving the 'new geography', and some development geographers did begin to map out the optimistic logic embodied in the linear stage theories of the modernisation school (Corbridge, 1986). However, this 'marriage of statistics and space' was short lived, as it became apparent that this perspective was 'blind to the historical reproduction of systems and processes of inequality' (Corbridge, 1986: 4–6). The ascendancy of alternative approaches was characterised by further work on the urban informal sector, the increased study of rural problems (a countenance to the idea of 'urban bias'), by a concern for decentralisation of planning and population and by theoretical work on regional development policy. There was often a policy and planning relevance to some of this work and an emerging focus on small groups, communities and local issues (Forbes, 1984). As we shall see in Chapter 4, two major theoretical traditions initially included the 'basic needs approach' and work emanating from a Marxist and neo-Marxist perspective (Watts, 1993a, 1993b).

One account of the remaking of the subdiscipline of economic geography in the 1960s suggests that the conventional understanding of the emergence of spatial science 'is interesting for all the places not included . . . Africa . . . Asia or Australasia' (Barnes, 2000: 18). Thus standard histories of the discipline focus on the movements of texts, ideas and individuals between key nodes such as Cambridge and Bristol (England), Lund (Sweden), Washington, Iowa, Michigan and Chicago (USA), but focus much less on the making of the discipline in Ibadan or Durban, Singapore or Rangoon. What was being suggested by McGee in his anti-development critique was that the whole post-Second World War movement for the political 'development' and 'modernisation' of the 'Third World' was based on the assumption that this was 'right for everyone' and that other paths, trajectories and concepts must therefore be 'wrong'. In this sense we might ask what happens, for example, to the geography of Africa (conventionally scripted as the 'least developed' continent) when Africa is seen as *rich* in economies, connections, cultures and lives whose contributions, diversity, wealth and worth are not captured adequately by being imagined as more or less *developed* (Myers, 2001).

Alternatively why are poverty, racism, marginality and deprivation (or, for that matter, excessive consumption among the affluent) in Europe, North America or Russia and post-communist 'transition economies' not seen foremost as issues of 'development' (Wood, 1998; Jones, 2000)? In this sense we need to understand how geography and geographers created a space in which only certain things could be said or even imagined about tropical 'others' (Escobar, 1995). It is this notion of the unique and distinctive space(s) and processes of 'the Tropics' that is the ghost of development geography today. People in the West had supposedly 'mobilised' themselves into forming nations and building sovereign states around these nations; so too therefore must the lagging and 'backward' 'Third World'. Modernisation theories of development envisaged rational bureaucracies, legal systems and military institutions in addition to periodic elections (similar to those of Washington or Westminster, for example), but as Lummis (2002: 67) has argued: 'calling this "development", which would mean the unfolding of the possibilities latent in the indigenous cultures, was simply a way of concealing the violence being done to these cultures'.

Fieldwork in 'developing countries' is today being increasingly recognised as deeply problematic; yet for geographers such as Jim Blaut and Keith Buchanan, fieldwork was to be embraced as an opportunity to transgress the imagined geographies of tropical development and engage wider, more revolutionary ideas. Their work spoke of planning research *with* people rather than *for* them. In this chapter we have sought to highlight the value of recovering and reworking the diversity of fieldwork traditions in geography, such as those that characterise this domain of tropical/development geography. By confronting the problematic of fieldwork (and the uneven power relations that structure these experiences) geographers can engage more fully in ongoing projects of decolonisation and the wider deconstruction of development geographies. In this project of decolonising development geography we need to look beyond the idea of a master western prototype:

> We need to attend to how places in the non-West differently plan and envision the particular combinations of culture, capital, and the nation-state, rather than assume that they are immature versions of some master western prototype.
>
> (Ong, 1999: 31)

Figure 3.5 Objective Decolonisation 2000: the UN and freedom to choose
Source: UN Department of Public Information

This then was a key weakness of the geography of modernisation school (McGee, 1974) and its 'pre-Historians' that had sought to map and interpret the regions they encountered *by analogy* with other cultures. In addition, most of the academic researchers who went to study the 'problems' of 'the Tropics' located themselves in what we might refer to, following Yapa (2002: 35), as the 'realm of the "non-problem"'. In this way these researchers often located themselves in privileged epistemological positions outside the problems they studied. Thus few geographers saw themselves as implicated in colonialism and the underdevelopment they sought to write about and map. With the radicalisation of geography in the 1960s and 1970s many geographers began to adopt different and more critical ways of looking at colonialism and its legacies in the Third World. Blaut (1976) argued, for example, against the idea of the 'European miracle', showing that Europe was not superior to other regions prior to 1492 and that colonialism and the wealth plundered from the Third World had led to the rise of Europe rather than any scientific process of rationality (Peet with Hartwick, 1999).

Buchanan's later work on the Southeast Asian world also drew an important distinction between a concern for description and a concern for 'trends and forces that have shaped and are continuing to shape the turbulent and diverse nations and the region' (Buchanan, 1967b: 11). Buchanan also made a number of important connections between his research on the West Midlands and that concerning 'the Tropics', locating the major features of human geography in Southeast Asia within a wider sense of underdevelopment:

> by comparison with the developed nations of the West their [Southeast Asian] social and economic structures have been warped and retarded and these processes have resulted in an impoverished and marginal quality of life for the great majority of their peoples. Superficial observers have been inclined to explain away this backwardness as the result of a tropical environment or the alleged lethargy of tropical peoples. [These are] environmentalist-racist assumptions [of] little validity.
>
> (Buchanan, 1967b: 20)

What was important about the work of radical geographers such as Buchanan (and he was not alone in this) was that earlier, descriptive geographical approaches to tropical areas were taken to task for the way in which they sought to 'explain away' the supposed backwardness of tropical zones in an apolitical fashion which was, at core, deeply racist and shot through with the language of earlier imperial endeavours. Most of the other geographical work on issues of development at this time was either a more or less direct link to tropical geography, or part of the vanguard of positivism and quantification. Notwithstanding some work on geographies of modernisation (see Chapter 4) which gave this a special character, the legacy of descriptive tropical geography and its association with determinism and other passé modes of geographical thought/analysis has proved enduring. As the 1960s progressed and revolutionary pressures in the South accelerated (epitomised by the insurgencies in Vietnam and the Portuguese colonies and the lurch into Mao's 'cultural revolution' in China), Buchanan embraced them – some years before a wider development geography (or indeed wider human geography) was recast as 'radical'. Buchanan's renditions of the achievements of Maoism have not stood the test of time – or rather the exposé of the devastating impacts of Maoist politics on China's environment (Shapiro, 2001). Ray Watters' (1998) obituary appreciation of Buchanan details some of the difficulties that Buchanan's radicalism produced in the Cold War climate of New Zealand's universities in the 1950s, among them the cancellation of exchanges with China.

It is also useful to think about the ways in which geographers sought to provide *descriptions* of the spaces and places of (economic) development and, from the early 1960s, to combine these with *prescriptions* of what was involved in 'catching up'. The interventions of radical geographers were thus beginning to show that these prescriptions for 'tropical' and 'backward' countries were not a neutral, rational enterprise that was untouchable but rather something which was very much caught up with the strategic political interests of the USA during the Cold War. According to Peet with Hartwick (1999: 90), the 'modernisation' framework was a 'cold war attempt at legitimating US domination of the global system'. This chapter has tried to excavate the genealogies (or intellectual histories) of development geography, relating it to geopolitical, economic and social traumas of decolonisation. While interested in such a 'big picture', we can also approach this story in part through engagements with the works of a series of radical geographers, among them Keith Buchanan. Buchanan was one of several radical geographers working in the late 1960s and 1970s, and his ideas were formulated largely through conversations with other radical geographers and other disciplines. Studying the particular movements and writings of Buchanan, it is possible to reflect on the connections between local cultures, economies and philosophies of geographical field-work and teaching (Power and Sidaway, 2004).

Through these various paths, the (postcolonial) present of development geography can thus be investigated. More recently, Rob Potter (2001b) has argued that today geographers once again need to enhance their sense of responsibility to other geographies:

> What is needed is an enhanced *responsibility to distant geographies*. Such distant geographies are not lacking in intellectual challenge because they are encountered far away; they are not irrelevant to us because they emanate from other societies and cultures.
>
> (Potter, 2001b: 425–426, emphasis in original)

Although it can never be straightforward or simple to conduct, the point here is not to reiterate the argument that crops up from time to time about the focus of geography on examples from Europe and North America and the neglect/marginalisation of the rest of the world (a point that Buchanan made so fluently). Instead we can argue, in the light of Bayart (1993) and others, that it is precisely this western modernity and its economic, cultural and political geography that can be creatively 'deepened' by engagement with the norms of what are usually scripted as its margins. Thus the story of 'tropical' and later 'development' geography can be enhanced by considering the role of marginal or peripheral spaces and peoples in allowing geographers to represent the world and its changing political and economic systems. Typically this story has focused largely on white, male, western academics and much less has been said about the role of women or non-western geographers in shaping development debates in geography and in defining the nature of its methods and agendas. Thus there remains considerable scope for more sustained and direct engagements with feminist readings of the history of these sub-disciplinary specialities and with 'post-colonial' approaches to thinking about the collaborations and 'partnerships' that were formed between western and non-western geographers in the new universities of Africa, Asia and Latin America from the 1960s onwards.

In the spirit of Buchanan, and his moves between Asia, Africa, Oceania and Europe, it is possible to invoke an ongoing crisis of and therefore *a struggle for* new representations and understandings of global geographical difference. If we look, for example, at Buchanan's (1972) map of what he termed 'The cultural empire', what today might be obscured in analyses of 'globalisation' is indicated in quite interesting ways by the global geography of Coca-Cola bottling plants that he constructs. This was among a trilogy of articles on the geography of Empire focusing on the 'Vietnamisation of the world', the 'Intellectual pace of the Third World' and 'the Economic pattern of Empire'. Moreover, an examination of Buchanan's departures suggests the value of biography as a mirror to disciplinary (and other) histories and the study of movements, places and departures as a counterpoint to the more familiar big stories (metanarratives). There is a danger, however, of representing this radicalisation of geography as a straightforward and simplistic undertaking when actually it was a slow process given the 'ostrich-like' behaviour of some (liberal) geographers who buried their heads in the sand and refused to engage with politics or the politicisation of geography (Peet, 1977b: 26). In one of his later essays entitled 'Reflections on a "dirty word"' Buchanan captured how many geographers had begun to be critical of the 'magic' of development medicines:

> a decade or so ago, as the 'underdeveloped' were administered larger and larger doses of the magic medicine of 'development' [geographers] began to notice how this therapy always – and strangely – enriched the developers and impoverished the supposed beneficiaries of the process.
>
> (Buchanan, 1977b: 363)

Thus, writing in 1977, Buchanan noted how the term had become a 'dirty word' and how geographers were seeking increasingly to build critiques of the 'generation of smooth-talking, formula fixated and model-making technocrats' (Buchanan, 1977b: 364) based in the 'white North'. Their prescriptions were seen to be failing with monotonous regularity and were shown to be rejected increasingly by the 'patients' they were aimed at. Radical geographers thus sought to illustrate the limitations of assuming

BOX 3.3

Chapter-related website

See 'The Stamp Papers: an introduction' which is available at http://www.sussex.ac.uk/library/manuscript/lists/stamplst.html/. This includes an introduction to the collection of fifty-three boxes of materials collected by Stamp in his travels through Africa, Asia and Latin America. Acquired by the University of Sussex in 1966.

that 'pie-in-the-sky' levels of affluence (proposed by modernisation theorists) could be enjoyed in the periphery, calling for a fresh look at theory and practice. Some of these geographers sought instead to view development as liberation, rejecting orthodox development patterns and emphasising the development of people rather than the development of economies.

4

Development Thinking and the Mystical 'Kingdom of Abundance'

From the unburied corpse of development, every kind of pest has started to spread. The time has come to unveil the secret of development and see it in all its conceptual starkness.

(Esteva, 1992: 6)

[S]ocial scientific enquiry regarding social problems is deeply implicated as a causative agent in the very problems it is designed to address.

(Yapa, 2002: 33)

INTRODUCTION: KNOWLEDGE AND THE ERA OF MODERNITY

The origins of development theories and ideas about progress and modernity are an important part of understanding the 'invention' of development as an arena of enquiry and state practice. For many commentators, this begins with the postwar creation of ideas and discourses about 'underdeveloped areas' (e.g. in Truman's 'bold new program') but we can trace this emergence back further still, to the seventeenth and eighteenth centuries of European Enlightenment rationality in particular. Development studies itself must be seen as a legacy of the Enlightenment in its rejection of artificial disciplinary boundaries and its search for common and collective approaches to problems and dilemmas of development. What we are interested in here is the extent to which deep-rooted 'western' and 'Eurocentric' traditions have cast their 'dominant shadow' over development thinking. The idea of progress forged in the Enlightenment era remains an article of faith in development thinking, while the idea of development has since taken on a 'quasi-mystical connotation' (Munck, 1999: 198). What emerges therefore is a kind of mystical belief that the project of modernity has not exhausted its capacity to bring about 'positive change'. Enlightenment ideals seem to have exerted a powerful influence on postwar conceptions of development and are also related to the wider question of whether modernity is still an unfinished project in the South. To what extent do theorisations of development become caught up in a 'western' perception of reality?

One discussion of 'development thinking' (Hettne, 1995) has argued that 'development' principally involves three main parts: *development theories*, *development strategies* and *development ideologies*. The first grouping (the subject of this chapter) refers to sets of logical propositions about how the world is structured which explain past and future developments. The second grouping discusses strategies of development adopted by actors and agents, ranging from the grassroots to central states. Different agendas reflect different development *ideologies* which constitute the third grouping that Hettne (1995) identifies. Various different ideologies of development have been formed and have come to shape a number of struggles around the world. Interestingly, Hettne also talks about the 'tendency of social science paradigms to accumulate rather than fade away' (Hettne, 1995: 64). Thus different streams of development thinking are not always entirely distinct, with one simply replacing another, but rather have often overlapped and sought to interact with one another in a kind of chessboard of philosophical positions. Where the idea of development comes together most often is around the notion of 'organized intervention in collective affairs according to a standard of improvement' (Pieterse, 2001a: 3). Each theoretical approach therefore has its own vision of 'collective affairs' and what types of intervention are necessary to bring about an improvement locally, nationally or globally. Development theory is thus partly about the negotiation of what constitutes 'improvement' and what 'appropriate' intervention means. Different epistemologies, or 'rules' of what constitutes knowledge have come into play as a result of this and the relationship between knowledge and power has emerged as a central issue. A key thematic in these debates is the contestation and conflict around these 'rules' for improvement: '[D]evelopment is struggle. To be precise, development is struggle over the shape of futures, a dramatic and complex struggle' (Pieterse, 2001a: 1).

A range of approaches to development thinking have been relevant since 1945 and it is not possible to cover each in equal depth here, but these have included the regional tradition (of *chorology* or dividing global space into a plethora of unique and unrelated chunks), quantitative spatial science, Marxism, humanism, structuration theory, realism and postmodernism (Cloke *et al.*, 1991). The subthemes associated with these streams of thought have included: economic geography as influenced by neoclassical economics, radical geography founded on Marxist geography, sustainable development, political ecology and recent work on post-development. The intention of this chapter is to focus on a small group of some of the most influential and enduring perspectives, moving towards a discussion of the post-structuralist approach developed by scholars such as Foucault and Derrida, which have been drawn upon in this book. A central objective here is to question the very framework from which the two terms 'more' and 'less' developed derive their authority: namely the economistic logic of development with its assumptions about hierarchies and continuums. Post-structuralism seeks to repudiate grand or master narratives of history and geography, focusing on 'local' narratives and stories while drawing attention to the gaps and fissures in these stories and narratives.

To do this, we need to trace the 'genealogy' or intellectual history of development over several centuries, not just in the imaginary 'West' but also around and along its fictitious borders. This chapter begins with a broad brushstroke history of 'Enlightenment' thinking in the eighteenth and nineteenth centuries, examining the important conceptions of progress and forms of knowledge that emerged from this era. We then move on to examine the beginnings of modernity and the rise of modernism and the social sciences, looking at how these ideas were then institutionalised and professionalised. It is necessary here to examine the implications of the Eurocentrism of these early theorisations. The latter part of the chapter explores three central approaches to the theorisation of development, discussing the modernisation, dependency and post-development perspectives. We need to ask how these different schools of thought and bodies of theory imagined very different scales and spaces in their conception of development's geographies. In addition, each envisaged quite different political and economic mechanisms in their understandings of progress, imagining different spatial scales and assuming particular kinds of links between peoples and places. Finally, this chapter seeks to outline the theoretical foundations for *rethinking* development geography, arguing for alternative and more radical approaches.

THE ENLIGHTENMENT AND THE THEORISATION OF DEVELOPMENT AS PROGRESS

The term 'Enlightenment' usually refers to a period in European intellectual history which began in the mid-seventeenth century and continued through most of the eighteenth century in which new attitudes to work and capital were formulated (Rist, 1997). It emerged during a century of commitment to enquiry and criticism, of a decline in mysticism, of growing hope for life and trust in effort (Hampson, 1968). Enlightenment thinkers were concerned to place the idea of progress on a more rational footing. Underlying these debates was a concern for social reform and the idea of a progression and development of societies built around an increasing secularism and a growing willingness to take risks (Gay, 1973). It is difficult to summarise these ideas and writings when they actually comprised quite a diverse and heterogeneous group, but there was in many ways an interconnected set of ideas, values, principles and facts which ran through them and provided both an image of the natural and social world and a way of thinking about it. In its simplest sense the Enlightenment was the creation of a new framework of ideas about the relationships between humanity, society and nature, one seeking to challenge some of the existing traditional worldviews which were dominated by Christianity. In this way, Europeans expanded their sense of knowledge as power:

> In the century of the Enlightenment, educated Europeans . . . experienced an expansive sense of power over nature and themselves: the pitiless cycles of epidemics, famines, risky life and early death, devastating war and uneasy peace – the treadmill of human existence – seemed to be yielding at last to the application of critical intelligence.
>
> (Gay, 1973: 3)

Geographically centred in France but with varied foundations in many European states, 'the Enlightenment' was thus a sort of *intellectual fashion* which held the attention of many European intellectuals. This was not some kind of singular or unanimous viewpoint shared by all in the eighteenth century; indeed there were many differences and disagreements. According to Black (1990: 208) the Enlightenment is therefore best regarded as 'a tendency towards critical inquiry and the application of reason' rather than a singular or coherent intellectual movement or institutional project. Again we see that the historical (as well as geographical) beginnings of the idea of development can be traced not to some singular movement or project but rather to this tendency to form wider critical enquiries about the organisation and structure of societies and to apply reason and rationality to such social scientific enquiries. This persistent metaphor of the 'light of reason' shining a powerful torch into the dark recesses of ignorance and superstition in 'traditional' societies was a powerful and influential one at this time. In some ways the term 'Enlightenment' refers to a secular intelligentsia that had emerged across Europe and that was powerful enough to challenge the clergy and the authority of the Catholic Church (Porter, 1990). In this sense what distinguishes the thought of the Enlightenment from earlier intellectual approaches (in addition to anti-clericalism) was a belief in the pre-eminence of empirical, materialist knowledge (with scientific method as the model), an enthusiasm for technological and medical progress (with scientists, inventors and doctors depicted as the cure for society's 'ills') and a desire for legal and constitutional reform (Hall and Gieben, 1992: 36). In addition, this intelligentsia expressed a search for new forms of political organisation which could establish new 'civil liberties' and freedoms in these rapidly changing societies. The Enlightenment thinkers sought then to redefine what was considered as socially important knowledge, to bring this outside the sphere of religion and to provide a new meaning and relevance.

There is a risk, however, of applying the term 'Enlightenment' too loosely or too widely, as if it had touched every intellectual society and every intellectual elite of this period equally (Power, 2001). In referring to this period and these ideas a process of simplification is at work which denies that the Enlightenment was an amorphous, hard to pin down and constantly shifting entity (Porter, 1990). There were, however, many common 'threads' to the evolving 'patchwork' of Enlightenment thinking such as the championing of new freedoms, the primacy of reason/rationalism, the concept of universal science and reason, the idea of progress, a belief in empiricism, the ethic of secularism and the notion of all human beings as essentially the same (Hall and Gieben, 1992: 21–22).

The idea of development can thus be seen partly to have emerged from the crucible of these early debates, particularly the Scottish Enlightenment and the work of Adam Smith, that had postulated a 'theory' of development embodying a series of 'natural' or 'normal' stages of human activity and progress. Smith's writings in particular were to become highly influential in the articulation of neoliberalism after the Second World War, where the basic liberal notions of free trade and the liberating potential of free, self-regulating markets have come to dominate development institutions such as the World Bank. Cowen and Shenton (1996: 13) remind us that the beginnings of the modern 'sub-discipline' of development economics are located as much in the 'rough and tumble of early industrialism' as in the work of Scottish Enlightenment writers such as Adam Smith. Thus these texts were themselves a product of their times and must be seen as emerging from the 'rough and tumble' of industrialisation in a particular region of the world which they attempted to naturalise and make sense of.

The Enlightenment writers adopted a very clear position on some of the important transitions underway with European societies at the time, particularly the transition from the tradition and mysticism of the past to the modernity of Enlightenment thought and the potential it offered for progress in the future. They sought to make their writings relevant and contemporary, linked closely to the political and economic challenges of the day. This darkened past traditional social order was often counterpoised to the bright progressive future promised by scientific understanding which meant that by concentrating on the future as a realm of unrealised possibilities there was a 'corresponding depreciation of the past' (Gay, 1973: 92). These new writings and ideas thus profoundly challenged the traditional role of the clergy as the keepers and transmitters of knowledge, constructing distinctively 'modern' approaches to thinking and knowing the world and an individual's place within it. The emancipatory potential of this knowledge turned out to be limited however, in that it was conceived of partly as quite abstract and utilitarian, as a kind of mastery over nature which thus becomes characterised by relations of power. As we saw in Chapter 3, many geographers subscribed to this idea of knowledge as a form of mastery over nature which was based around binaries, hierarchies and illusions of power over natural change. About the Enlightenment writings, Doherty argues that:

> Knowledge is reduced to technology, a technology which enables the *illusion* of power and of domination over nature. It is important to stress that this is an illusion. This kind of knowledge does not give actual power over nature. . . . What it does give in the way of power is, of course, a power over the consciousness of others who may be less fluent in the language of reason. . . . Knowledge thus becomes caught up in a dialectic of mastery and slavery.
>
> (Doherty, 1993: 6; emphasis in original)

One of the main Enlightenment concepts related to this question of harnessing nature and natural resources for social change came in the late eighteenth and early nineteenth-century debates about 'trusteeship'. In this way, many Enlightenment thinkers viewed the remedy for the disorder brought on by industrialisation as related to the 'capacity' to use land, labour and capital in the interests of society as a whole. Only certain types of individual could be 'entrusted' with such a role (Cowen and Shenton, 1996). Property, for example, needed to be placed in the hands of 'trustees' who would decide where and how society's resources could be most effectively utilised. The changing social orders brought about by the making of European modernity and the transition from feudal to capitalist modes of social organisation could thus be managed by 'trustees' who had the power to harness these capacities for societal good. Interestingly, some degree of attention was focused on the role of banks and bankers that might be made fit for trusteeship, using their knowledge to support industrialists.

In eighteenth-century France, the changing social organisation of the country was represented as three 'Estates' – the Clergy, Nobility and the 'Third Estate', which comprises everyone else, from wealthiest bourgeois to poorest peasant (Hall and Gieben, 1992: 33). This gap between enlightened thinkers (who were often members of the second Estate) and the peasantries of European eighteenth-century societies is an important part of the historical context of Enlightenment thinking. Although they appeared to represent a threat to the established order, these ideas and writings sought evolutionary rather than *revolutionary* change, arguing that progress and development could come about within the existing social order through the spread of ideas among 'men of influence' (Hall and Gieben, 1992).

It also follows that Enlightenment notions of truth and objectivity (which continue to underwrite much contemporary development thinking) actually mask underlying power relations. The claim to speak legitimately for others has to be viewed in this context. Various different forms of development have thus carried an implicit assumption about their right and authority to represent and speak for others. The Enlightenment is also important here because for some critics of developmentalism, the post-1945 development project is 'the last and failed attempt to complete the Enlightenment in Asia, Africa and Latin America' (Escobar, 1995: 2–4). In addition, many postcolonial theorisations of development (see Chapter 8) have sought to critique this Enlightenment project generally and the failures of national development projects that it has led to more specifically (Berger, 2001). Perhaps the clearest indication of the continued importance of the Enlightenment I can offer here comes from the following quotation from a speech by World Bank

President James Wolfensohn in 1996:

> Knowledge is like light. Weightless and intangible, it can easily travel the world, enlightening the lives of people everywhere. Yet billions of people still live in the darkness of poverty – unnecessarily.
>
> (Wolfensohn, quoted in Patel, 2001: 2)

Once again, development philosophies such as that of the Bank see knowledge as something which travels easily (business class perhaps?) with limited baggage, enlightening wherever it shines. For the Bank, the 'darkness' of billions of people's lives demands that that this light should not be extinguished. It also requires that the Bank be established as the organisation 'entrusted' with the task of harnessing this light for the improvement of societies in the South. The foundations of many social sciences were therefore intimately bound up with the Enlightenment's concept of progress and the idea that development could be created through the application of reasoned and empirically based knowledge. Science would improve the practice of agriculture and industrial organisation, harnessing natural forces for human interests and cushioning the path through the 'treadmill of human existence'. The Enlightenment had forged the intellectual conditions in which the application of reason to practical issues could flourish through such 'modern' institutions as the academy, the learned journal and the conference. In turn social and political ideas could be disseminated to a 'modern' audience that was constituted in relation to a class of intellectuals writing about future progression (Hall and Gieben, 1992).

Thus what begins to emerge here is the arena of 'development' constituted through a group of professional, 'learned' individuals who seek to disseminate their knowledge of progress and freedom. This emerged directly from a growing concern about the formations of modernity being made in Europe at this time. The very idea of modernity is central to the foundation of many disciplines such as geography and sociology (Spybey, 1992). In terms of the links between Enlightenment ideals and modernist thought there were a number of important legacies here. Knowledge continued to be seen as a kind of technology based around the language of reason and rationality. The highest forms of knowledge were scientific and rational; disembodied 'science' would shine the 'light of reason' into all the darkest 'underdeveloped' corners of the world, where they would replace the backwardness of tradition. Modernist thought also envisaged a process of Enlightenment, of becoming more modern and less traditional and also saw a group of enlightened western scientists 'guiding' the paths to progress of distant others.

One of the major implications of the Enlightenment was that it firmly rooted in the popular imagination the notion that the 'West' was superior to other societies and that there were 'stages' of progress with all nations travelling the same road to the modern. No society, however, could hope to match the pace of western modernity, where 'scientific reason' was seen to play a more dominant role (Rist, 1997: 40). The emergence of an idea of 'the West' was also important to the Enlightenment in that it was a very European affair which put Europe and European intellectuals at the very pinnacle of human achievement. There has been a commonly held view that sees 'the West' as the result of forces largely internal to Europe's history and formation (Hall, 1992) rather than as a 'global story' involving other cultural worlds (Said, 1978). As we saw in Chapter 3, the making of nineteenth-century European 'modernity' established a sense of cultural and economic difference from other worlds and geographies which shaped the ways in which other (e.g. 'tropical') spaces were viewed as belonging to distant, uncivilised and immature stages in the progress of humanity. The establishment of modern modes of scientific enquiry, of modern institutions and the modern 'development' of societies in nineteenth-century Europe, was based around a contrast with the 'savage' and 'uncivilised' spaces of the non-western world, a feature of 'development thinking' that has proven particularly enduring. The notion of 'civilisation' is crucial here in that there are many similarities with the idea of 'development':

> Like 'civilization' in the nineteenth century, 'development' is the name not only for a value, but also for a dominant problematic or interpretive grid through which the impoverished regions of the world are known to us. Within this grid a host of everyday observations are rendered intelligible and meaningful.
>
> (Ferguson, 1990: xiii)

This interpretive grid was very clearly shaped by Enlightenment thinking and languages. The paradigms and philosophies of 'the Enlightenment' influenced Hegel's nineteenth-century writings, for example, on the history of human civilisation and progress. Hegel was also particularly interested in the question of mastery over nature and placed European civilisation as the furthest advanced along a scale of world historical development. By contrast, Hegel saw Africa as a dark, 'unhistorical and undeveloped land', a singular space far removed from the awakenings of self-consciousness among 'learned' Europeans. Like the philosophers of the Enlightenment, Hegel articulated the view that there was an underlying unity or seamless continuity to human history and believed firmly in the possibility of an ambitious, all-encompassing or 'totalised' human history (Doherty, 1993). History was written 'from above' by enlightened philosophers then, according to this view, rather than from below in the everyday or popular realm. For Hegel, this unifying thread that ran through all history was related to the 'principle of development' which dominated his thinking about world historical truth and freedom in self-determination.

Karl Marx disagreed with the fundamental premises of Hegel's approach to world history but none the less drew upon the Hegelian 'principle of development' to explain the genesis from feudalism to capitalism and his notion of historical materialism. Marx reread history and was also interested in the laws behind the 'rough and tumble of early industrialism' in Europe, beginning to raise important questions about the structure of societies and the organisation of industry and agriculture. Marx's narrative of emancipation of the working classes, oppressed by the extraction of value from their labour, also operated like Enlightenment reason in abstracting meaning from diverging local histories and traditions and translating them into the terms of a meta-narrative or 'master code'. Marx's attitude towards the bourgeoisie was interesting in that on the one hand it was full of admiration for its 'civilising energies' and yet on the other it was critical of its 'incipient barbarous tendencies' (Doherty, 1993: 11).

Modernist reason was not as inherently good as the 'enlightened' thinkers believed and has been used for a wide variety of purposes. Chaturvedi (2000) argues, for example, that the British started constructing 'their India' during the later nineteenth century 'as an integral part of the larger Enlightenment project' which attempted to control non-European peoples and order them within a colonial nation. Independent India has inherited this nation, and the British ideology of difference has also survived and flourished after independence (Chaturvedi, 2000). Reason can be imperialist and racist (as in the making of the idea of 'the West'), taking a specific form of consciousness for a universal, a standard that all must aspire to reach. Reason was also a potent weapon in the production of social normativity during 'the Enlightenment', driving people towards conformity with a dominant and centred 'norm' of behaviour (Doherty, 1993). Modernist reason was structured around a process of 'othering' of nonconformists, cultures and societies that were not informed by this reason and these social norms, and were thus banished to the lower echelons of humanity, defined as 'undeveloped' or 'uncivilised'. The emergence of new ideas about social, political and economic development was therefore bound up with these pressures to conform to particular rules of what constitutes knowledge, reason and progress and with the making of a 'Third Estate' or 'Third World' of nonconformity as the alter-ego of a developed 'West' (see Chapter 5). The rise of the social sciences is one aspect of that broad set of changes to which we often refer as the 'rise of the modern world'. During the nineteenth century in particular, the rise of the social sciences took place in the context of accelerating military and economic colonisation of the non-western world, of nation-building in the First World and more generally the development of modernity (Schuurman, 2001). This context heavily influenced the theorisation of progress as something which centred upon nation-states, on new forms of collective national identity and on national markets and transfers. The work of social scientists was thus concerned with understanding and ordering the many different changes which had been set in motion. The shift from agrarian feudalism to industrial capitalism in Europe was accompanied by a growth in the social sciences – many of the individuals and organisations that brought about this shift were also enthusiasts for social change. The growth in the social sciences was thus closely bound up with the emergence of the biological sciences – which had clear and lasting implications.

Metaphors of organic change in the natural world were adapted for the explanation of social change. The growth from an acorn into a tree or from a seed into a plant implied biological and organic changes that were used to explain the growth of industrialisation and other *man-made* social changes. Evolutionary theory also had a key impact on the formulation of theories of social change. Thus social development or the development of societies was likened to the process of evolution that Darwin outlined – with its natural succession or progression through a series of stages. What is interesting is that few analysts followed these metaphors through to their conclusion, which is decay, decline and eventually death and finality (Wallerstein, 1994). This kind of developmentalism sees non-western spaces and their inhabitants as essentialised, homogenised entities and pointed up the 'makeability' of their societies. This was to be an enduring legacy of the organicist and evolutionary thinking of the Enlightenment.

Marx's interpretation of capitalist modernity in particular has continued to be relevant to the theorisation of 'development' well beyond 1945 (see below), and ideologies of Marxism have been drawn upon in development strategies in a variety of different historical and geographical settings.

THE 'SINATRA DOCTRINE' AND THE COLD WAR

> Before development, there is nothing but deficiencies. Underdeveloped areas have no history of their own, hardly any past worth recalling, and certainly none that's worth retaining. Everything before development can be abandoned, and third world countries emerge as empty vessels waiting to be filled with the development from the first world.
>
> (Abrahamsen, 2001: 19)

'Development thinking' (Hettne, 1995), or the sum total of ideas about development theory, ideology and strategy, went through a variety of twists and turns in the twentieth century. With the end of the Second World War, 'machineries of decolonisation were set in motion by colonial states' (Sylvester, 1999: 704) and, with the establishment of the United Nations, this gave a decisive stimulus to theorisations about the political and economic trajectories of non-western others. Growth theory and development planning had thus become ever more contested theoretical domains. Many development approaches, however, share a belief that 'development' is a process which has industrialisation as its culmination, its end-point or ultimate outcome. Resource distribution has also often tended to be seen primarily as linked directly to the state, although the ways in which people should benefit from 'trickle-down' or 'redistribution' is often much less clear (Sylvester, 1999). This is partly a consequence of the emergence of these approaches at particular historical conjunctures and geopolitical moments.

In the twentieth century, specific western theories of development first began to emerge as a result of the great recession in trade in 1930s Europe and North America and again at the end of the Second World War, which had illustrated the need for reconstruction and development as very important in a Europe ravaged and devastated by conflict. In the 1940s, 1950s and 1960s political 'winds of change' were sweeping across many regions of the South and many new and exciting ideas about social and economic progress were in the air. None the less, many of the theories still carried implicit assumptions about western economic, social and cultural superiority. However, they were still taken up by many nationalists struggling for the development of a new nation and for a new sense of national identity. As we will see in Chapter 5, many of these movements and struggles for national independence in the 1960s and 1970s drew strength and unity from the idea of the Third World and its underdevelopment by 'westerners'. The resulting forms of nationalist developmentalism that emerged illustrate further the various ways in which development ideas became a common ideological space in which different movements (on both the left and the right of the political spectrum) came to fight their battles and to justify their struggles. In discussing these different conceptions it is important to think about where and when they emerged. Most conceptions reflect some of the priorities of development thinking characteristic of their era and the particular time and space in which they were written. According to Allen and Thomas (2001) the practice of constructing development theories generally involves an attempt to respond to different *perceptions* of 'development challenges' at different historical moments.

The simple 'task' of development, for some modernisation theorists, was to provide 'an ethos and system of values which can compete successfully with the attraction exercised by Communism' (Watnick, 1952: 37). Morris Watnick called for the modernisation of 'underdeveloped areas', painting a picture of 'underdeveloped peoples' (Watnick, 1952: 22) confined to 'backwardness' but somehow *torn* between the appeal of communism and the prospect of western modernisation. This was a key characteristic of what has become known as the modernisation school, which was often dualistic, opposing 'traditional' to 'modern' lifestyles and 'indigenous' to 'westernised' cultures, as if nobody could belong to both categories. Hirschmann (1958) was another key proponent of modernisation ideas, voicing the optimistic view that the forces of concentration ('polarisation') will 'trickle down' from the core to the periphery at national, regional and global levels. In some of his work Hirschmann explained how development economics might 'slay the dragon of backwardness' (quoted in Rist, 1997: 219). Governments were advised not to intervene in this polarisation process despite the fact that Hirschmann (1958) was advocating 'a basically unbalanced economic growth strategy' (Potter *et al.*, 1999: 46). Gunnar Myrdal (1957) adopted a more pessimistic view, arguing that capitalist development produces deepening regional and social inequalities. This happens when population migration, trade and capital movements come to focus on the key growth points of the economy, such as the largest urban areas. As we have seen, geographers contributed to the mapping of a 'modernisation geography' (Peet with Hartwick, 1999), a spatialisation of these visions of growth and a geopolitical imagination of development as pro-capitalist. Geographies of modernisation (Gould, 1960, 1970; Friedmann, 1966) would seek, therefore, to map modernisation 'surfaces' and the spatial diffusion of progress as something 'cascading down urban hierarchies, and funnelling along transport systems' (Peet with Hartwick, 1999: 84).

Another modernisation theorist who identified stages of growth in the development process was Walt Rostow (1960), who outlined them in his book *The Stages of Economic Growth*. According to André Gunder Frank (who later became one of the most vociferous critics of this approach), Rostow had lofty objectives for his anti-communist vision, confiding to Gunder Frank that since the age of 18 he had made it his 'life mission to offer the world a better alternative to Karl Marx' (Frank, 1997: 2). First drafted as a series of lectures presented at Cambridge University in 1958, this notion of development displayed a strong organicism, speaking of laws applicable to all societies and degrees and stages of development. The theory advanced by Rostow permeated the practice of international relations and the international structure of development in various ways (Rist, 1997). Again, in the

"Join us. It's only a step."

Figure 4.1 Join us: it's only a step! Just a small step to progress?
Source: *South: The Third World Magazine*

approach favoured by Rostow, the focus was on a top-down 'trickling' of capitalist development from urban-industrial areas to other regions (Stöhr and Taylor, 1981). Rostow (1960) predicted that nations would 'take off' into development, having gone through five stages which he likened to the stages an aeroplane goes through before take-off. Ranging from stage one, 'traditional society', to stage five, the 'age of high mass consumption', the theory assumes our faith in the capitalist system since Rostow argues that all countries will be in a position to 'take off' into development. Radical geographer Keith Buchanan was particularly critical of the formula-fixated, model-making technocrats such as Rostow whom he saw as largely irrelevant:

> the Rostovian comet (*The Stages of Economic Growth*) has flashed across our horizon, finally to vanish into the academic haze of some US university – and the poor are still with us, poorer and

> more numerous, while the economic systems which were to offer the magic escape from poverty have been crippled by the war which Rostow himself helped to shape and their imperial nakedness is finally exposed for all to see.
>
> (Buchanan, 1977b: 364)

What is particularly interesting about Buchanan's reading of the 'Rostovian comet' is the sense in which this is seen to have encoded a notion of finding a 'magic' escape while masking imperial nakedness and the motives of Cold War geopolitics. The fifth stage of 'high-mass consumption' seemed to be characterised by American Fordism (Rist, 1997). In addition, the role of the western technocratic 'expert' was crucial to modernisation theorists such as Rostow. The analysis that Rostow provided was heavily taken up by the United States Agency of International Development (USAID) and other international aid agencies in the early 1960s (USAID, 2002a). Many expert 'missions' were sent to Saudi Arabia in the 1950s and 1960s, for example, to 'guide' the creation of national-level policy agencies and practices so that the country could harness the potential of its oil resources (Vitalis, 2002). Here, oil companies (many of them US-owned) constructed themselves as 'engineers' of modernisation in the Saudi kingdom and foreign interventions of all kinds were prescribed to help transform the Saudi economic landscape. By following the 'rationality' of modernisation:

> [A] country afflicted by underdevelopment could hope to move briskly into the modern tempo of life within a relatively few years, perhaps a decade. The state would be the key monitor of development, economic growth and macroeconomic policy its main concerns.
>
> (Sylvester, 1999: 705)

States therefore could work together with oil companies, for example, to help this brisk quickening of pace towards a modern tempo of life. Modernisation has also to be understood partly as a response to the failures of aid programmes in the 1940s, in the first cases of decolonisation such as India (Leys, 1996). Enshrined in the World Bank and the USAID, modernisation discourses therefore assumed a brisk and instant leap into the modern. Like many other modernisation approaches, Rostow's model devalues and misinterprets 'traditional societies' which represent the lowest form of development. Traditional society for writers such as Rostow was a kind of 'degree zero of history corresponding to a natural state of "under-development" ' (Rist, 1997: 95). Progress was to get away from this 'natural' state. USAID also later drew on expertise from the Adam Smith Institute and has recently promoted privatisation in the South of the national industries that were once viewed (by the same agency) as the leading lights of modernising forces (George, 2001).

The advanced state of modernisation was nearly always represented as 'western modernisation' and so non-western traditional societies seem like distant poor relations. Modernisation interventions, despite the desire to create modern capitalist economies, also often produced a variety of modernities and of non-capitalist classes and activities which were in many cases closer to 'traditional' modes of social organisation (Gibson-Graham and Ruccio, 2001). Thus there were a variety of economic landscapes created in the wake of colonial capitalist development and not always a universal reduction to homogeneity in the wake of an all-conquering capitalism or modernisation. In fact this was precisely the 'mission' of development thinking during the first decades of the Cold War, to produce and achieve the same kinds of capitalist relationships and 'steady state' institutions in the periphery as were obtained in the 'core' of the world system. What soon began to emerge was an ideology based upon the 'mission to secure a "fit" between structural, normative and behavioural components' of political development (Apter, 1987: 16). This enabled a mutually reinforcing vision of core–periphery relations and statehood so that the West could seek to achieve the same 'steady state' in the 'periphery' as obtains in the 'metropole' (Apter, 1987: 16).

This mission to secure a 'fit' between core and periphery is crucial to the very formation of development thinking. In one sense, then, modernisation theory (and the various attempts to apply it to 'Third World' contexts) involved the idea of replicating and mimicking the development of others. As Gunder Frank has argued, they can also be understood as reflections of the 'Sinatra Doctrine':

> Do it my way, what is good for General Motors is good for the country, and what is good for the United States is good for the world, and especially for those who wish to 'develop like we did'.
>
> (Frank, 1997: 13)

Geographies of modernisation, as we saw in Chapter 3, were a popular product of geographers working in Africa, Latin America and Asia during the 1950s and 1960s. Geography was thus quite an important question in modernisation theory. Slater (1976), writing in *Antipode*, argued that this represented a kind of 'bourgeois social science' which led to what he called the 'hitherto banal orthodoxy of Anglo-Saxon geography' (1976: 88). The modernisation school also often talked about the 'diffusion' of development from what were known as 'spread effects'. Thus a factory might move to a particular location and this might in turn spur development elsewhere. This notion of 'trickle down' was very important to the modernisation writers. The core–periphery model which came out of this (Friedmann, 1966) represented the city as constituting a very important development region – diffusing modernisation and modern social organisation to peripheral areas (especially those with traditional societies). Geographies of inequality and development cannot be summarised neatly as a set of prescriptive stages however, and neither was the city always the development Mecca it was frequently constructed as. Urban areas are themselves subject to uneven development and inequality and did not always disseminate development. Modernisation approaches also failed largely to address the importance of gender and gender relations, assuming that men and women occupied equal positions in terms of power relations and decision-making within the household. The 'trickle-down' effect has often failed to materialise among those who have been the subject of modernisation projects. Modernisation discourses also assumed that all human beings/agents were the same and that the interests and needs of men and women were essentially the same. Important questions were also not raised about where the 'core' ends and the 'periphery' begins or about the wider links to the geopolitics of liberal cold warriors: 'modernization theory was so popular [in the United States] in the aftermath of World War II that it approximated a civil religion championed by liberal Cold warriors' (Nashel, 2000: 134).

Figure 4.2 Women in urban Africa
Source: *Environment & Urbanization* (2001)

It is difficult to forget that the background of the modernisation school is dominated by the USA. The Second World War had undermined the global importance of a number of European countries, while bringing about the rise of countries such as the USA and Japan. After 1945 the USA became the centre of power in the world system, and this continued to be the case through much of the twentieth century (the 'American century') and had important implications for the shaping of development thinking about the 'Third World' (Sidaway, 2000). The USA sought to address the concerns of what it called the 'Free world' which would be furthered by a crusade against colonialism and later communism. The USSR also became a major centre of power after 1945 and its rivalry with the USA in the Cold War is an important feature in the making of development theory. Thus in the 1950s, when modernisation theory first emerged, the 'model of the modern' was the image of the USA writ large (Luke, 1991). Modernisation ideas were most popular and at their zenith throughout the 1960s (Escobar, 1995), which was also a decade of considerable upheaval and conflict around the world. Rostow's book was subtitled *A Non-Communist Manifesto* and comprised an anti-communist focus with the USA as the base of modernisation theory. In the 1950s poverty was thus widely regarded as a breeding ground for the spread of communism as many western policy-makers worried that continued material deprivation 'would drive third world countries in to the hands of Moscow' (Abrahamsen, 2001). Also during this time western countries supported and armed some of the world's most oppressive dictators, sponsored anti-socialist guerrilla movements as in Angola, Afghanistan and Nicaragua and intervened directly to defend dictatorships as in Iran (1953), Guatemala (1954), Brazil (1964) and Chile (1973).

Traditional societies were, however, infinitely more complex than was acknowledged in some of the areas about which geographies of modernisation were produced. These societies were not as 'back-

Figure 4.3 Peace demonstrations in Luanda, Angola, after decades of war (April 2002)
Source: Humana People to People (Angola)

ward' and irrational as they had been portrayed. As Abrahamsen (2001) argues, in the modernisation schema there is nothing before development that is seen as worth retaining or recalling, only a series of deficiencies and absences. The approach was also in a sense very much based around a 'top-down' rather than a 'bottom-up' approach, implying that the process could be brokered by states or development institutions rather than emerge from the 'grassroots' struggles of 'Third World peoples' as had been called for in some more radical approaches. In terms of common criticisms, the division of the world into modern and traditional has often been seen as problematic (Pletsch, 1981). Modern societies were much more fractured and were divided by race, class and politics, and were not as united and responsive to the blueprints of planners as was often assumed. The scale of modernisation programmes was also often a problem in that they assumed that 'big is beautiful' (e.g. large dam-building and irrigation projects). Further, the school and its practitioners often depoliticised development, making few if any references to history and culture. Rostow in particular was able to say little about the 'final stage' of the organic model of development since its underlying principle was that growth has no limits (Rist, 1997); instead he simply discussed a number of very generalised scenarios. Like so many theorisations of development that followed, it ended with a creed, a set of principles about what was to be done and heavily invested faith in the goals of mass consumption and westernisation.

DEPENDENCY: JUST ANOTHER NARRATION OF TRADITION VS. MODERNITY?

The dependency approaches that emerged in the 1960s and 1970s set out to oppose the modernisation approach point by point to such an extent that Rist (1997) argues that in a way they ended up offering only a variation of the tradition/modernity dichotomy of earlier approaches. Although there were important variations in focus on internal and external factors, both approaches saw an imperialist centre and a (post)colonial periphery, reflecting and reproducing long-standing historical divisions. Theorising the manipulation of the periphery by the core was an important process given that at this time a variety of state socialisms had begun to appear (e.g. in Cuba, Angola, Mozambique, Tanzania and Vietnam). The dependency approach has important roots in the USA, Brazil, Chile and Colombia and later opened out into a variety of regions including Africa, the Caribbean and the Middle East. The economic programme pursued by Iran in the 1980s, for example, reflected many of the ideas on 'de-linking' and self-sufficiency propagated by dependency theorists such as Samir Amin and André Gunder Frank (Halliday, 2000). This work emphasised the growing dependence of Third World economies on the structure of international capitalism and occasionally drew upon Marx's writings about the unevenness of capitalist development, calling attention to the manner of incorporation that each country had into the world capitalist system as a cause of this exploitation. Latin American countries had been independent for nearly a century, yet many Latin Americans remained poor despite the multiple promises of modernisation and its mirages. Questions were asked about how colonies had been stripped of their resources and had their land systems reorganised and labour relations reconfigured by colonialism, and the 'parasitic' and exploitative relations it had put in place between core and periphery. None the less, the dependency school continued to focus on the nation-state and the international economy, on the roles of states in preventing inequalities and urging states to reform land structures and on redistributing wealth. Towards the end of the 1960s a growing disillusion with the *laissez-faire* and diffusionist approach of modernisation theory began to emerge as it became clear that there had been a failure to deliver the promised 'nirvana of material benefits' (Sylvester, 1999: 707). A more wide-ranging critique of development theory had evolved which was also firmly rooted in the 'Third World' and in certain traditions of Thirdworldism (see Chapter 5).

The dependency school contended that dependency on a metropolitan 'core' (e.g. Europe, North America) increases the 'underdevelopment' of satellites in the 'periphery' (e.g. Latin America, Africa). Unlike many modernisation approaches, dependency theorists sought to view development in a historical context, arguing that colonialism had helped put in place a set of dependent relations between core and periphery. These peripheral satellites, it was argued, were encouraged to produce what they did not consume (e.g. primary products) and consume what they did not produce (e.g. manufactured/industrial goods). The promised altitude that was the fifth stage of Rostow's model envisaged an urban-based, western lifestyle of consumption, but the dependency scholars showed that the planes of the South had been stalled on the runway by unequal relations and a history of colonialism, denying them a chance of ever being airborne or 'industrialised'. Unlike the modernisation theorists, the dependency scholars focused more on the international and global scales and spaces of development, examining the structural relations of nation-states to the world economy. Just as the modernisation approach was adopted in a variety of ways by international institutes and bilateral donors, the dependency school was made up of all those opposed to US policy and by groups of what were called 'Thirdworldists' (Rist, 1997).

Towards the end of the 1960s a growing disillusion with the *laissez-faire* and diffusionist approach of modernisation theory thus emerged as it became clear that modernisation had failed to deliver the promised flowering of modernity. The dynamism of the Chinese socialist state had encouraged a number of new countries (especially in Africa, Latin America and Asia) to adopt a more socialist agenda for change. New countries such as Tanzania, Mozambique, Vietnam and Cuba all tried socialist experiments. Moreover, many Marxist social scientists in the West had begun to show that development planning and aid was not a neutral process as it had been portrayed in modernisation theory but actually involved exploitation of the Third World in favour of the West. In Latin America, André Gunder Frank made the relations between 'North' and 'South' a key point of focus in the study of *Capitalism and Underdevelopment in Latin America* (Frank, 1969). Frank argued that the relations between what he called the Metropole (North) and the satellite countries (South) were exploitative, pointing out that any surplus generated in the satellite countries was siphoned off to the North, breeding conditions of underdevelopment. Frank saw this exploitation happening on a variety of spatial scales (national, international, provincial, regional, etc.).

Theoretically, the dependency debate was an assault on the conventional wisdom concerning the relationship between international trade and the development process. The neo-Marxist aspects of its critique offered a revolution against capitalism as a way out, highlighting the weakness and vulnerability of western capitalist economies and their dependence on the labour and resources of others as well as focusing on the political role of a local (comprador) bourgeoisie in the process of underdevelopment. In terms of its core versus periphery, a focus on the centre would benefit the wealthy, while the peripheral countries suffered and were actively impoverished as a result of this. The development strategy of the dependency school was formed partly by an institution set up in the 1950s, known as the Economic Commission for Latin America (ECLA, or CEPAL in Spanish). ECLA emphasised industrialisation by import-substitution (ISI) as well as planning and state interventionism in general and the need for regional integration in particular. For a few years, as Leys (1996) has shown, dependency approaches held the initiative, and eventually even the international development community was obliged to accommodate at least some of its critique: the International Labour Office called for 'redistribution with growth', for example, in 1972 (Sylvester, 1999).

Key criticisms directed at the dependency approach were that the theory represented a form of 'economic determinism' and overlooked social and cultural variation within developed and underdeveloped regions. The term 'dependence' had also been used immoderately and had led to oversimplification (Rist, 1997). It may have said much about the origins of underdevelopment but a clear statement of what 'development' itself might be was obscured by a rigid core–periphery model which some have read as a simple inversion of the tradition–modernity binary of modernisation discourse. Another point of contention was that the dependency theorists seemed to be calling for a de-linking from the world capitalist economy at a time when the world economy was undergoing further globalisation. Furthermore, the notion of underdevelopment in a way endorsed concepts of First World–Third World or core–periphery rather than seeking to fundamentally challenge this schema and begin a search for alternative ways of differentiation (Palma, 1981). In this sense the dependency school 'preserved the dualist simplification of the dominant discourse' (Sharp, 1988: 120) and reinscribed a sense of 'us' and 'them' rather than acknowledging that the world does not involve an undifferentiated 'us' and a uniform 'them'. The main propositions of the dependency scholars could also be counterpoised 'point by point to Rostow's theory' (Rist, 1997: 110).

While Rostow's was a history 'from above', a philosophy even, the dependency scholars raised questions about the making of history by real people in definite circumstances. Where Rostow saw colonialism as part of an 'awakening' of modernity, the dependency approach highlighted how colonialism underdeveloped the periphery (Rist, 1997). Its final aim was also to modernise and industrialise and, as with modernisation approaches, 'solutions' to problems were often rather vague and thin on the ground. There is also a sense in which the dependency framework left the impression that there was an 'evil genie who organizes the system, loading the dice and making sure the same people win all the time' (Rist, 1997: 122). In a way this idea of an 'evil genie' is very simplistic. Arguably the break with Enlightenment ideas and modernist rationality was therefore incomplete, in that there remained a focus on industry and industrialisation based around long-established binaries and dualist divisions of the world. In addition, the economy (rather than the culture or politics of individual spaces and places) was still seen as of primary importance by the dependency scholars in a way which lacked nuance and verged on the deterministic. Despite its appeals dependency also did not inspire many policies of development except for short periods in Chile and Cuba.

In the early 1980s some authors began to perceive an impasse in development theory, the topic of several books (Booth, 1985; Schuurman, 1993) from where it has since become a 'landmark in the teaching of development studies' (Munck, 1999: 199). Development pessimism had thus set in by the early 1980s with the realisation of the ever-increasing gap between rich and poor. Essentially, the dependency theorists had stood on its head and inverted the long-standing supremacy of modernisation approaches, which meant that there was some confusion and misrepresentation in early exchanges between these approaches partly because modernisation and dependency 'seemed to checkmate each other' (Schuurman, 2001: 6). There was also confusion around the links that the *dependistas* had to Marxism and a misguided belief that what was being offered in their critiques represented a radical departure from the existing theoretical and political co-ordinates of modernisation approaches (Munck, 1999). The development of Marxism in these debates and its links to anti-colonial struggles deserves much wider recognition, however (Laïdi, 1988). In many parts of Africa, anti-colonial nationalisms found in Marxism a code of legitimation against imperialism, a justification of power, the promotion of new structures of rule and the possibility of 'catching up with history' (Laïdi, 1988: 11). By the early 1980s many commentators noted the diminishing returns of the dependency critique and pointed to an impasse because there seemed no way to go beyond the theoretical coordinates of these previous approaches. While some writers have since sought to 'go beyond the impasse' (see Schuurman, 1993), it has been more common to see a return and retreat to liberal orthodoxies (Hettne, 1995; Munck, 1999).

IMAGINING A POST-DEVELOPMENT ERA

In some ways much of the relevant literature groups writings about 'anti-development', 'post-development' and 'beyond development' together as if they were all exactly the same in their signalling of the end of development. Pieterse (2000), for example, groups all three together as 'radical reactions' to standard and conventional development rhetoric and practice. These are not three fully individual or separate theorisations of development however; neither do they represent some singular reactionary stance. The mindset, intentions, worldview and results of development are rejected by these radical reactions, starting from the basic theme that attaining a middle-class life of high mass consumption for the majority of the world population is impossible (Pieterse, 2000). These writers have also sought to problematise 'poverty' and the portrayal of development as westernisation alongside a critique of modernist, scientific reasoning. Poverty is seen here as a culturally, geographically and historically variable notion (Rahnema, 1992), while an attempt is made to show how people are often misrepresented through distorted images portrayed by global poverty alleviation campaigns. The economics of development is seen as truly pauperising in this regard (Pieterse, 2000). Development is thus seen as the 'new Religion of the West' (Rist, 1997).

A key starting point is the idea that a western, middle-class lifestyle is not a realistic or desirable goal for the majority of the world's population (Pieterse, 2000: 175). These writings could also be read as a critique of the idea of 'three worlds' which continues to shape 'western hegemony' in Africa and other world regions (see Chapter 5). This and some of the other representational practices we have considered so far continue to confer upon western countries the right to administer development and democracy to the 'South', on behalf of the (already) developed/democratic world of freedoms. Thus, it is often argued, Africa can be 'delivered' from its 'current underdeveloped stage' only by First World modernisers. As we shall see in Chapter 7, this good governance agenda is not a neutral humanitarian effort to promote development and growth/democracy but rather a particular discourse which has tapped into larger discursive practices through which global power and domination are exercised (Abrahamsen, 2001).

Modernisation is thus still an important idea and is very influential among a number of senior world leaders, political officials and their discourses. Modernisation was at the very heart of the UN system when it was formed and it still is in a way, dominating the ways in which UN organisations view the transfer of knowledge and technology between regions, for example. Despite all the empty promises of development, G8 states still account for 40 per cent of the voting power in the World Bank, compared to just over 4 per cent for the whole of sub-Saharan Africa. Politicians like Tony Blair in richer countries such as Britain still have the power to influence and define the ideological contours of development policy within the IMF and World Bank and there is now an increasingly close correlation between the policies of the major G8 countries and those of Bretton Woods institutions (Abrahamsen, 2001). Moreover, positivist and empiricist epistemologies still predominate within these global institutional sources of power and the 'knowledge bank' they have assembled is still seen as 'objective':

> Instead knowledge comes to represent the objective 'truth' that the World Bank and others concerned with development have discovered and accumulated about the third world and underdevelopment.
>
> (Abrahamsen, 2001: 13)

These key questions about knowledge and 'truth' in development are associated very much with the post-development school. Post-development writings have 'more than a whiff of counter-modernism about them' (Munck, 1999: 200) and tend to be sceptical towards grand and all-encompassing stories and narrations of development, while the theoretical frames it brings to examine these are quite novel (Sidaway, 2001: 17). For post-development writers there are no easy answers and the whole question of development needs to be further problematised or rejected altogether. In addition, a key thematic in these debates has been the focus on the role of human beings as agents or 'subjects', exploring how their 'subjectivity' is socially constructed in development discourses. 'The subject' is a term that refers to individual human beings as agents and the ways in which they think their place in the world. There are a variety of theories of subjectivity, each imagining the subject in different ways, for example, as bounded, unique and contained or self-knowing. There is no subject prior to knowledge according to post-development theorists; rather subjects are produced by power and discourses. The post-development writers have examined the ways in which subjects are used to justify discourses or to produce stable grounds of knowledge. This was itself quite a major point of departure from earlier streams of development thinking and the ways of knowing constructed in previous approaches which had seen people as the 'objects' of development, rather than as its active subjects and agents. Thus the kinds of analyses of development geographies being produced in such perspectives were very different in that they re-centred attention (to a degree) on people, places, subjectivities and identities rather than assuming homogeneity and downplaying difference.

Post-development writings also seek a critique of the key assumptions of progress and modernity which are formulated alongside a concern with the technocratic measures and definitions of previous approaches. Development is seen here as a particular vision, a set of knowledge, interventions and worldviews or as discourses which are related to the power to intervene, to transform and rule over 'others'. Thus, rather than seeing knowledge and science as disembodied and all powerful, this approach recognises knowledges of development as situated and partial (Yapa, 1996, 2002). Arturo Escobar, a Colombian anthropologist and one of the leading contributors to these debates, argues that different discourses of development have created a 'regime of truth', an accepted way of speaking about and acting towards 'developing' or 'Third World' countries (see Escobar, 1995). The work of Escobar looks at the 'kingdom of abundance' promised by theorists and politicians in the 1950s, highlighting the failure of developmental benefits to materialise. Escobar (1995: 4) argues that the discourse and strategy of development has produced its opposite 'massive underdevelopment and impoverishment, untold exploitation and oppression'. In order to see the 'achievements' of development (if we agree there are any) Escobar argues we must also explore the 'dark side of domination', looking at how subjects become the objects of development, 'normalised' in a global push for rationality.

Following the work of post-structuralists such as Michel Foucault, Escobar views development as a discursive field or system of power relations which produces domains of objects and rituals of truth. His work focuses mainly on the post–1945 period in which the objects of development internalise and come to define themselves in development's terms and looks in particular at how the 'Third World' became a global domain ripe for developmental intervention. Escobar also raises questions about the relations between knowledge and power in development, asking by whom and for what purposes is development knowledge produced and how does this knowledge put in place particular power relations? For post-development writers, conditions of deprivation are understood partly as a form of socially constructed scarcity:

> Discourse is deeply implicated in creating poverty insofar as it conceals the social origins of scarcity. Although the experience of hunger and malnutrition is immediately material 'poverty' exists in a *discursive materialist formation* whereby [the] material, discourse and power are thoroughly intertwined.
>
> (Yapa, 2002: 36, emphasis in original)

Thus the geographical discourses of tropical development produced from area studies centres (see Chapter 3) could be seen as examples of this implication of social scientific discourses in the creation of poverty. This is not to suggest that poverty is not also and in a very real sense a material process (like development itself), but rather to signal that it is also simultaneously a discursive process, involving an enunciation through discourse of what 'problems' or 'opportunities' a country or region faces. This also means that it is necessary to acknowledge the situatedness of post-development writings as well, to be reflexive and to understand how such post-development critiques emanate from certain (sometimes privileged) positions grounded in a particular time and place. As Sidaway notes of Arturo Escobar's work, it is worth remembering that this work offers a *particular rendition* of contemporary Colombia that cannot be generalised easily, reflecting his experience as an anthropologist and his take on Colombia as 'a society of ongoing violent civil war and foreign intervention, whose main export (by value) is cocaine' (Sidaway, 2001: 18). Escobar's critique of development is suggestive, but there is a

Figure 4.4 Post-structuralism and geographers
Source: © David Griffith and Rob Kitchen

risk that it assumes that all experiences of development are as traumatic and problematic as that of Colombia (Sidaway, 2001: 18).

These particular ways of thinking and speaking about the 'Third World' have in turn made possible and legitimised certain practices and interventions towards these parts of the globe. Most usefully for our purposes, Escobar (1995) argues that thinking about development in terms of discourse enables us to maintain a focus on power and domination, while looking also at how that discourse is made possible and what its effects are. In President Truman's famous speech of 1949 which spoke of 'bold new programs' and 'scientific advances', post-development readings of this discourse have focused on the portrayal of 'underdeveloped areas' as passive victims of disease, poverty and stagnation. As Abrahamsen (2001: 16) has argued, the inertia of underdevelopment was seen in modernisation approaches to stand 'in sharp contrast to the dynamism and vitality of the "developed areas" and the USA in particular'. Happiness can be delivered to underdeveloped areas that live in 'misery' by the advancing and expanding western world, through its better technical knowledges and scientific rationalities. By coupling development and underdevelopment, Truman thus justified the possibility and necessity of intervention despite the fact that underdevelopment appeared in his speech to have no cause and poverty was seen as a 'handicap' that produces 'victims' (Rist, 1997).

Poverty was a 'handicap' and a 'threat' both to poor people themselves but also to the prosperous and prosperity itself. Truman's interventions meant that it was impossible to do nothing. Development then has often been considered a humanitarian and moral concern by the wealthy to help and care for the needy and less fortunate, an idea contested in post-development writings. Post-development writers such as Gustavo Esteva have shown that on that important day in 1949 of President Truman's speech about a 'bold new program' some 'two billion people became underdeveloped . . . transmogrified into an inverted mirror of others' reality' (Esteva, 1992: 7). It seemed to matter little if these people were African, Asian or Latin American since they were all 'underdeveloped' at the end of the day. Post-development seeks to focus in part on the 'social technicians of the development apparatus' (Escobar, 2002: 198) and looks to map the complex of knowledge–power and the work of those 'doing the developing' (Ferguson, 1990). The suggestion here is that the the poor were denied a chance to define themselves, changing their names and identities such that all backward peoples were united by being labelled and lumped together as 'underdeveloped'. Their identities were forfeited (along with their autonomy) and they were now 'forced to travel the "development path" mapped out for them by others' (Rist, 1997: 79).

Instead of asking the question 'Why are poor people poor?', post-development writings suggest it is better to ask why particular groups in specific places experience hunger, homelessness and so on. The answers to the latter question are very different from those to the question 'Why does poverty exist?' (Yapa, 2002: 36). When studying the history of development geographies and writings about 'the Tropics', we are interested in how these works stood as focal points for the thinking of a wider scholarly community of geographers. There is clearly much of value then in Escobar's critique and his suggestion that through North–South relations, 'development' has facilitated control of the 'Third World' (and intervention therein). For Roxanne Doty, this has important implications for the issues of representation, in that the 'Third World' was often seen as a space with a 'formless' population of poor and destitute people (Doty, 1996). Both Escobar and Doty contest the idea that all Third World countries are homogeneous and that all are therefore 'primitive', poor, illiterate and lacking in development. As Escobar's work shows, the structuring of a series of absences and deficiencies reinforces the idea that this world needs to be reformed and improved by regions that are not lacking or deficient in some way. Escobar also recognises that even those opposed to conventionally defined development have remained until recently trapped within its language and imagery. In this sense as objects of Third World development, governments have internalised these categories and seen themselves in these terms.

These identities are thus not forced on to passive recipient countries; they can find leverage in them, access to international aid resources as well as a common sense of shared experiences and trajectories. Images and hierarchies in development discourse are also reinforced continually by the media, where stereotypical images of poverty overshadow alternative representations of non-western societies. In the same way that colonialism sought to construct the colonised as 'degenerate' and in need of instruction and administration, contemporary development discourses have produced knowledge and institutions which support (and normalise) western intervention in 'developing countries'. The post-development thinkers in particular have shown that development, for all its power to control other worlds and peoples, is not immune to challenges and resistance. Much attention centres on the way in which development discourses have depoliticised the meaning of these debates and privileged the role of states and state institutions in what Ferguson (1990) refers to as the 'anti-politics machine':

> [Development] agencies present the country's economy and society as lying within the control of a neutral, unitary and effective national government, and thus almost perfectly responsible to the blueprints of planners. The state is seen as an impartial instrument for implementing plans, and the government as a machine for providing social services and engineering growth. 'Development' is, moreover, seen as something that only comes about through government action; and lack of 'development' by definition, is the result of government neglect.
>
> (Ferguson, 1990: 226)

In some ways, Ferguson's work was more 'grounded' than some of Escobar's later contributions. Ferguson had looked at the specific discourses of development that had enframed Lesotho in certain ways. Focusing in particular on agencies such as the World Bank, his work provided strong examples of how this approach could be drawn upon in order to understand power relations, subjects and discourses as they are constructed in various localities (see also Box 4.1). Many institutions, it was argued, positioned the state as the neutral and natural arbiter of progress within national territory in ways which played down the importance of politics and struggle. Ferguson (1990: 256) thus suggests that poverty is depoliticised by the 'hegemonic problematic of "development" '. Post-development writers thus seek to remove the mask of neutrality that is cast over multiple developers and development agencies. These perspectives attempt in particular to 'unveil the shroud of humanitarianism that obscures the way the Third World has been produced and controlled by the discourses and practices of development' (Van Ausdal, 2001: 578). They also explore the power relations of development, the construction of its categories, codes and truths. Truth is a key question here, in that the creation of a 'politics of truth' about the 'Third World' has conditioned a considerable portion of the debate about 'development'. Escobar argues that underdevelopment is a fictitious concept, leading to a discourse which instilled in Third World countries 'the need to pursue [development] and provided for them the necessary categories and techniques so to do' (Escobar, 1988: 429). A particular reality is thus made visible through this fiction and discourse, leading to specific 'treatments' and 'solutions'. Some post-development writers have shown that the material and discursive aspects of development are thus thoroughly intertwined and have sought to show how the labels, images and goals of development have infiltrated local mentalities, philosophies and epistemologies (Marglin and Marglin, 1990; Van Ausdal, 2001).

One idea that runs through much of this work is that there has been (in Escobar's words) a 'triumph of the Western economy' and the creation of an economic rationality, with subjects subordinate to capitalism, which is seen to spread globally through the 'guise of development'. Escobar also refers to the need for a 'discursive insurrection' (1995: 17) which is capable of disrupting the powerful blinders of development theory and practice, to displace its categories and therefore open up a space for alternatives. Much attention has been focused in this regard on the role of minority cultures and social movements, which are discussed in more detail in Chapter 9. One of the major problems with this work, however, is that it often creates an image of development as based on an exogenous model of the industrialised world, and this has been taken by critics to mean a prejudicing of the local over the global, a kind of return to ISI days. We need also then to think here about the multiple centres and regions that shape development discourse (including those in the South) – not just about 'westernisation' (and obvious icons or centres of consumption). What about the East Asia 'miracles? What of the role of Japan or the Middle East? We may also question the assumption that the "worship of progress" (Rist, 1997) is simply a western preserve. Resistance to western discourse (as we shall see in Chapter 9) occurs within as well as between countries.

Development has none the less been characterised by post-development writers as a kind of 'Frankenstein-type dream', an 'alien model of exploitation', and what is needed (especially for Escobar) is not more development but rather a 'new

BOX 4.1

Expectations of modernity in Zambia

> Throughout the 1960s and 1970s, we must remember, Zambia was not reckoned an African 'basket case' but a 'middle-income country' with prospects for 'full' industrialisation and even ultimate admission to the ranks of the developed world.
>
> (Ferguson, 1999: 6)

Development expertise and knowledge in the 1950s and 1960s spoke of an 'emergence' in the economy of a forward progression towards modernity that has not been forthcoming. If these discourses of development are geared so closely towards the future, then how can they accommodate the realities of stagnation and decline? The Zambian Copperbelt existed in theory as a resource. Ferguson's book *Expectations of Modernity: Myths and Meanings of Urban Life on the Zambian Copperbelt* (1999) spells out the trajectories of the economic declines that followed and their impact on people's ways of understanding their lives. The book refers on several occasions to the 'modernisation myth' and what happens when it is 'turned upside down, shaken and shattered' (Ferguson, 1999: 13). This work illustrates how the collapsed hopes and continuing poverty of an area once thought of as at the vanguard of Africa's industrial development belies a simplistic faith in the benign effects of globalisation.

> The mythology of modernization weighs heavily here. Since the story of urban Africa has for so long been narrated in terms of linear progressions and optimistic teleogies, it is hard to see the last twenty years of the Copperbelt as anything other than slipping backward: history, as it were, running in reverse. How else to account for life expectancies and incomes shrinking instead of growing, people becoming less educated instead of more, and migrants moving from urban centres to remote villages instead of vice-versa.
>
> (Ferguson, 1999: 13)

The myth of modernisation was more than just an academic myth; the idea that Zambia was destined to move ahead and join the ranks of modern nations was shared by many national and international policy-makers and many ordinary Zambians. Modernist metaphors did not disappear with the demise of modernisation theory in the 1960s and (as we shall see in Chapter 5) are at the heart of contemporary debates about post-colonialism (Ferguson, 1999: 17). As Ferguson shows, modernisation refracted colonialism and development 'through the looking-glass', where modernity was seen as the object of nostalgic reverie, and 'backwardness' became the 'anticipated (or dreaded) future'.

The breakdown of those narratives of modernity and associated ways of understanding the world has had a variety of impacts on Zambian livelihoods, experiences and interpretations of development today. A very complex and differentiated global political economy was simplified around a race for economic and political 'advance'. Ferguson also explores the myth of the 'modern family' on the Copperbelt and instead looks at the diversity of actual trajectories of men and women who move across rural and urban space. The conclusions of his book explore the meaning of decline (calling for new ways of theorising this) and the breakdown of modernisation for Copperbelt residents. Today the state's ownership of the Zambian copper industry has been privatised and faces severe problems, with the possibility of mine closures and unemployment for the country's 15,000 copper industry workers. The product remains central to Zambian exports but privatisation has not been without its problems. Zambia receives about US$1 billion a year in foreign aid, much of which was conditional on the state's accepting privatisation. In 2000 a debt-relief package of US$3.8 billion was agreed, making Zambia one of the highest recipients of official aid in Africa (*The Economist*, 1 June 2002).

regime of truth and perception'. Post-development ideas more generally have also been criticised for ignoring discontinuities in the World Bank's discourse over time and for drawing upon dependency approaches very closely, focusing on external relations, but now taking those ideas further to a view of development as an international power–knowledge regime (Pieterse, 2000: 181). The priority given to local and grassroots initiatives is seen as a weakness by Pieterse and other critics of this approach. Many focus on the way in which Escobar confers enormous power on the external world system, privileging certain localities and cultural identities. In this sense, as Watts (2001: 171) has argued, Escobar replicates the weaknesses of the dependency scholars' arguments in that the nature of external power is 'crudely articulated in bold outline'. While appearing to be concerned with a variety of spatial scales and with rooted, embodied knowledge and place-specific practices, critics have argued that there is a simplification and homogenisation in seeing development in this way, which aggregates development in ways that prioritise the foreign and the exogenous and see this as uniform or monolithic. Post-development is seen to have constructed a rather basic model of development as *social engineering*, as the ambition to shape economies and societies in interventionalist and managerialist ways. As Pieterse argues, for writers such as Escobar:

> It [development] involves telling other people what to do in the name of modernisation, nation-building, progress, mobilisation, sustainable development, human rights, poverty alleviation and even empowerment and participation.
>
> (Pieterse, 2000: 182)

The post-development approach has thus been seen as essentialising development, misrepresenting its history (and its claims to progress) and underestimating the complexity of 'motives and motions in modernity' (Pieterse, 2000: 183). Some even say the approach is centred on 'language games' rather than on any form of critical analysis. In terms of ways forward, Escobar mentions three major discourses – democratisation, difference and anti-development – which can serve as the basis for radical anti-capitalist struggles (Escobar, 1992). These struggles, Escobar has argued, need to be expanded and articulated, although the agenda for this is not especially clear. Escobar's work is seen by some to romanticise local/social movements and the wider process of resistance which is dismissed by Pieterse as 'quasi-revolutionary posturing' (Pieterse, 2000: 186). Other critics of the approach also argue that it presents 'a conventional and narrow view of globalisation, equated with homogenisation' (Pieterse, 2000: 188). Development is seen to be ascribed a single and narrow meaning and a simplistic conception of power which ignores its 'polysemic realities' (Pieterse, 2000, 2001a).

To some extent a few of these reservations about post-development/anti-development writings are shared by Van Ausdal (2001) who tries to use the framework to explore development discourse and practice among the Maya in Southern Belize. Examining a project by the International Fund for Agricultural Development (IFAD), Van Ausdal found that far from there being a 'depoliticisation' of development through agencies such as these, the IFAD project inspired a *greater* politicisation of land claims. In this way the attention of post-development writers to discourse and alternative visions is welcome but may be problematic in that '[m]ore often than not, however, their conclusions are overstated' (Van Ausdal, 2001: 580). Interestingly, Van Ausdal concludes with a warning of the dangers of generalising from the Belizean case.

CONCLUSIONS: DECONSTRUCTIONS AND RECONSTRUCTIONS OF DEVELOPMENT

> The theoretical criterion [for rethinking development] is to enable observers to transcend the immediacy of events and see alternatives and possibilities not visible on the ground or to the naked eye. Such a goal is itself a political matter. It assumes that by 'rethinking' development, we can treat it as a fund of knowledge, a form of intellectual capital. . . . Its potential value as a constituted perceptual universe is a reordering of experience, the revaluation of case materials, and the derivation of new comparative hypotheses. In the last analysis we are talking about the power to appreciate the larger picture.
>
> (Apter, 1987: 15)

> Theorising about development is . . . a never-ending task.
>
> (Hettne, 1995: 15)

It is impossible to cover so many centuries of intellectual history in such a short space but it is possible to draw out some of the major notions of growth, progress and development that have been formulated. There have been important continuities and breaks between Enlightenment thinking and what followed, but what does come through loud and clear is the persistent influence of the ways in which western societies have conceptualised their relationship to the past and future and how this has closely shaped contemporary ways of envisaging history and geography (Rist, 1997). In addition, all theories and all forms of geographical writing about development have, to different degrees and in various ways, assumed some theory or notion of subjectivity of the role of the individual human being as an agent of this process. Each also articulates certain visions of power and a sense of the spatiality of power and development, and each is based on particular epistemologies or senses of what constitutes knowledge. These are key themes in rethinking development. In Chapters 1 and 2 we recognised the possibilities of deconstruction as a method of thinking about development, and what this movement of ideas and writings shows is that critical discourses can themselves easily slip into and adopt the assumptions, postulations or positions that they seek to contest (Manzo, 1991, 1995). A good example of this process at work comes from our analysis of the dependency approach which shared the same theoretical and discursive spaces as modernisation approaches, prioritising the national scale and national territory and foregrounding state intervention. These two approaches have been singled out here because they remain 'inextricably intertwined in one another's assumptions' (Munck, 1999: 203). As Slater (1993) has argued, however, dependency approaches represented an important attempt to 'theorise back', as a response to the assumptions of modernisation theorists about growth and the diffusion of progress.

The extent to which alternative (postmodern or post-structuralist) approaches can go beyond these discursive spaces and these familiar theoretical co-ordinates remains to be seen. Another key question concerns the extent to which deconstructions of development can be drawn upon to formulate *reconstructions* of development and attempts to build another or alternative strategy of resolving material scarcity and inequality. As Munck (1999) points out, one of the main reasons to be positive about anti-/post-development discourses is that they give a voice to the excluded and they seek to break with the dominance of western rationality and some of the legacies of Enlightenment ideals. Geographers in particular, concerned with the aftermath of colonialism, might welcome the new emphasis on identities and place-specific practices. The emphasis on national communities and national identities is now being opened up to consider a wider range of spatial scales and wider forms of identity formation. These writings are also useful in that they highlight the universalist pretensions of the Enlightenment (and its reinstatement through postwar development). They seek a radical break with existing mainstream global development discourse, but just because this break is incomplete or still operates within older binaries and frameworks, it does not imply that these writings are not relevant. Rather we must seek to understand how their conclusions should be qualified and their knowledge situated and located in particular spaces and places. Post-development writings do not seek to deny globalisation and modernity but instead aim to find new ways to imaginatively transcend them (Hoogevelt, 1997).

It is important therefore to recognise that these various schools of thinking never had a singular or simplistic 'theory' but rather a series of prescriptions or recipes for change which was never uniform among its proponents. They have also emerged from different spaces and different times, which, in tandem with changing geopolitical relations, changes the meaning and theorisation of development over time across spaces and 'worlds'. None the less, it is important to point out that in the specific case of modernisation theory one of the most important unifying factors that brings the school together is Eurocentrism. Modernisation theory implied that the history of Europe or of the West more generally could be *read on to the future history* of the Third World. Dependency and post-development writings attempt to disrupt and dislodge this overwhelming theorisation of development from 'the North'. However, even here some traces of Eurocentric thinking did still resurface. In some ways this is not surprising given that the predecessor of modern development economics was colonial economics (Pieterse, 2001a). Radical dependency theory turned attention 'from top to bottom' and shifted it towards attacks on authority and power relations rather than on how to establish and maintain them while arguing 'against the state rather than for it' (Apter, 1987: 48). In Middle Eastern geographies, the politics of development and underdevelopment have also been debated widely: 'Writers from predominately Islamic countries (most notably Iran) saw the obsession with development as part of a misplaced "intoxification" with the West' (Dabashi, 1993).

Marshall (2001) also examines dependency debates in a Caribbean context, noting the particular connotation that these ideas acquired in that setting over time. Geographers even applied dependency theory to an understanding of Britain's relationship with the Republic of Ireland (Regan and Walsh, 1977) and Wales (Buchanan, 1977a). Thus an important question here is the extent to which different development practices and theories have sought to map and represent 'developing

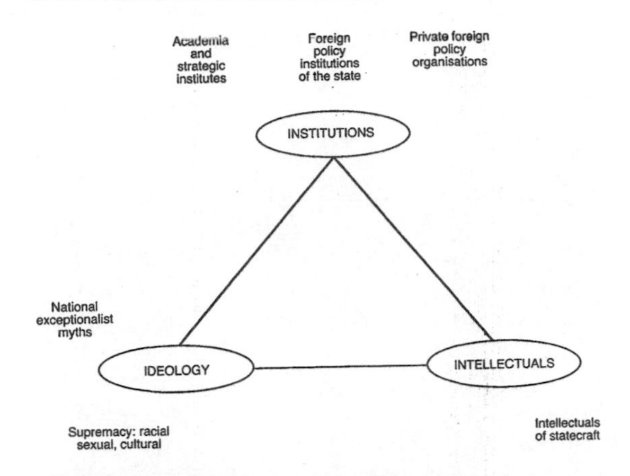

Figure 4.5 Geopolitics as power/knowledge
Source: Gearoid Ó Tuathail

regions'. Mitchell (1995) argues that many development agencies tend to portray a region they have targeted in ways that seemingly justify predetermined sets of prescriptions. With respect to Egypt, Mitchell argues that many development texts invariably begin with a kind of opening snapshot emphasising the country's natural resource limits or an image of the population squeezed by the desert into a narrow river valley. Through such imagery, he argues, USAID-led modernisation proposed to overcome the country's physical geography through technology and management. In countries such as this, however, the experience of postwar development has been highly contested and has been littered with unintended and often disastrous consequences. For this reason, many critics of the post-development approach are perhaps somewhat premature in the dismissal of what they term as its 'quasi-revolutionary' postures. As Esteva and Prakash (1998) have pointed out, not all post-development writing involves a simplistic romanticisation of the indigenous or a privileging of the local over the global.

The important point to consider is why development's discrepancies have been so common and so widespread. Discrepancies between policies and concrete results have been so great and experiences so often disastrous that 'another round of [development] thinking is called for, this time in transdisciplinary rather than interdisciplinary terms' (Apter, 1987: 14). Post-development writings do build on previous streams of development thinking and this should not be seen as a weakness but rather as an important attempt to re-engage development's past. The 'postmodern' foundations of post-development contributions have also been seen by some as Eurocentric and therefore not relevant. For Simon (2001: 122), '[f]ew theoretical propositions have elicited such strongly polarised views' as postmodernism. Simon (2001) also asks, if postmodern literally means 'after the modern', then how relevant is postmodern thought in the global South where the majority of people are 'still poor and struggling to meet basic needs' (Simon, 2001: 123)? Indeed this is precisely where post-development writers seek to intervene, in this social and geographical construction of the 'poor majority' and their struggle to meet their needs. This approach

TABLE 4.1 DEVELOPMENT THEORIES AND GLOBAL HEGEMONY

Development thinking	*Historical context*	*Hegemony*	*Explanation*
Progress, evolutionism	Nineteenth century	British Empire	Colonial anthropology, social Darwinism
Classical development	1890–1930s	Latecomers, colonialism	Classical political economy
Modernisation	Post-war boom	US hegemony	Growth theory, structural functionalism
Dependency	Decolonisation	Third World nationalism, NAM, G77	Neo-Marxism
Neoliberalism	1980s>	Globalisation, finance and corporate capital	Neoclassical economics, monetarism
Human development	1980s>	Rise of Asia and Pacific economies, 'emerging markets'	Capabilities, developmental state

***Source*: Adapted from Pieterse (2001b: 9).**

***Notes*: Non-Aligned Movement (NAM) Group of 77 (G77) countries favouring non-alignment with Cold War powers.**

BOX 4.2

Disability and the theorisation of development

Few discourses of development actually recognise the subjectivity of people with physical disabilities. The complex relationships between space and disability have received increasing attention from geographers in recent years as it has become necessary to explore how social and spatial processes can be used to *disable* rather than *enable* people with physical disabilities. Gleeson (1999) talks about the 'long disciplinary silence' in geography and writes that geographers were 'absent without leave' from the broader intellectual campaign around disability issues: 'A failure to embrace disability as a core concern can only impoverish the discipline, both theoretically and empirically' (Gleeson, 1999: 1).

Debates about how space informs experiences of disability expanded considerably in the 1990s but largely urban, Anglophone, western societies remain the predominant focus of attention. Gleeson (1999) has referred to the need to bring about 'enabling environments and inclusive social spaces'. Instead, many development organisations arguably construct elaborate 'landscapes of dependency' (Power, 2001).

As geographers interested in development, it is absolutely crucial to play our part in bringing an end to these disciplinary silences through an illustration of the discipline and power of development and dependency and by exploring the possibility of alternatives. There are a number of common themes and experiences shared by disability movements within and between countries, and regional and international co-operation remains an important objective for the future. It is important then to explore the 'commonality of disability', since, on almost every indicator of participation in 'mainstream life', disabled people come out extremely badly; for example, in employment statistics, income levels, suitable housing, access to public transport, buildings, information (newspapers, radio and television) and leisure facilities. It is also useful to ask questions about the limited extent to which development theories are inclusive of disability issues and the voices of those with disabilities.

The UN often talks of disability issues as a 'silent crisis' affecting societies of the South with UNICEF, UNESCO, the ILO and the WHO as the major agencies involved in developing responses. According to the UN, 'it is imperative that planners remain sensitive to the disability dimension early and throughout the development process' (UNDP, 2000). Again, however, there is the notion of a singular disability dimension that needs to be appended to an untransformed process. Many development agencies silence disabled people in their representations, in their disabling politics and in their desire for impairment-specific and technological solutions. This may be partly a consequence of their continued affinity with problematic discourses of development, which homogenise subjects and genders.

Figure 4.6 UN Year of the Disabled, 1981 (UNESCO poster)
Source: UN Department of Public Information

The World Bank's neoliberal position is that disability issues are central to their mission but they talk of disability-reduced productivity within the workforce and view disability very much within a kind of 'cost–benefit' analysis. As a creator of poverty in Africa, where does the Bank's moral authority and expertise come from (BWP, 2001a)?

BOX 4.2 (contd.)

Solutions to disability oppression are often impairment-specific and frequently technological, however. This kind of disability focus is disabling in that it silences what are undoubtedly shared meanings (Ingstad and Whyte, 1995). Sentimentality and patronage are important, with disabled people portrayed as powerless and the victims of violence (Priestley, 1999). There remains none the less an 'urgent need for development organisations and funders to take disability on board as an equal opportunities issue (as with gender and ethnicity)' (Stone, 1999: 10). Disability is much more than just a singular issue to take on board however, while the co-option of disability movements by development organisations is not unproblematic. To add disability to a development agenda as if it was some kind of cumulative list of needs means that the underlying ablest assumptions of development(ism) remain unchallenged. Disability is profoundly relevant to the theorisation of development and has been acknowledged very infrequently in the literature.

seeks to focus on the agents, methods, objectives and values of different discourses of development. Postmodernism refers in part to an intellectual movement which rejects periodisation, the search for origins and, most significantly, the idea of progress – all of which were important modernist problematics. Postmodernism also celebrates fluidity, ambiguity and difference and rejects the normalising tendencies of Eurocentric discourses with their grand, universalising claims.

Although many critics are sceptical of the relevance of postmodernism in the South, feminist geographers have been attracted to the way in which reason is seen as gendered and ethnocentric rather than disembodied and neutral (McDowell and Sharp, 1999). Simon (2001) shows that postmodernism has been considered by some as 'yet another Northern paradigm', a kind of 'Northern intellectual toolkit', mentioning how postmodern writers often also fall into the trap of exclusivist modes of argument (Simon, 2001: 125). None the less, it is the contention of this book that post-development writings are critical to any attempt at rethinking development geographies and offer crucial new perspectives about power and knowledge which deserve much further engagement rather than casual dismissal as 'utopianism'. The approach to development issues adopted in this book is closely informed by post-structuralist thought, in particular the work of Jacques Derrida. In his writings, Derrida claims that western philosophy is founded on the programmatic logic of opposites which he suggests we should not seek to reverse or invert but rather to destabilise, upsetting the simplicity of binary divides which structure much contemporary development thinking (see Chapter 8). Escobar's work also seems to suggest that it is necessary to take a step back and make room for other ways of knowing, other kinds of conceptual mappings of the world: 'imagine moving away from conventional Western modes of knowing in general in order to make room for other types of knowledge and experience' (Escobar, 1995: 216).

In Chapter 5 we will be looking at the idea of 'three worlds', an important conceptual mapping in a variety of debates about the theory and practice of development. Various debates have focused on the utility of this term in allowing collective projects of development and in encouraging debate and discussion. Over time, international development organisations have formulated newer terms and languages however, and have occasionally co-opted progressive-sounding ideas into their conceptions of development, using terms such as 'endogenous', 'sustainable', 'popular' or 'human', integrated' or

TABLE 4.2 GENERAL TRENDS IN DEVELOPMENT THEORY OVER TIME

From	*To*
Macro-structures	Actor-orientation, agency, institutions
Structuralism	Constructivism
Determinism	Interpretative
Homogenising, generalising	Differentiating
Singular	Plural
Eurocentrism	Polycentrism

***Source*: Adapted from Pieterse (2001b: 3).**

TABLE 4.3 MEANINGS OF DEVELOPMENT OVER TIME

Period	Perspectives	Meanings of development
1850>	Latecomers	Industrialisation, catching up
1870>	Colonial economics	Resource management, trusteeship
1940>	Development economics	Economic (growth) – industrialisation
1950>	Modernisation theory	Growth, political and social modernisation
1960>	Dependency theory	Accumulation – national, autocentric
1970>	Alternative development	Human flourishing
1980>	Human development	Capacitation, enlargement of people's choices
1980>	Neoliberalism	Economic growth – structural reform, privatisation, deregulation
1990>	Post-development	Authoritarian engineering, disaster

***Source*: Adapted from Pieterse (2001b: 7).**

'comprehensive' development (Munck, 1999) partly as a way of masking the lack of any real transformation of core assumptions. For geographer Neil Smith 'the paradigm of modernisation has been reinvented as globalisation, but they both issue from the same mouth' (Smith, 1997: 174). Similarly, Dirlik (2002) refers to the replacement of the paradigm of modernisation with the paradigm and models of (neoliberal) globalisation.

There are, however, important differences between modernisation approaches and neoliberalism and it is important to recognise that one consequence of the perceived and imagined 'impasse' in development thinking was that modernisation orthodoxies were reinscribed, dusted off and dressed up in new clothes but were never copied and mimicked in their entirety through neoliberal ideologies. This is an important contention none the less, that the promised kingdom of abundance envisaged by modernisation has been reinvented and reconstructed into contemporary neoliberal discourses which issue from the same sources, roots and voices as those of the modernising agents of the early postwar period. If modernisation was a Cold War attempt at legitimating US domination of the global system, then (as we shall see) neoliberalism is a continuation of this project in the 'era of market triumphalism' (Peet with Hartwick, 1999: 90). In this way the contemporary world order of international development involves an important geo-economic dimension in that it is based upon this doctrine or ideology of transnational liberalism or neoliberalism. The fundamental principle of such an approach is that economic liberty must be preserved and that an economy must be free from the social and political 'impediments', 'restrictions' and 'fetters' placed upon it by states that seek to regulate markets in the name of public interest (Routledge, 1998). The connections to the Enlightenment and to the writings of Adam Smith are thus enduring, and continue to shape development thinking and practice today in a number of ways.

BOX 4.3

Chapter-related websites

http://www.daa.org.uk/
Disability Awareness.

http://www.dpi.org/
Disabled People's International.

http://www.worldbank.orb/dpnet/
World Bank.

http://www.pha2000.org/
People's Health Summit.

http://www.thelandminesite.com/
Landmines.

http://www.landminesurvivors.org/
Landmines.

5 Thirdworldism and the Imagination of Global Development

INTRODUCTION: A CRITICAL GEOPOLITICS OF THIRD WORLD DEVELOPMENT

> At times the 'West' excludes Latin America, which is surprising since most Latin Americans, whatever their ethnic origin, are geographically located in the western hemisphere, often speak a European tongue as their first language, and live in societies where European modes remain hegemonic. Our point is not to recover Latin America – the name itself is a nineteenth century coinage – for the 'West' but only to call attention to that *arbitrariness of the standard cartographies of identity* for irrevocably hybrid places like Latin America, sites at once Western and non-Western, simultaneously African, indigenous, and European.
>
> (Shohat and Stam, 1994: 13, emphasis added)

> It is strange to interpret a political world through cardinal points.
>
> (Williams, 1983: 200)

Where and what exactly is 'the West'? This question puzzled Christopher Columbus and in a way remains quite puzzling today. It is important when thinking about 'development' to consider the extent to which countries aspire to become 'western' – at least in terms of achieving 'western' standards of living. Although it should have become clear to Columbus that the New World into which he had stumbled was not the 'East' at all (with its many sources of wealth and richness in resources), he none the less still peppered his reports with a range of fantasies and myths about the mystic 'East' (Hall, 1992). Our ideas of 'East' and 'West' have never been free of myth and fantasy and even to this day they are not primarily ideas about place and geography. These terms or cardinal points have important and complex lines of descent (genealogies). The contrast between 'East' and 'West' is very old, but it has changed its content repeatedly and has been transformed into the contemporary contrast between 'North' and 'South' (Williams, 1983). Some commentators have argued that in a way we have no choice but to use phrases such as 'Western' or 'Third World', but it is also important to remember that these terms represent very complex ideas and have no single or simple meaning. The idea of the West then is as much *an idea* as a fact of geography (Hall, 1991). These terms are not some universal shorthand but rather must be viewed as dynamic and complex representations related to important and changing geopolitical discourses and languages. In order to understand the use of terms such as 'the West', 'North'/'South', 'First World'/'Third World', it is necessary to develop a critical geopolitics of development, which raises questions about states, territories and the geographical nature of power relations.

Political divisions between Western and Eastern Europe at the end of the Second World War were soon generalised into what was then offered as a larger 'universal contrast, between political and economic systems of different types' (Williams, 1983: 201). Thus where once there had been a long-standing division of 'East' and 'West', after 1945 these were in a way transformed into a kind of North–South opposition. According to Raymond

Williams, the 'East' is generally perceived negatively in this binary framework and it is generally an advanced capitalist economy that is seen to qualify a country as 'western':

> Thus the North–South contrast is either: (a) an absolute contrast between industrially developed (rich) and underdeveloped (poor) economies; or (b) an extension of this to the terms of the East–West polarisation, as this primary competition extends itself to the 'South', which can go either way; or (c) an ideological form of relations between 'the West' (the capitalist world) and the exploited 'South', with the 'East' (the socialist world) as the 'South's' natural ally.
>
> (Williams, 1983: 202)

In some ways it could be argued that this image of an 'exploited South' relates to a wider tendency to see what Willliams (1983) refers to as a 'blocked and generalised' image of poverty outside the West or the 'first world', one which plays down complex differences between and within regions. This chapter seeks to highlight the Eurocentrism of many systems of ordering the world and also aims to show that these systems of ordering are partial and frequently negative truths which deny cultural differences and stereotype the 'Third World' as a kind of overpopulated world of problems. Further, the chapter illustrates that the terms 'First World', 'Second World' and 'Third World' all reflect important Cold War conceptualisations of space and systems of ordering the world that themselves quickly became important (geo)political projects. The labelling of certain regions of the world is a complex and contested process which has involved the formation of various kinds of important alliances. More importantly, this chapter seeks to show how discourses surrounding the term 'Third World' have in many ways represented, embodied and eventually reproduced many of the inequalities present in (and characteristic of) colonial relationships of power.

The word 'geopolitics' is now familiar to many people but remains quite difficult to define. The term refers largely to the way in which space is important in understanding the constitution of international relations. A critical geopolitics of development may offer some useful ways forward in understanding the history and geography of the 'Third World' as a political idea. As a school of thought this approach focuses on political meanings and representations and explores the contestation of the geopolitical world while examining the discourses of geopolitics which underpin the production and reproduction of geopolitical spaces such as 'Third World' (Johnston et al, 2000). Political geographers have identified patterns in the development of international politics (which have been called *geopolitical world orders*) and have thus tried to understand the geopolitical transition from one order to another (e.g. before and after the Cold War). Adopting a critical geopolitics perspective, it is necessary to avoid major generalisations across 'Third World Politics' (and the assumption that 'Third World' political systems are essentially the same) and thus we should begin by distinguishing between nations and states. Nations are defined on the basis of culture, religion and ethnicity, for example, whereas states represent political units whose authority is bounded by other political spaces. The 'critical geopolitics' literature explores the rights and abilities of each state to control the territory encompassed by its boundaries or to gain international recognition for its territorial authority. Relationships between power and territory can be observed at all geographical scales:

> The study of geopolitics involves considerations of territory, power and conflict between nations and states. In international relations, control of territory usually increases power, while increased power can expand control of territory. More powerful states exercise direct or indirect territorial control over weaker ones. In many cases, power implies direct, formal political sovereignty over designated territories.
>
> (Braden and Shelley, 2000: 9)

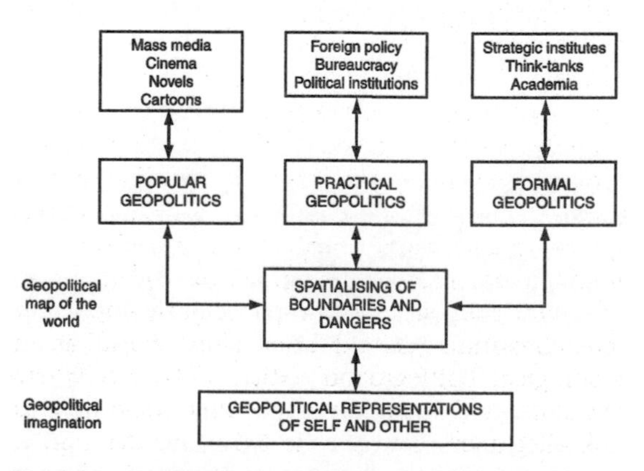

Figure 5.1 A critical theory of geopolitics as a set of practices
Source: Gearoid ÓTuathail

It is thus necessary to examine how international political relations have been *geographically produced.* Rather than assuming a fully formed state system and state-delimited territories, the concern of a 'critical geopolitics' perspective is thus with the power struggle within and between different societies over the right to speak sovereignly about geography, space and territory (Ó Tuathail, 1996: 11). What is particularly interesting and important about this perspective is that it argues that territory can never be understood in isolation from a wider complex of state power, geography and identity (Ó Tuathail, 1999: 140). This complex interaction between geography, power and identity shows how everyday life can and has been defined and delimited around concepts of spatial divides, borders and boundaries and at different spatial scales. Through such processes as class, wealth, gender, race, citizenship and political power, borders are being produced continually (Ó Tuathail, 1999: 150) in ways which have important implications for the changing spatiality of global inequality. In South Africa, for example, territorial control was central to the privileging of whiteness and certain versions of national identity during the apartheid era. As Luke (1993) has argued, states have often sought to establish their power by *in-state-ing* themselves in space, imprinting a mark of their territorial presence.

The form of in-statement of power and authority through political space varies considerably, reflecting the variety of historical struggles that have gone into the creation and maintenance of states as coherent territories and identities seeking international legitimacy (Ó Tuathail, 1996: 12). The 'scramble for Africa' in the late nineteenth century is one important example of this surrounding and enclosure of political space. More recently, in Mexico, this process of in-state-ing power in space has been resisted strongly by the Zapatistas in the state of Chiapas (see Chapter 9). The way in which geopolitics works 'from below' is a crucial theme in this book which can allow for an understanding of how grassroots resistance serves as a response to development theories and practices and can change or modify these theories and practices. In addition, Cold War geopolitics is also particularly important here, constituting as it did the 'Third World' as an 'ideological battleground' (Rist, 1997: 80) where new states and liberation movements would have to seek allegiances between the USA and the USSR, between capitalism and communism. Examining the 'Third World' as a geopolitical space, we can explore further the relationships between 'otherness' and territory or the ways in which an 'inside' and an 'outside' have been created across the world political map. It is also possible to focus on the historical and geographical patterns of inclusion and exclusion after 1945 and to point to the ways in which the world political map was manipulated in order to enforce internal/external boundaries during the Cold War. The USSR in particular played on its ambiguous position as both 'inside' the international system and 'outside' it, presenting itself as the natural 'midwife' for completing the independence of newborn states (Laïdi, 1988). With the accelerating pace of decolonisation and the creation of independent states in the South, geopolitical questions were addressed from a set of new or 'Third World' perspectives, a kind of North–South geopolitics which emerged from the legacies of colonialism and the perception that 'underdeveloped' countries had distinct geopolitical considerations from those of western societies. The Cold War was an important backdrop to all this, as non-western countries debated their relations to Cold War geopolitical discourses emanating from the USA and the USSR and sought to position themselves in relation to these discourses. In addition, as we have already seen, debates about the role of geopolitics, foreign aid and development remain central to the post-Cold War geopolitical world order.

Many questions have been posed here about the role of the state as the primary actor in geopolitical relations, leading to enquiries about whether this is shifting in the context of globalisation (Braden and Shelley, 2000). In sub-Saharan Africa, states have

Figure 5.2 Photo of Havana skyline
Source: Alix Wood

been described variously as 'failed' or 'collapsed', as 'patrimonial' or 'clientelist', 'kleptocratic' or 'predatory'/'extractive' and even 'vampire-like' (Frimpong-Ansah, 1991; Englebert, 1997; Goldsmith, 2001). The notion of a 'vampire-like' state dramatically illustrates the idea of the state as fundamentally extractive, draining away the very lifeblood of the nation and its people. To attach a particular label to a state is to ignore the variations of ideas and opinions that can exist within the state apparatus: there are often conflicts within the state and opposition to corrupt practices, for example. In focusing on geopolitics in the so-called 'Third World' there are important questions to be asked about the role of a variety of states in 'Third World development' and about the representation of 'failed' or 'weak' states in particular as they have been represented by global development agencies since 1945.

Politics in the 'South' is fundamentally about conflict, territory and boundaries, and about access to power structures which benefit some groups and disadvantage others. States can govern partly because their people recognise their own identity as *citizens* of that state and identify with the government of national territory as *their* government. This recognition of statehood, both internally and externally, entitles governments to act on behalf of the people in internal and external transactions. The lack of correspondence between national territories and state boundaries is a significant issue in many non-western societies since boundaries were often drawn up 'at conference tables in European capitals by diplomats with little or no knowledge of local conditions' (Braden and Shelley, 2000: 8). All over the world there are peoples who do not recognise the authority of the state that claims them. Governmental legitimacy therefore does not always exist (e.g. in countries ruled by military regimes). In this chapter we can begin to raise questions about how 'Third World' states sought to build this popular legitimacy in conjunction with other states or found their territorial authority undermined by national and international conflicts. In referring to a 'state' one is often talking only about the group which owns or dominates its apparatus at that particular historical conjuncture. What we need to examine here therefore are the aims and objectives of groups which sought to control the state and its 'technologies of domination' (if we follow Foucault's notion of power/knowledge) in each 'Third World' country.

This assumed moral relationship between peoples and the states that control them is often extremely weak (e.g. in terms of ethnicity). Colonialism created artificial territorial boundaries and African states in particular have to deal with the consequences of their lack of territorial legitimacy today. The international recognition of state authority is important here, but it does not depend on a consistent set of criteria but rather on fluctuating and historically dynamic international criteria of legitimate statehood. Although African states, for example, appear to be like states everywhere else, the reality is not so simple. The state in Africa is thus partly the result of protracted historical processes (of colonialism or nationalism). While all African states seem to look alike, their genesis differs markedly (Bayart, 1993). The political communities which formed in each colonial territory and the principles of legitimacy established between the governed and government in that territory (in colonial/postcolonial times) is therefore *historically specific* and difficult to generalise.

Political parties, governments, parliaments and ideologies have tended to dominate our understanding of politics outside 'the West', suggesting that this is much more state-centred than in reality. The term 'civil society' is often used to point to the interactions between states and societies and to groups that are somehow supposedly outside the state, but also aggregates together all those who are seen to have become excluded, disenfranchised and powerless. Although often without formal power structures however, civil societies do have informal power and can respond to the power of the state in a variety of ways. Although a European construct, if reworked and seen as case-specific, this concept is useful in allowing us a partial understanding of geopolitics in non-western contexts. Because political independence was a goal shared by many peoples in the South, nationalist politics has often been seen to dominate the politics of civil society. Indeed a 'postcolonial' order in many societies has been constructed on the assumption that nationalism was the only unifying force in civil society (Ahluwalia, 2001). When ethnic politics flashes up in postcolonial Africa it is often understood as part of the contest between state and civil society (see Chapter 9). In relation to notions of the Third World and traditions of 'thirdworldism' a key issue thus concerns the extent to which new kinds of 'Third World' elites came to power after decolonisation and relates to the ways in which these elites sought to interact with and represent their (often divided) societies as well as issues of national identity and belonging.

The number of sovereign states has increased steadily since 1945 as a result of the independence of many former colonies, with important implications for the spatial imagination of global development. The breakup of a 'second world' of states after 1990 with the collapse of state socialism is also very relevant here. Geographers have explored the relations between societies and spaces quite frequently and are also now beginning to explore the relations between power and spatiality. Indeed, spatiality is a central component of state power and space is used strategically in relation to the use of the concept of 'territoriality'. The definition of the word 'territory' relates to attempts by an individual or a group to 'influence, affect or control objects, people and relationships by delimiting and asserting control over geographic area' (Braden and Shelley, 2000: 9). States are thus defined partly by their territorially based claims to sovereignty. State power is then exercised across this territory, giving a very geographical focus to the state's activities. Geopolitics is thus relevant to the formulation of specific plans for territories undergoing 'development', such as, for example, Mexico, India and Brazil (see Chapter 9). This chapter seeks to examine in more detail the geopolitical imagination of development in particular contexts, exploring how the definition and representation of the worlds of global development (the Third World, the South, and the Rest) are imbued with geopolitical discourses and meanings. Formal geopolitical theories often ignored non-western societies, assuming that these societies were of limited overall importance to the changing nature of the world political map. In one study of perceptions of geopolitics among US undergraduate students it was found that the USA, Russia, China, Japan and the countries of Western Europe were ranked as very important compared to states such as Mozambique, Uganda, Sri Lanka and Burma (Myanmar), which were ranked near the bottom of this global order (Braden and Shelley, 2000: 36). In this chapter we will be examining the geopolitical spaces and positions of so-called 'Third World' countries, exploring the views of geopolitics and the forms of geopolitical reasoning that have emerged from these countries and the idea of Third World territories as geopolitical spaces.

The first section of this chapter examines the invention of ideas about the West and the Rest, to which the concept of three worlds is intimately related. Through the age of Empire the idea of East/West divisions (as we have already seen) was a crucial one and raises critical questions about being inside/outside world systems and the mutuality of relations between insiders and outsiders, coloniser and colonised. The second section focuses more specifically on the concept of three worlds, examining its emergence and varied connections to geopolitical struggles and to movements against neocolonialism. The sense in which this term acquired a revolutionary vision of internationalism is important here but did this bring with it other associated problems of representing people and places? The penultimate section goes on to explore the possibility of alternatives to the three worlds schema and looks at the politics and practice of what has been called 'worlding', focusing on how it might be possible to go beyond the arbitrariness of standard cartographies of global development and identity.

'THE WEST' AND 'THE REST'

In many ways 'the West' is a historical rather than a geographical construction. By this we mean that that 'western societies' first emerged at a particular phase in world history – closely associated with the rise of the 'modern world' and the making of 'modernity' in 'western' or European/North American contexts. According to Stuart Hall (1992), the idea of 'the West' has four main functions:

1. It allows for characterisation and classification of societies into different categories – i.e. 'eastern', 'non-western' and so on. It is a kind of 'tool' to think with.
2. The West is an image or set of images. It condenses a number of images in our mind's eye (e.g. the multiplicity of different societies, cultures, histories, peoples, places). It works in conjunction with other images and ideas (e.g. urban, developed, industrial) which back it up and reinforce it.
3. The West provides a standard or model of comparison. It allows comparison between different societies; it permits an explanation of economic or political difference or how and why countries differ in terms of development.
4. It provides criteria of evaluation against which other societies are ranked and around which

powerful positive and negative feelings cluster. It functions as an ideology, imbuing certain parts of the world with good attributes (West = good, non-West = bad), while denying them to others.

Keywords such as condense, criteria, characterisation and comparison seem to dominate the way in which the Rest are represented through images and ideas, always in relation to other (western) places and spaces and often through a simplification of complexity. It is important to remember then that the idea of the West did not simply reflect an already established western society – rather it was essential *to the very formation* of that society. The idea of the West since then has had real consequences – organising global power relations and enabling certain kinds of knowledge – like the theory and practice of international development. The so-called uniqueness of the West was, in part, produced by European contact and self-comparison with other, non-Western societies ('the Rest') which were very different in their histories, ecologies, patterns of development and cultures from the European models of modernity. It was in this context that the idea of the West took shape and meaning (i.e. in comparison and contrast to its other, its alter-ego), and the non-western societies that therefore lay outside the map of global modernity. A key part of this process involved ranking regions and countries and the establishment of hierarchical structures and oppositions. This chapter examines how these hierarchies work alongside others such as First World/Third World in establishing the *spatiality of contemporary development theory and practice*.

In order to think about these issues in more depth it is necessary to ask further questions about the nature of European expansion and imperialism. It is difficult to summarise some five hundred years of contact, conquest, settlement and the colonisation of 'new worlds' in which overseas territories were first annexed to Europe as possessions, or harnessed through trade. What tends to emerge much more prominently in the literature is the period in which permanent European settlement, colonisation or exploitation was established (e.g. through plantation societies in North America and the Caribbean, mining and ranching in Latin America, or the rubber and tea plantations of India, Ceylon and the East Indies) (Hall, 1992). With a concern to show the way in which capitalism emerged as a global market and the shape of wealth concentration, the dependency scholars focused in turn on the period during which the scramble for colonies, markets and raw materials reached its climax. This was the 'high noon of imperialism' and led into the First World War at the beginning of the twentieth century. As we shall see in Chapter 6, there are also important debates about the continuing relevance of colonialism in a supposedly 'postcolonial' world when much of the 'South' remains economically dependent, even when formally independent and decolonised.

These different phases and periods of Empire produced different representations of non-western peoples which had an important bearing on the scripting of postwar development in the 'Third World'. These representations (along with associated ideas about the potential of 'the Tropics') have not disappeared entirely or been decolonised fully and continue to shape popular imaginations of 'other' regions in a variety of ways. In the earliest phases of colonialism non-European peoples were regarded initially as exotic cultural equals (Hall, 1992). Later, in the period of the Enlightenment (see Chapter 4) as trade and commerce grew, non-Europeans were taken to represent innocence, and ideas of the 'noble savage' were presented. In the nineteenth century, as European colonial holdings in the Third World grew and became tied more closely into European systems of trade, investment and commerce, the imagery of the colonial world and of colonial peoples shifted once more. Ideas of 'uncivilised savages' were represented just as European countries began to have more direct and practical responsibility for the supposedly 'less advanced' people of the world. By the beginning of the twentieth century this concern for the colonies and their peoples expanded into a wider sense of responsibility for the development of colonial territories.

Some of the initial voyages of discovery generated images of non-European peoples as fascinating simply for curiosity value, as people untainted by the distortions of 'civilised life'. As Europe's interests in the non-European world began to grow, so the imagery surrounding colonialism began to change. Non-European peoples began to be portrayed as 'uncivilised' or 'backward' peoples, thus justifying the control that incoming Europeans exerted over them. In this sense it is important to remember that European images of the non-European world were often deeply racist. Frantz Fanon, an important and well-known writer about

European colonialism in Africa, once argued that 'Europe is literally the creation of the Third World' (Fanon, 1967). This seemed to suggest that it was impossible to understand 'First World' histories or cultures *without* reference to their experiences with colonialism, but Fanon also meant that Europe has been built up partly with the blood, sweat and toil of Third World peoples. Having an underdeveloped, savage or backward world gave to Europe and Europeans a sense of (and confidence in) the idea of Europe or the West as 'modern' and fully developed spaces of identity and economy.

As the Second World War approached, more and more colonial powers began to see their mission as one of bringing colonial peoples towards independence. In many ways the Second World War destroyed colonial systems around the world, ushering in a post-war period of quite rapid decolonisation or the withdrawal of European colonial powers from colonial territories. The UN-centred system of nation-states facilitated a construction (or reconstruction) of nation-states and national identities then in a variety of locations. Between 1945 and 1981, UN membership rose from 51 to 156 nation-states (Berger, 2001a: 216). The USA in particular took on the mantle of 'leader of the free world' and applied pressure to European colonial powers to speed up the process of decolonisation. The USA wanted to rid the world of Empire-centred trading blocs so free trade could rule in a free world. This was in turn related to the emerging importance of trusteeship as a model for ushering the newborn states towards progress.

One important theme that this book tries to explore is the extent to which it is possible to think beyond the West and the power of western hegemony. The inextricable linkages of West and non-West are difficult to disentangle however, especially in regions such as the Middle East, where there has been something of a collective discontent against an imaginative construction called 'the West' which has animated many revolutionary movements as in Iran (Dabashi, 1993). The leader of the Iranian revolution, Ayatollah Khomeini, demonstrated an exceptional ability to build and maintain a broad base of support for the revolution, including the young and poor whom he called the 'disinherited'. The Islamic ideology promoted in the Iranian revolution included important texts on 'Westoxification' (*Gharbzadegi*) which had argued that alternative political systems to those developed in the West were needed and could be based around political ideas rooted in Islamic texts and traditions rather than in western canons and doctrines (Dabashi, 1993). The 'Islamicity' of the revolution was a crucial question here then, providing a 'theology of discontent' by calling on Islam and Allah in the drafting of a political agenda. This 'Islamic ideology' was the 'inevitable off-spring of an unwanted marriage between "Islam and the West" ' (Dabashi, 1993: 499). Both are important contemporary cultural constructs and are products of the universal need to produce a self and an 'other', an inside and an outside:

> As 'the West' created 'the Orient' to complete its 'Self' imagination, Muslims, in collaboration with their European and American counterparts, invented 'The West' for precisely the same purpose. While in 'The West', 'The West' was the self-congratulatory pronouncement of all things good and admirable, for Muslims it became the symbolic construction of corrupted excellence, an object of discrete adoration and manifest hatred.
>
> (Dabashi, 1993: 500)

For many parts of the Islamic world, 'the West' has also served as a stark cardinal point of contrast to something else, to the self-imagination of non-westerners. Thus the West became an imaginative geography from which radical ideologies spread to 'conquer the world, Islam and all' (Dabashi, 1993: 500). A repulsive rejection of the West animated the revolution in Iran as this construct became the epitome of moral corruption, of illegitimate global domination and of plundering the wealth and sovereignty of other sovereign states. In this way the 'West' was the 'cause of all ills, the mother of all corruptions, the condition of all despair, the father of all tyrannies' (Dabashi, 1993: 507). It is necessary to be conscious of these kinds of fictitious animations of western cultural or economic difference. Similarly the idea of 'westernisation' is not without its own limitations. There are implicit assumptions about the values of western modernity that run through a number of contemporary discourses of development, some of which point to the supposed total exhaustion of viable systematic alternatives to western liberalism. The 'triumphs' of western modernity in the so-called 'Third World' are also said to illustrate this point, allowing another backdrop to the stories and dramas of western progress and enlightenment.

THE CONCEPT OF THE THREE WORLDS

> The division of the planet into three worlds is based on a pair of very abstract and hardly precise binary distinctions. First the world has been divided into its 'traditional' and 'modern' parts. Then the modern portion has been subdivided into its 'communist' (or 'socialist') and 'free' parts. These four terms, underlying the idea of three worlds may be thought of as an *extremely general social semantics*.
>
> (Pletsch, 1981: 573, emphasis added).

> We did not try to represent 'the real' (of the third world). This was everybody else's project, and part of the problem from the post-development perspective.
>
> (Escobar, 2000b, 13)

Europe has been seen as the world's centre of gravity, as ontological 'reality' to the 'rest of the world's shadow' (Shohat and Stam, 1994: 2). Eurocentric thinking attributes to the 'West' an almost 'providential sense of historical destiny' which illustrates that colonial and Eurocentric discourses are thus intimately intertwined. Eurocentric discourses are complex, contradictory and historically unstable, but we are left with a kind of 'composite portrait' with the idea of linear trajectories towards progress and democracy in the West. Thus in rethinking development many have been tempted to simply invert this ordering of the globe and posit Europe or America as the source of all social evils in the world. In Derrida's words this is Europe 'exhibiting its own unacceptability in front of an anti ethnocentric mirror' (Derrida, 1976: 168). 'Third World elites' beyond the West have also been an important part of the subordination and impoverishment of Third World peoples. As we shall in Chapter 6, this was a common feature of debates about neocolonialism which focused on the connections that capitalism made locally with Third World economies and politicians. Eurocentrism works by mapping the world in a cartography that centralises and augments Europe while literally 'belittling Africa' (Shohat and Stam, 1994: 2). The important point to take on board here is that in thinking about geographies of development we need to make connections in spatial and geographical terms, to place debates about representation in the 'broader context of the Americas, Asia and Africa' (Shohat and Stam, 1994: 5).

More recent attempts to assess and describe national differences on a world scale have tried to recognise the problem of cultural prejudice implied by the term 'Third World'. This comes back to questions that Fanon raised about guilt and innocence. The important question here is not how far Europeans have been 'guilty' and Third World inhabitants 'innocent' but rather how guilt and innocence are measured by certain criteria and determined and constituted historically (Asad, 1993). The concept of the three distinct worlds has a long historical pedigree. Not surprisingly (given the imprecise nature of the term itself) there are several competing explanations of the term's origins. Fieldhouse (1999) has claimed that the term can be traced back to a British Labour Party pamphlet written in 1947 which had suggested that a 'third force' was needed to avoid polarisation in world politics between eastern and western, socialism and capitalism. Worseley (1979) credits Claude Bourdet who in 1949 had described the French political left as a 'third force' separate from communist parties but opposed to the capitalist right. The majority view seems to be that the term was first used in France in the 1950s ('Tiers Monde'/ *Tiersmondisme*) to refer to the poorest elements of French society (Hadjor, 1993; Mason, 1997). It was coined by French demographer Alfred Sauvy in the 1950s by analogy to the revolutionary 'Third Estate' of France – that is, the commoners, in contrast with the First Estate (the nobility) and the Second Estate (the clergy). Once again the history and culture of non-western worlds were interpreted by analogy.

The term posited three worlds: the capitalist First World of Europe, the USA, Australia and Japan; the Second World of the socialist block (China's position within the scheme was much debated) and the rest of the world that remained, the 'Third World' (Muni, 1979). What is interesting about the term is its continued usage and popularity, in spite of its over-simplification (Roy, 1999). The term is also often used in conjunction with other terms such as 'poor', 'non-industrialised', non-white/non-western, with cultural categories like 'backward' or geographical frames of reference like 'East' or 'South', all of which are constructs. Rethinking development geographies involves an awareness of how these terms are deployed in different contexts and acquire different meanings. A wide variety of debates began between about 1955 and 1975 – from the Bandung Conference (see Box 5.1) to the call by

BOX 5.1

The Bandung era and the Non-Aligned Movement (NAM)

The Bandung Conference was a meeting of representatives from twenty-nine African and Asian nations, held in Bandung (Indonesia) in 1955. Its aim was to promote economic and political co-operation and to oppose colonialism. The Conference was sponsored by Burma (Myanmar), India, Indonesia, Ceylon (Sri Lanka) and Pakistan. What it tried to do was cut through the layers of social, political and economic difference that separated nations of the South or the 'Third World' in order to think about the possibility of common agendas and actions. South Africa, North Korea, Israel, Taiwan and South Korea, however, were not invited. Richard Wright (1908–1960), an American writer who attended the Conference and wrote a book about his experiences, read the aims of the twenty-nine nations that attended as follows:

> (a) to promote good will and co-operation among the nations of Asia and Africa, to explore and advance their mutual as well as common interests and to establish and further friendliness and neighbourly relations (b) to consider social, economic and cultural problems and relations of the countries represented (c) to consider problems of special interest to Asian and African peoples, for example, problems affecting national sovereignty and of racialism and colonialism (d) to view the position of Asia and Africa and their people in the world of today and the contribution they can make to the promotion of world peace and co-operation.
>
> (Wright, 1995: 13)

Bandung was in many ways the 'launching pad for Third World demands' where countries distanced themselves from the 'big powers seeking to lay down the law' (Rist, 1997: 86). Richard Wright argues that it was not hard for countries with common histories of colonial exploitation to find something in common since the 'agenda and subject matter had been written for centuries in the blood and bones of participants' (1995: 14). For Wright, there was something almost 'extra-political' about this meeting since it had not been called with any particular ideology in mind. In his opening speech to the conference on 18 April 1955 President Sukarno of Indonesia urged participants to remember that they were all united by a common 'detestation' of colonialism and racism (Sukarno, 1955: 1) and pointed out that colonialism was not dead or in the past but also had its 'modern' (neocolonial) forms. Sukarno also talked about the global arms race and the contrasting capacity of Asia and Africa to mobilise millions of human beings to demonstrate the will of the 'majority world' for peace. The Conference was especially successful in hastening the arrival of new international institutions explicitly dealing with 'development' (Rist, 1997). The term 'Bandung Era' was coined by Samir Amin in a discussion of the national developmentalisms that emerged from this period (whose dynamics and objectives were of course varied and complex) (Berger, 1994, 2001; Amin, 1997a). The Non-Aligned Movement (NAM) was then founded (among twenty-nine states, most of which were former colonies) in 1961 on a principle of non-alignment with either the USA or the USSR in the Cold War. The NAM emerged principally from India, Indonesia and Yugoslavia which were its principal advocates. Its members did, however, include countries that were in fact aligned during the Cold War with one or other superpower. A year after the NAM was set up, the 'Group of 77' was established (made up largely of the attendees of previous conferences) to press for changes from UN agencies in the ways they viewed 'developing countries'. The NAM has continued to function as a space through which nations of the so-called 'Third World' could meet and discuss common issues and objectives and it now has some 115 member nations. The Conference was led initially by several key statesmen of 'thirdworldism': President Sukarno (Indonesia), Prime Minister Jaharawal Nehru (India) and Prime Minister Nasser of Egypt. The organisation has been restructured since the

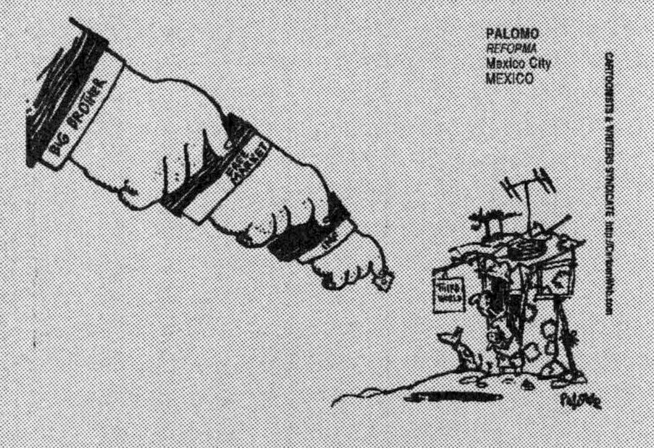

Figure 5.3 Big Brother and the Third World
Source: CWS and Palomo

BOX 5.1 (contd.)

end of the Cold War and continues its commitment to its founding principles, while also now incorporating a focus on reforming the international economy. South Africa took over as Chair in 1992 when the organisation met in Jakarta to decide on future policy directions, while today the NAM co-operates with NGOs, the UN, the EU and the G8. At its fortieth anniversary in New York in November 2001 the NAM restated its commitment to work with the multilateral agencies to help bring about a reform of the HIPC initiative concerning debt as well as more support for NEPAD (The New Partnership for Africa's Development), a new economic partnership for African development (see Chapters 6 and 10). The NAM site is available at http://www.nam.gov.za/

the non-aligned movement and the Group of 77 for a New International Economic Order (NIEO). Development was still seen as necessary here within a perspective of economic co-operation, integration and modernisation. This was a period of extraordinary global change and confrontational political realignment. Most of the original participants at Bandung in 1955 were from Asia, as many African countries were still enduring colonial rule, but it marked the beginning of collective political demands in the fields of development and politics (Rist, 1997). During this time connections were made between (post)colonialism and the idea of political struggle on behalf of the 'Third World':

> [T]he only recently constituted 'Third World' became the site of intense debates regarding options for the 'development' and the early 'Bandung regimes' . . . (Nehru's India, Nasser's Egypt, Sukarno's Indonesia, Nkrumah's Ghana) the stage for arguments of what came to be called . . . a 'non-capitalist path to socialism.' . . . Moreover, from the late 1960s through the 1970s there emerged a number of radicalizations of the Bandung project – Salvador Allende's Chile, Julius Nyerere's Tanzania after the Arusha Declaration, Michael Manley's democratic socialism in Jamaica, Maurice Bishop's Grenada and Sirimavo Bandaranaike's Sri Lanka. In each of them 'socialism' was the name of a variously configured oppositional idea of political community defined largely in terms of anti-imperialism, national self-determination, and anti-capitalism.
>
> (Scott, 1999: 144)

Figure 5.4 Building a socialist administration in Inhambane, Mozambique
Source: The author

In the 1950s the term was deployed increasingly by governments and movements and their sympathisers to generate unity and support in the face of the USA and a handful of former colonial powers, and gained popular currency as African and Asian countries became independent (Berger, 1994). By the 1960s *Tiersmondisme* also had its Soviet and Chinese varieties, establishing Moscow or Peking oriented worlds of global struggle. 'Thirdworldism' and the non-aligned movement forged as a result of this therefore often involved a kind of 'revolutionary ideology', one which 'aimed to contest and overturn the terms of economic engagement between North and South' (McGrew, 2000: 356). The Vietnam War exemplified for many the wider struggles against US imperialism, and, as we have already seen, anti-communism (and the threat of anti-capitalist revolution) was closely linked with the invention of development, offering a kind of counter-discourse to the spread of communism. This also allowed western countries to intervene in the name of Third World 'modernisation', an idea which still remains hegemonic at the popular and policy levels:

> What began in the 1950s as an attempt to forge a political and diplomatic alliance ostensibly outside the capitalist and socialist 'camps' has now become an all-encompassing category reducing the governments, economies and societies of Africa, Asia, Latin America and Oceania to a set of variables distinct from and inferior to the 'First World'.
>
> (Berger, 1994: 270)

This notion of being distinct from and inferior to the 'First World' is important here. The way forward, according to Berger (1994), is to try to locate the politics of development in historically particular places but also (simultaneously) to examine the connections to global processes. Whenever the term has been used a certain kind of history and geography is created which suggests that the countries of Latin America, Asia and Africa have failed to become the idealised versions of modern industrial democracies predicted by the modernisation school of development thinking in the 1950s and 1960s. The use of the term thus continues to encourage a homogenised understanding of a considerable part of the globe and its diversity of cultures, histories and peoples. What is particularly amazing about the longevity of the concept is the simplicity of its categories:

> It is an extremely rudimentary system of classification, however, and certainly not a natural or necessary one. With the possible exception of the political categories of left and right, the scheme of three worlds is perhaps the most primitive system of classification in our social science discourse. One wonders how it could have assumed such authority.
>
> (Pletsch, 1981: 565)

The continued usage of this system of classification, both by academics and in everyday parlance, testifies to the resources, time and efforts that are deployed in maintaining its boundaries and the implicit hierarchies that underscore it (McDowell and Sharp, 1999: 298). In some ways then the use of this term seeks to reinforce the continued spatial exclusion of Third World others. A very general image is given by the idea of three worlds, which divides up the globe and its inhabitants for study by various social scientists. Intimately bound up with Cold War geopolitics and the scripting of the USSR as other, the original use of the term also emerged partly from western anxieties about the emergence of a second world of socialist nations in Eastern Europe. The role of the Soviet Union in shaping the co-ordinates of 'Third World' geopolitics is a neglected theme in the literature and little is known about the instrumentalisation of the Soviet model in development practice (Laïdi, 1988). For much of the twentieth century geopolitics was crucial to the international construction of 'development' as a specific kind of alternative to communist planning. Indeed, experiences with national development planning in the first three decades of the twentieth century in post-revolutionary Russia preceded the beginning of debates in America about the possibility of 'modernisation' (Engerman, 2000). Economists and policy-makers in North America between the two World Wars had become increasingly interested in the Russian 'version' of the national development project (which also saw a strong role for the

Figure 5.5 Child soldiers aboard a Russian Mig fighter in Angola
Source: Alex Hofford

state). Many of the pioneers of development economics were also from East Central Europe (Berger, 2001) and wrote about their experiences of economic progress in this region.

President Truman's doctrine and speeches in the United States at the end of the 1940s launched a campaign for 'backward peoples' in the 'less developed' parts of the world. This was, at least in part, based heavily around the 'popular myth of America itself as Empire's antithesis' (Vitalis, 2002: 191). Many countries were just reaching independence in the decades of the 1960s, at the 'zenith' of international development and, as a result, both the USA and USSR drew upon the idea of development in a variety of African, Asian or Latin American contexts, constituting the Third World of newly independent states as an arena of global political and ideological struggle. It has even been argued that since the end of the Cold War the Third World has ceased to exist since it was defined largely by the geopolitical structure and discourses of the Cold War (Roy, 1999).

Pletsch (1981) looks at how different branches of area studies were developed in the Cold War concerned with communist areas, coinciding with various policies of containment that were adopted by the USA. An international division of labour emerged around this time within the social sciences based upon the idea of three separate worlds which excluded other kinds of participation and narration (Pletsch, 1981). In the First World therefore it follows that the practice of economics, geography, politics or sociology supposedly corresponds closely with our best ideas of what science should be, the most exact. What is important here is that a great variety of social scientists (including geographers) suddenly found the three worlds schema useful for organising their thinking about the international order after 1945. Said (1978: 255) argues, however, that area studies 'expertise' was not always as sought after as it was during the Cold War:

> Because we have become accustomed to think of a contemporary expert on some branch of the Orient, or some aspect of its life, as a specialist in 'area studies', we have lost a vivid sense of how, until World War II, the Orientalist was considered to be a generalist (with a great deal of specific knowledge of course) who had highly developed skills for making summational statements.

Alfred Sauvy (credited originator of the term) was also very critical of the world to which he applied it and argued that the Third World had implications for the 'social evolution' of European and American societies. Sauvy's prescription was that 'Third world' countries must 'modernise', a dominant idea in the 1950s and 1960s and intimately linked to this notion of three worlds. Further scientific/detached observation by social scientists such as Sauvy was needed. The Third World was therefore a residual category of unaligned objects for the competing empires of the first two worlds, but there were also much deeper and wider connotations. Thus the 'Third World' represented what was left when it had been subtracted from the world as a whole and the industrialised West in particular. The term interlinks with words such as 'traditional', 'modern' and 'communist/socialist' in some important ways. The Third World was seen as the world of tradition, culture, religion, irrationality, underdevelopment, overpopulation, political chaos and so on. The First World on the other hand is purely modern, free, efficient, technological/scientific and natural. Some have even assumed the coherence of the Second World while setting out to attack the notion of the 'Third World'. Since the free world is taken to be natural (guided by invisible hands) it seems to imply that the socialist countries should slowly but surely come to approximate the free or First World. The West is also, however, a collective heritage of cultures that were constituted by non-European influences (Fanon, 1967).

Aijaz Ahmad (1992: 308) has emphasised that the 'Third World' does not come to us 'as a mere descriptive category' but carries within it contradictory layers of meaning and political purpose. Thus in many instances the 'third force' ascribed to this world often only existed in rhetoric or in speeches and political texts, while in reality many so-called Third World countries erred towards one of the other two 'worlds' of political influence (Mason, 1997). Ahmad's work distinguishes between the theoretical use/abuse of the term 'Third World' and its role in common, everyday parlance to describe 'developing countries' and their issues or needs. Some still believe in the revolutionary *oeuvre* of the term and its capacity to mobilise global coalitions (e.g. around environmental issues) (Williams, 1993; Berger, 1994). While many books and articles are still produced which claim to rethink the 'Third World', many often hold back from an explicit or sustained questioning of the validity of talking about a 'Third World' as such. Can we still retain this notion with the rise of the Asian tiger economies and amidst its growing cartographic complexity – how can we map this world amidst the escalating mobility of labour, capital and elites? It is also possible that those who claim to speak for the Third World may also be deeply implicated in the prevailing international discourses and power structures which seek to discipline and 'manage' Third World peoples and places (Berger, 1994).

In a related sense, Roxanne Doty (1996) has argued that many texts about Third World sovereignty and statehood in international relations scholarship share a fear or sense of danger regarding the entry of the 'Third World' into the international society of states. Doty (1996) argues that in trying to make sense of 'Third World' sovereignty scholars drew on 'a whole array of hierarchical oppositions' (Doty, 1996: 149) with 'weak states' in the South needing to live up to the western ideal model. Development is again split into positive/negative, to invoke images of identity once used in many past imperial encounters between 'North' and 'South'. The image of 'Third World' corruption, incompetence, the lack of modern 'thinking' and 'attitudes' were blamed by many for these 'Third World failures' (thus legitimising intervention in the affairs of those states). The benevolent, democratic 'international community' thereby replaces the superior 'West/white man' of earlier imperial encounters (Doty, 1996). In this sense many social scientists participated in the production and legitimating of 'world ordering possibilities' (Doty, 1996: 157). In attempting to rethink geographies of development today we need to understand where and how these possibilities operate. This involves a more nuanced interpretation of how the three worlds schema works through specific institutions, practices and ideas. For some, 'Third World' states are a source of radical potential: 'Third World' states have radicalised international society by 'introducing collectivist ideologies and goals that challenge classical positive sovereignty doctrine' (Jackson, 1990: 114).

The important point here is that international development forms the common base of knowledge around which 'Third World' societies and polities are understood and 'the take-off point, so to speak, regarding the place of the third world in international society' (Doty, 1996: 158). The point of entry is crucial here; it defines and limits the ways in which those countries are enframed and understood and creates a role for experts in 'international society'. The 'extremely general social semantics' (Pletsh, 1981) that are deployed around the term have important implications. None the less, over time an 'amazing conceptual arsenal' (Said, 1978) has been deployed by development 'experts' and planners around these and associated notions. Increasing attention to the problems of poor countries, however, has altered the meaning of this concept over time. Shohat and Stam (1994: 26) talk of 'Third World Euphoria', referring to a time 'when it seemed that First World leftists and Third World guerrillas would walk arm-in-arm toward global revolution'. Not all that was associated with the Third World was revolutionary, however, and the Cold War geopolitics of the 1960s and 1970s, which required a degree of solidarity and alignment, has clearly changed.

The term is now stuck in a kind of swirling, permanent terminological crisis, an 'inconvenient relic of a more militant period' (Shohat and Stam, 1994: 26). The 1960s was also a decade of 'fundamental shifts on the local, national or global level [which] resonated with and grew out of each other' (Fink *et al.*, 1998: 2). In 1968, student movements protested the 'imperialist war' against Vietnam, in China students and workers were lodged in the vanguard of the Great Proletarian Cultural Revolution instigated by Mao to block the 'capitalist roaders', and in Mexico students protested the political monopoly of the Institutional Revolutionary Party (PRI) (Watts, 2001). As a decade of strategic reverberations within the global geopolitical system, the 1960s helped to shatter the Eurocentric idea that 'the advanced proletariat of the West [brings] socialism as a "gift" to the "backward" masses of the periphery' (Amin, 1974: 603). As Mike Watts (2001: 170) has put it:

> Third worldism corroborated not only a sense of revolutionary internationalism, but also confirmed that there were models of revolution and liberation outside, and beyond, both the Communist and Social Democratic traditions.

Key writings such as Aimé Césaire's *Discourse on Colonialism* (Césaire, 2000) thus offered something of a 'Third World manifesto' (Kelley, 1999: 1) which consisted of polemics against the established (neo)colonial order and sought to mobilise support for new kinds of postcolonial internationalism (see Chapter 6). In a way, therefore, the uneven

BOX 5.2

A Third World atlas?

The notion of the 'Third World' has occasionally found expression in the production of atlases of 'Third World' development (Crow and Thomas, 1983; Unwin, 1994). These have included geographies of Islam or the global arms trade, for example, and have tended to focus on what are perceived to be 'Third World issues', while being (to differing extents) conscious of the limitations of this term, raising questions about maps and their implicit projections of ethnocentricity. Crow and Thomas (1983: 8) point out that a major connotation of the term 'Third World' is the idea of being 'underdeveloped' or simply 'poor'. They report on two Open University surveys of students which aimed to take a sample of student perceptions about the 'Third World'. Thirty-six students were interviewed on six broad questions, including what they understood by the terms 'underdeveloped countries', 'Third World' and 'developing countries'. Subsequently, 200 students were given a list of countries and asked to rank and divide them into three groups, of which the 'Third World' is the third group 'as you see it'. These researchers found that two kinds of conception were in common use. The first and most frequent was to rank and order countries according to their levels of industrialisation or of development.

A second conception was dubbed 'West'/'East'/'Rest' and was often a much less clear-cut picture of global difference. Many students felt that 'underdeveloped', 'Third World' and 'developing' all meant essentially the same thing (namely 'poverty' and various kinds of 'problems'), seeing the Third World as a distinct geographical area or as exactly one-third of the Earth's surface. The exploitation of 'Third World peoples' and dependence on aid were seen as uniting problems for the Third World among many students. Generally, the term 'Third World' was seen as having very negative connotations. One student referred to 'them over there who are not developed'; another pointed out that 'When you say Third World I tend to think of the unpleasant aspects . . . Third World to me suggests that they have been out of the running altogether, that there's nothing that can be done about them' (cited in Crow and Thomas, 1983: 9). Popular conceptions of the 'Third World' involve similar negative images of 'Third World' difference, producing a powerful negative conception that has 'coherence, resolution and definition' (Bell, 1994). Rethinking development geographies today involves understanding how these powerful negative images operate and an attempt to build more radical and alternative geographies. It also involves getting away from the assumption that 'something can be done' about Third World others (see also Sutcliffe's (2001) *100 Ways of Seeing an Unequal World*).

development of the world system conditioned the composition and character of the diverse anti-systemic movements which began to appear (from the Middle East to Mexico). In Africa, the first generation of independent states (some of which had descended quickly into military dictatorship) became Fanonite postcolonial movements with nationalism and institutionalised elite politics as the focus of their critiques (Watts, 2001). Jean-Paul Sartre argued that the year 1968 demanded 'the enlargement of the field of the possible' (cited in Watts, 2001: 175), enlarging and opening up the very constitution of the political and political struggle in particular. Within the new kinds of social movement emerging at this time (which took multiple and hybrid forms, for example, around Marxism) specific political questions to do with feminism, the environment, race and disarmament began to be raised which seemed to cut across these broad movements, making for enormously complex political hybrids, cross-overs and connections (Watts, 2001: 176).

LOST FOR WOR(L)DS?: WORLDING AND OTHERNESS

This theme of 'cross-overs and connections' made across and between worlds is an important one in retracing the themes and issues of development debates. What united many struggles was a 'growing questioning of Western consumer-urban-industrial models' (Jolly, cited in Arndt, 1987: 108). The rise of consumerism in many 'western' countries began to be increasingly questioned therefore as many associations were made between these conditions and those of other countries (such as Vietnam). As Hardt and Negri (2000: 394) have put it:

> Far from being defeated, the revolutions of the twentieth century have each pushed forward and transformed the terms of class conflict, posing the conditions of a new subjectivity, an insurgent multitude against imperial power.

This question of an insurgent multitude against imperial power is a theme that we will return to later in the book, but it is clear that this was a notion which ran through many revolutionary conceptions of Thirdworldism. There were many important questions raised by these revolutions about subjectivity and identity (not just national identity but class and gender identity as well). None the less, there have been many critics of 'Thirdworldism' and the idea of three worlds who have argued that identity and difference were sometimes played down and over-simplified in these struggles. Shiva Naipaul, for example, views the term as symptomatic of a 'bloodless universality that robs individuals and their societies of their particularity' (Naipaul, 1997). Iran and Turkey, two countries which were never colonised directly, fit uneasily into the tripartite schema (as do Middle-Eastern geographies of development more generally), and although these countries are economically peripheral, they have different histories and experiences of European intervention and domination. Thus the three worlds idea is sometimes seen to flatten heterogeneity and masks contradictions as well as obscuring some similarities. As Jenny Robinson argues, much of the developmentalist literature on 'Third World cities' is quite problematic in that it clings to a category that sees such cities as essentially very similar, constructing an image of all poor cities as infrastructurally poor and economically stagnant. The literature on 'world cities' reinforces this distinction, assuming that investigations of world cities and Third World cities cannot inform one another (Robinson, 2002). As Robinson (2002: 533) argues: '[T]he persistent use of the category of "third world city" [imposes] substantial limitations on imagining or planning the futures of cities around the world.'

Understandings of 'city-ness' are thus drawn from a very select group of countries and others are measured according to this pre-given standard of world 'city-ness'. Similarly 'developmentalist' accounts bring into view only the poorly serviced parts of cities. Many cities around the world are thereby consigned to structural irrelevance and somehow 'fall off the map' (Robinson, 2002). Contemporary urban life in Britain and the United States of America intertwines these worlds; thus struggles between worlds take place within as well as between nations. For David Rieff (1990) Los Angeles is today the real capital of the Third World. In reality these lines and spatial demarcations between worlds are incredibly blurred, becoming fictions, geographical imaginings or mental maps rather than actual territorial demarcations. There is no first world without a third 'other' to compare and contrast it with.

At the end of the 1960s a common exclusion from decision-making and an oppressive experience of global development and industrialisation had led many countries to discuss and explore their common obligations to the 'advanced capitalist countries' (Jalée, 1969: ix–x). Although this notion of 'advanced' countries is problematic in some ways, Jalée's point about obligations and oppressive experiences remains relevant. Shohat and Stam (1994: 26) argue that 'Third World' is still a relevant term and better in geopolitical and economic terms than referring to North/South polarity (since the 'South' has occasionally included Australia!). In particular they examine the rich tradition of 'Third World cinema' and the notion of 'third cinema' which emerged from the Cuban revolution. Despite these imbrications of first and third worlds, the global distribution of power 'still tends to make First World countries "transmitters" and to reduce most third world countries to the status of receivers'. It is not a simple case therefore of an active First World universally forcing its products on passive Third World peoples, since global and local cultures can co-exist. There are also powerful reverse cultural currents that we must examine here (from countries such as Brazil, Mexico, Egypt and India). We can be cautious, however, about how useful the idea of the 'Third World' ultimately is:

> All these terms, like that of the 'Third World', then, are only schematically useful; they must be placed 'under erasure', seen as provisional and only partly illuminating.
>
> (Shohat and Stam, 1994: 26)

What is needed then is a more flexible conceptual framework. Shohat and Stam (1994) chose to retain the term since for them it remains capable of signalling 'the dumb inertia of neo-colonialism and the energizing collectivity of radical critique' (Shohat and Stam, 1994: 27). It is also useful to remember that in a way the paradigm of the 'Third World' has been eclipsed in recent years by the term 'postcolonial' (see Chapter 6), which evolved during the 1980s to designate work which examined colonial relations and their aftermath: '[t]he new term arrived with a magnetic aura of theoretical prestige, in contrast to the more activist aura once enjoyed by the phrase "Third World"' (Shohat and stam 1994: 38). The key question to explore here is if this represents a move 'beyond' obsolescent discourses or the closure of something old and the passage towards something new. The 'Third World' is most problematic for geographers in that it has an ambiguous historicity and spatiality. In addition, the term 'postcolonial' is seen by some to be less useful than the idea of the 'Third World' in that it appears to posit no clear domination and calls for no clear opposition, suggesting that thinking about postcolonialism is not relevant to a critique of the unequal distribution of power and economic resources. There is the assumption here then that this newer term leads to a blurring of the assignment of perspectives or political positions which was seen to come from 'Thirdworldism' and to be based on a sense that colonial experiences were shared (albeit asymmetrically). For some critics, the term 'Third World' referred to something in particular but it is not clear what the 'post' in postcolonial refers to: the ex-colonised, the ex-coloniser, the ex-colonial settler or the displaced migrant living in London, Lahore or Luanda?

What was often understated in Thirdworldist gatherings was the big difference in the timing and nature of anti-colonial struggles across Latin America, Asia and Africa (where the history of colonialism in these regions was very different). For some authors, 'where colonialism left off, development took over' in terms of the means of legitimating interventions in the affairs of 'Third World' states (Kothari, 1988). The idea of 'neocolonialism' was central to the notion of Third World people and is also still seen as relevant today in that it emphasises a repetition and regeneration of colonial relations and designates important forms of contemporary hegemony in the world economy. The point perhaps to take from these debates and varied opinions and perspectives is that it is not a simple case of declaring one conceptual frame to be 'wrong' and the other to be 'right', but rather that each is only partly illuminating. In rethinking development geography, this does not mean throwing the baby out with the bath water, but rather that we can use a variety of terms 'as part of a more mobile set of grids, a more flexible set of disciplinary and cross-cultural lenses . . . while maintaining openings for agency and resistance' (Shohat and Stam, 1994: 45).

The idea of three worlds on its own does not necessarily provide these kinds of 'mobile grids' and cross-cultural lenses that allow us to understand difference between geographical regions. It is thus also very important to recognise the hybridity – of peoples, places and cultures – that resulted from colonialism and to be open to various interpretations and forms of politics, centred on communities and refusing the 'ghettoising discourse' that comes with some representations of the Third World (Shohat and Stam, 1994: 47). This would be 'polycentric' in that Europe or North America would be displaced from their position at the centre of global debates about development. Concerning research on non-western areas in social science disciplines today, these connections are not always made and there is a kind of privileging of other spaces and places:

> The privileging of the Anglo-American cultural world, and the tracing of cultural studies' pedigree only to London or Birmingham, prevents dialogue with Latin American, Asian and African studies, whatever does not belong to the Anglo-western world is peripheralized as 'area studies'.
>
> (Shohat and Stam, 1994: 6)

What would happen then if Latin American, Asian and African studies in any discipline were not always othered and peripheralised as 'area studies' (since many of these centres were created at the height of the Cold War)? In this sense it is crucial to break down the spatial metaphors and languages of representing distant and different geographies, constructing alternatives. From about 1975 onwards many debates began to emerge about the question of precisely how many worlds there are or were. The argument was made by a number of researchers interested in development that the 'Third World' contains enormous cultural, social, historical and political diversity and that it could not possibly account for the range of this diversity between countries. Many of the discussions that have taken place in the Islamic world around the three worlds schema have rested on the assumption of an essential Islam. This was the case in the 1960s and 1970s in the debate about whether Islam favoured capitalism or socialism (Halliday, 2000). Some researchers then began to identify other 'worlds' within the 'Third World'. This expanded the numbers of 'worlds' recognised in development to as many as eight in some cases. A 'fourth world', for example, was identified by some authors referring to one made up of some of the world's poorest indigenous peoples. The point here is not that there are more than three worlds, that there are in fact 'X' number of worlds but rather that perhaps the whole system of dividing the globe up into 'worlds' is erroneous to begin with. We need to be careful in using these and other terms, to have a sense of their historical emergence and usage and all that it implies about the economic or political order of the world. We have thus to realise that the 'Third World' has been used as a tool with which to think. The 'three worlds' schema has therefore enabled various commentators to summarise and 'skim over' diversity and difference.

The extent to which the three worlds schema can help us to understand all this diversity and differentiation is thus questionable. We have to see that in some ways people and institutions 'world' Africa, Latin America and Asia *away from themselves*. Other worlds are made to appear as distant places outside the universe of immediate moral concern, as 'out there' and as 'satellite' or 'peripheral' concerns. Thus many students travel away from home communities to collect 'Third World places' (Desforges, 1998). The point here is that it is vital to be careful in forming our understandings of development so as not to create the impression that the Third World is miles away, worlded away from our own immediate universes of concern. In some ways Africa in particular has been 'overworlded' (Paolini, 1997: 93) by a variety of development discourses, occupying a disjunctive space and seen as separate and disjoined from other worlds. In critiquing development therefore we can understand this process of worlding and constructing a disjuncture between regions which conveys the impression that continents such as Africa are somehow outside or beyond modernity. This is an interdependent world however, and the extent and value of its interdependence also need highlighting and rethinking. Elsewhere it has been argued, following Bell (1994), that more radical development geographies will require liberation from the tyranny of these kinds of dualisms and hierarchical structures (Power, 2000). From them flows a particular construction of geographical difference as otherness which needs much more careful scrutiny:

> What is the geography of the Third World? Certain common features come to mind: poverty, famine, environmental disaster and degradation, political instability, regional inequalities and so on. A powerful and negative image is created that has coherence, resolution and definition. But behind this tragic stereotype there is an alternative geography, one which demonstrates that the introduction of development into the countries of the Third World has been a protracted, painstaking and fiercely contested process.
>
> (Bell, 1994: 175)

This then is one of the major objectives of this book, to try to formulate a more radical and alternative geography of development, one which is liberated from the simplistic dualities and negative images that tend to crystallise around the 'three worlds' scheme. In this sense we need to bring out the extent to which 'development' has become a 'protracted, painstaking and fiercely contested process'. The term 'radical geography' was associated very much with debates about development and the 'Third World' in the 1960s, at a time when anti-poverty and anti-imperial movements where beginning to emerge in a variety of countries. In the first issue of the radical journal of geography *Antipode*, Richard Peet listed the concerns of radical geographers as the war in Vietnam, apartheid in South Africa, Israeli occupations, urban poverty, US imperialism and defence spending. Today it is possible to widen the compass a little to include postcolonial and postdevelopment critiques of development, while retaining the focus on poverty, conflict, ethnicity, statehood and contemporary forms of imperialism. In so doing however, we need to move beyond the range of myths and stereotypes that emerge from the three worlds schema, such as the idea of 'Third World women' who have been seen by many western feminists as 'frozen' in time and space and as exotic 'others' (Mohanty, 1988). Similarly, Chaliand (1977) once cautioned about the 'revolutionary myths of the Third World' and the mythical images of socialism and national liberation that were offered to Third World peoples in the course of various revolutions in the periphery.

Power relations in 'Third World' politics must thus be understood as dispersed and contradictory. Internal contradictions within socialist revolutions

Figure 5.6 Photo of the revolutionary Che Guevara appears on a Cuban building
Source: Alix Wood

were therefore sometimes played down in these mythologies of transformation. The myth of 'Third World women' also denies women the right to be active, experiencing subjects. Representations of the agents, individuals and subjects of development in the imaginary 'Third World' thus frequently paper over this diversity and difference rather than seeking to accommodate and celebrate this as part of a radicalisation of development geographies. For Williams (1983) it is also necessary to listen to the spokesmen and spokeswomen of the 'Third World' and to ask to what extent they are representative or can act in the general interest of their peoples. The problem remains, however, that 'these are very hard questions to resolve within the terms of undifferentiated cardinal blocks' (Williams, 1983: 204). It is thus difficult to distinguish different kinds of leadership and different political regimes in a framework that does not really seek to differentiate. Colonel Qaddafi of Libya is one example of a ruler who maintained power partly by undemocratic means but who was none the less widely seen as a champion of various 'Third World causes'. As Sharp (1988: 120) writes with respect to the use of the First World/Third World terminology in South Africa (see also Box 5.3):

> In both 'first' and 'third' worlds there are many gradations of wealth, prestige, status and power. Therefore all claims to speak on behalf of one or other of these 'worlds' as a whole are suspect.

One useful way of exploring the contemporary use and significance of the term 'Third World' is to think about the work of the Third World Network (TWN) based in Penang, Malaysia. The TWN describes itself as an independent non-profit international network of organisations and individuals involved in issues relating to development, the Third World and North–South linkages. The organisation campaigns on a range of themes including trade issues and rules, the WTO, global economic and financial crises, UNCTAD (United Nations Conference on Trade and Development) and trade developments, biodiversity, biotechnology and indigenous communities, the environment, gender issues and rights and health issues, among others. The TWN's objectives are 'to conduct research on economic, social and environmental issues pertaining to the South, to publish books and magazines, to organise and participate in seminars and to provide a platform representing broadly Southern interests and perspectives at international gatherings such as those of the UN conferences and processes, (TWN, 2001: 1). The organisation publishes daily SUNS bulletins (South–North development monitors), the fortnightly *Third World Economics* and the monthly *Third World Resurgence*. It also has a collaborative partnership with the 'South Centre' in Geneva. The international secretariat is based in Penang, Malaysia, but it has offices in Delhi, Montevideo, Geneva, London and Accra and is affiliated to other organisations across the world. The organisation's website is also now an important means of disseminating this idea still further. One interesting component of the TWN website deals with the 'Seattle debacle' surrounding the Third WTO ministerial talks: 'While the immediate cause of the collapse was the inability of the US and EU to bridge their differences, the most important factor was the refusal of developing countries to be bullied' (TWN, 2001: 1).

This brings us back to the question of a 'platform for broadly Southern interests' and the question of Third World 'others' being bullied. Is it possible to represent this diversity of opinion and perspective within one voice and to respond collectively through resistance? Solidarity between countries has been organised around the notion of three worlds for many years but is it possible that movements like the TWN will be able to organise themselves around alternative identities if we place the term under erasure? What is particularly interesting about the TWN is its use of the term 'Third World' in relation to debates about 'globalisation'. In many such debates (see Chapter 7) the Third World is seen to exist on the 'outside' of modern and global forces, as an 'other' that forces the global and the

BOX 5.3

South Africa as a combination of 'First World' and 'Third World'?

In describing the use of the terms 'First World' and 'Third World' in South Africa during the 1980s, Sharp (1988) notes how many people came to use these terms after they were drawn upon in the South African Broadcasting Corporation (SABC) *Comment* programme. On the one hand, reference was being made to the high-rise buildings, hi-tech industries and the efficiency and infrastructure of cities, and to the rudimentary dwellings, pitted roads and inefficient bureaucracies of the rural areas and 'homelands' of South Africa created by the Apartheid regime on the other. As a result people ended up saying a great deal more than was meant when they first used the terms 'First World' and 'Third World' as a 'convenient, descriptive shorthand' (Sharp, 1988: 111). The daily transmission of the SABC *Comment* programme in the early 1980s (along with the widespread use of the terms by other South African media such as the *Financial Mail*) played a major role in shaping public opinion. From about 1983, references were made to Africans 'outside' the 'national states' and to a vision of a 'Third World' population in rural areas beginning to develop modern values. Identifying African people as a part of the 'Third World' sector within South Africa justified their continued separation from the people who made up the First World.

In this way, international discourses of development were drawn upon to replace the outdated and controversial racial categories, as the First/Third World structure replaced the earlier reference to 'black' and 'white'. Presented and assimilated as 'common sense', the terminology accorded white people a high status since they were seen to represent the 'First World', the highest stage in a meritocracy, one that 'Third World' black South Africans should aspire to (Sharp, 1988). In 1983, the Development Bank of South Africa (DBSA) was also formed, which modelled itself on the World Bank, and sought to position South Africa as part of international endeavours to bring development to 'less developed' countries and peoples, just as the World Bank had done for many years. Williams (1981) has shown that in Africa generally, the World Bank's activities were animated by a view that people in 'Third World' countries are incapable of their own development and that this is a process that must therefore be instigated by 'First World' experts and their agencies. The world of the 'Third World peasant' was, by contrast, seen as static and irrational. If people in the 'Third World' reject free enterprise, it was seen to be because they are deluded by the ignorance of their traditions:

> Despite ostensible revisions to its programmes over the years, the World Bank remains firmly attached to the vision of a static 'third world' which is unable to help itself, and a dynamic 'first world', which, by virtue of its own state of development, has both a duty and a right to tell the other segment of humanity what is good for it.
>
> (Sharp, 1988: 116)

The point about these terms in South Africa during the apartheid years is that they were used in a fashion prescribed by official and SABC usage, thereby lending support to official ideas (such as that of the World Bank) about the relationships between these words and worlds. The terms also offered a way out of the growing doubts which beset the older apartheid languages of domination, naturalising a racist social order and apartheid ideologies of division. Thus the dominant vision of a benign 'First World' carrying the benefits of progress to a benighted 'Third World' had its own specific utility in South Africa under apartheid. There were numerous problems, however, with viewing black South Africans as 'outside progress', as incapable of their own development and as universally 'backward'. The white population of 'First World' South Africa was, by contrast, seen as glorified, as the dynamo driving the economy forward.

modern to encounter themselves and interrupt their assumptions about the non-western world (Paolini, 1997). For the TWN the term has a variety of meanings however. The organisation clearly recognises that the term 'Third World' refers to a past historical period and does not effectively map the diversity of social, cultural and economic difference at the global level (assuming that it ever did), but it also seems to recognise in the term 'Third World' the potential for creative imaginings of new global communities that are capable of contesting and resisting inequality. Globalisation, as we shall see in Chapter 7, fosters politico-economic interconnections and cultural hybridities that do not neatly correspond to a division of the world into three separate or distinct parts.

CONCLUSIONS: COLONIALISM, DECOLONISATION AND THE PURSUIT OF DEVELOPMENT

> It [the Third World] delineates nothing that really exists. It is a flabby western concept lacking the flesh and blood of the actual . . . it is a form of bloodless universality that robs individuals and societies of their particularity.
>
> (Naipaul, 1985: 10)

> Replacing 'Third World' with 'postcolonial' has drawbacks as well as advantages. 'Third World' still evokes a common project of (linked) resistances, and has served to empower intercommunal coalitions of peoples. . . . 'Third world' implies that the shared history of neocolonialism and 'internal' racism form a sufficient ground for alliance.
>
> (Shohat and Stam, 1994: 40)

Figure 5.7 The Cold War legacies of US and USSR rivalry
Source: CWS and Paresh

Following the collapse of the Soviet Union and the end of the Cold War, the USA responded by reducing or eliminating its military aid to Kenya, Somalia, Liberia, Chad and Zaire, and reduced its presence in Africa by closing nine aid missions and fifteen intelligence posts, redirecting aid personnel to new assignments in Eastern Europe and the former Soviet Union (Bratton, 1994). As we have seen, development is an industry, but it is also very much something which has been grounded in Cold War geopolitical orders and discourses, producing a particular set of images and stories about 'Third World' spaces and places. From the early 1980s, development assistance has increasingly been tied to evidence of democracy and 'good governance' that each state in the periphery can produce before its western creditors. Abrahamsen (2001) argues that the good governance agenda developed for Africa in the 1990s represents a kind of rediscovery of politics in the international development arena (after it had been seen as irrelevant for years) but none the less one which represents a continuation of western hegemony over the continent and a perpetuation of three world hierarchies. The good governance agenda may thus be regarded as a 'discursive transformation' that, while claiming to liberate the poor, enabled the West to continue its hegemony in Africa under the changed circumstances of the new world order(s): 'It reproduces the hierarchies of conventional development discourse, whereby the third world is still to be reformed and delivered from its current underdeveloped stage by the first world' (Abrahamsen, 2001: 44).

The 'Third World' and 'Third World chaos' have in some ways come to be perceived as the major threat to the prosperous global North. The main discourse and ideology of northern intervention in conflicts of the South is 'development', but even the logic for development interventions is now increasingly being questioned across the globe. The Cold War provided an important rationale for this war on global poverty with anti-communist discourses used to justify development spending on peoples in faraway places (Gendzier, 1985). A whole range of representations, languages and images thus surround the geopolitical imagination of the 'Third World'. Arnold's (1989) *Third World Handbook* and Hadjor's (1993) *Dictionary of Third World Terms* even offer readers a dictionary or handbook guide to the 'Third World' and the use of its associated languages. For some observers the question of whether the Third World is best considered as a geopolitical or analytical category belongs to a distant era, a question that has become an 'anachronism' (Stallings, 1995: 349). Between the two World Wars colonial systems around the world began to destabilise slowly as movements of anti-colonialism emerged and as international public opinion began to turn against the idea of having colonies. Ironically, European ideals of independent nation-states and nationalism have often provided the main organising principles. Colonial nationalisms took various forms, however, as elites looked to a series of sources of legitimacy for their struggles to build authority and territorial power in the newly formed nation-states (Wallerstein, 1994).

It is therefore necessary to examine the extent to which 'truths' about the First and Third World are constructed through a variety of discourses (such as those of 'Third World leaders' themselves). As regimes of truth, discourses are encased in institutional structures that exclude specific voices, aesthetics and representations. Here we are referring to discourse in the Foucauldian sense of something which transcends the individual and constitutes a multi-institutional archive of images and statements, providing 'a common language for representing knowledge about a given theme' (Shohat and Stam, 1994: 18). Thus when studying 'Third World development' it is important to think about how different voices, aesthetics and representations are locked out of the equation and the ways in which development is seen to transcend individuals and to provide a common language for representing knowledge about the 'Third World'. Racism differs from Eurocentrism in that any group can be ethnocentric or come to see the world through the lenses provided by its own culture. What is racist about many images of the Third World is that difference is stigmatised in order to justify an unfair advantage, just as had been the case under numerous forms of colonialism. In this way it becomes clear that development ideas, discourses and practices have often come to stigmatise economic and political difference, as a way of justifying the advantages one region has over another and of legitimating interventions of various kinds. Talk of the free world and use of the phrase 'saving American lives' has been invoked several times 'as a pretext for murderous incursions in Third World countries' (Shohat and Stam, 1994: 24). Whole lists of countries in this imagined 'Third World' have seen US intervention in their affairs (Vietnam, Cuba, Angola, Nicaragua, El Salvador, Chile, Liberia, to name just a few).

In so many instances the 'Third World' is enframed as a world of poverty and violence, and death and decay abound in these representations. A uniform and universalised global landscape of impoverishment is created, which 'lacks the flesh and blood of the actual' as Naipaul (1985: 10) has argued, robbing us of a sense of the particularity and specificity of spaces and places. The role of the media in shaping popular perceptions of 'other worlds' is important here since there is a 'media penchant for associating the Third World with violent, unnecessary, random death, or with disease and natural disaster' (Ignatieff, 1998: 292).

Figure 5.8 Demining training and a prosthetic limb factory in Angola after years of civil war
Source: Alex Hofford

This is problematic because it is precisely this sense of place specificity and this grasp of particularity that we should be trying to move towards in rethinking development geographies. As Spurr (1993: 24) puts it, 'the dead or dying body has become in itself the visual sign of human reality in the Third World'. Simplistic symbolism and imagery thus come to stand for a 'world of problems'. This is particularly important because we need to think about the ways in which terms such as 'developing' and 'Third World' have been internalised and about how people come to see themselves in these terms. Escobar (1995) argues that development sets up the world as a picture, so that the whole system can be grasped in an orderly fashion and is thereby seen as forming a structure or whole. Post-development writers suggest that we need to understand how this 'picture' was painted, depicted and imagined or even internalised by people, planners and politicians.

In this way it is important to understand further the 'toxic' keywords through which the language of development is constructed and seen to relate to particular spaces and provenances. Escobar argues that discourses of development have produced their other ('underdevelopment') as a condition that is manageable through normalisation and the regulation of knowledge but also through the 'technification' of Third World poverty. In this way, 'Third World poverty' becomes something that only technicians of poverty eradication can 'solve' rather than being something that the poor of the 'Third World' can change for themselves. A new space is invented, he argues, a field of power that can be dominated by the development sciences with their truth claims and assumptions of neutral, disembodied knowledge. This takes us back to the main themes of this book which are to explore how discourses about the 'Third World' and the 'West' create imaginary places that are imbued with certain sorts of moral, cultural and socio-political attributes. The concept of the three worlds thus has a significant impact in terms of delimiting the spaces and scales at which policy can (quite literally) take place and ultimately also determines in which spaces and places resistance and contestation are possible. The 'Thirdworldist' notion of linked resistance envisioned only certain kinds of mobilisation and privileged only certain types of ideology.

The definition of the 'Third World' thus flows from our discussion of colonialism, development and racism above, because the term refers to colonised, decolonised and neocolonised nations and minorities whose structural 'disadvantages' have in some way been shaped by colonialism and by the unequal division of international labour that the colonial process entrained. With the third World as a world of tradition, irrationality, overpopulation, disorder, chaos and so on, the idea assumes a racial character that perpetuates, both conceptually and actually, relations of domination, subjugation and exclusion (Goldberg, 1993). We therefore cannot continue to exclude this issue of representation from mainstream development debates. The term 'Third World' was formed and constructed against the backdrop of the patronising vocabulary which saw these nations as 'backward', 'underdeveloped' and 'primitive'. It also represented an important political coalition which specifically emerged from the 1955 Bandung Conference of 'non-aligned' African and Asian nations and their enthusiasm for anti-colonial struggles in Vietnam and Algeria. Doty (1996) asks if it is possible to recognise difference (between First and Third Worlds/North and South, for example) without invoking 'the hierarchical oppositions reminiscent of the superior/inferior classifications that have justified the practices characteristic of earlier imperial encounters' (Doty, 1996: 161). This is a crucial question in any attempt at rethinking development geography – the extent to which it is possible to move beyond the oppositions, binaries, dichotomies and classificatory schemas of imperialism. The North is constituted *vis-à-vis* the South as modern, efficient, competent, as benevolent and humanitarian. The South or 'Third World' becomes its absence, its deficiency, its other. This 'politics of representation' may well be so widespread and so pervasive that to some degree we cannot escape its confines, nor its production of 'us' and 'them' (whatever the urgency of a linked project of resistance). There is a continuing complicity therefore of colonial representations in North–South debates today which ranges from a politics of silence and neglect to 'constructions of terrorism, Islamic fundamentalism, international drug trafficking, and Southern immigration to the North as new threats to global stability and peace' (Doty, 1996: 170).

In terms of explaining the sudden ubiquity of this terminology we need to think again about colonialism and the postwar making of development (Pletsch, 1981). Whereas imperial and racist interests had often dominated representations of other societies, after 1945 came the search for a kind of 'cleaner' language of government and social science which talked of 'developing nations' rather than 'primitive peoples' – apparently neutral terms clothed in the garb of science. The three worlds schema did, however, have some value in helping social scientists make sense of the world, dividing it up into researchable proportions, suggesting which part to work on and when:

> It gave the various social sciences a systematic grounding in the new world situation and permitted the establishment of new disciplinary matrices . . . academic specialities invented out of whole cloth to correspond to the new areas of political and economic influence being sought by the United States.
>
> (Pletsch, 1981: 588)

Thus we need to shift attention towards the theoretical frames that these disciplinary matrices were built around, to ask how development, tradition and modernity have been defined in postcolonial times. Much has been written about the emergence of a 'Third Way' among many contemporary world politicians such as Tony Blair, defining alternatives in opposition to what they perceive to be dominant paradigms. Thus, in Britain, Blairites claim to support an economy that combines the individual choice of the market with a concern for the social opportunities of the welfare state (see also Giddens, 2000). For many critics, these ideas simply throw a rhetorical gloss over a new style of right-wing politics. As Petras (2000) has argued however, there are many other varieties of a 'Third Way' in the modern world besides the Euro-American brand. The idea of a Third Way has a long history, as we have seen in the 'Third World', dating back to the 1940s and 1950s, to Third World leaders such as Peron (Argentina), Nehru (India), Tito (Yugoslavia), Nkrumah (Ghana) and Sukarno (Indonesia). In addition, many Islamic movements and regimes (such as the Iranian clerical government which took power in 1979) have also claimed to be pursuing a Third Way, railing against western decadence, atheistic communism and the conceptual models that they produced. Similarly, many international NGOs claim a privileged realm for 'civil society' in their vision of development today, arguing that this represents an alternative, third path to progress (outside neoliberal capitalism or statism). This is odd, since many NGOs are financed directly by neoliberal institutions such as the UN, World Bank and IMF. Ironically, in a strange twist to the history of Thirdworldism, the Euro-American brand of 'Third Way' politics is now promoting the hegemony of capitalist organisation and domination in the Third World far more effectively than many of the 'old right' regimes of Margaret Thatcher or Ronald Reagan, for example (Petras, 2000).

The key question we must therefore consider is the extent to which these ideas are real in their consequences. Do the taxonomies we construct and impose on the world shape the way we think about the globe as well as the way we act in it, as Worseley (1979) suggests? One important example of this came with the beginning of the Iranian revolution in 1979 led by Ayatollah Khomeini. The Iranian revolution involved a mass revolution from below against an authoritarian modernising state and was supported by key clerics such as Khomeini. Founding ideological cornerstones of the revolution included the division of the world into two categories: the oppressed or *mostazafin* and the oppressors or *mostakbarin*, two Quranic terms (Dabashi, 1993). Khomeini's jihad appealed to the poor, excluded elements of Iranian society, dividing the world neatly into the camp of struggling oppressed Third World peoples and their enemies, the non-Islamic, 'western' powers. Iran represented itself until Khomeini's death in 1989 as a model for other oppressed peoples, seeking to export its revolution elsewhere (e.g. to Iraq, Lebanon and Afghanistan). What is important here is that many of the standard themes of 1940s and 1950s 'Thirdworldism' (anti-imperialism, dependency, hostility to monopolies, solidarity with the oppressed peoples of the world) recur in the statements of a number of Islamic leaders in countries such as Iran, Tunisia and Turkey. Several countries, Iran included, came to believe that they could make a 'revolutionary leap' into industrialisation and progress, that they could take off or leap-frog into the good life or take a short cut to that process.

All of this needed to be very carefully theorised if that leap was to be a successful one in every case, for 'underdevelopment' to become but a distant memory. As we have seen, dependency theorists' concern with this notion of underdevelopment did not always lead to a decentring of the three worlds

Figure 5.9 Cartoon of First World/Third World, World Summit
Source: CWS and Hajjaj

schema, seeking in some ways a reversal of the traditional/orthodox argument rather than completely challenging it. This chapter has sought to argue that Eurocentrism still pervades many systems of ordering the world, and that systems of ordering (and the pictures they paint) are partial and frequently based around negative 'truths' which deny or efface cultural difference and stereotype the 'Third World' as a world of problems. We have also seen that some systems of ordering are ahistorical, simplifying the complex political struggles which have brought about change in many parts of the world. None the less, all systems of dividing and differentiating the world involve the formation of certain kinds of alliances and are important (geo)-political projects in themselves.

BOX 5.4

Chapter-related websites

http://www.globalissues.org/
Global Issues site.

http://www.twnside.org.sg/
Third World Network.

6
Postcolonial Geographies of Development

INTRODUCTION: WHAT IS POSTCOLONIALISM?

> [P]ostcolonialism is a much needed corrective to the Eurocentrism of much writing on development.
>
> (McEwan, 2001: 130)

> Why is it that most development academics and practitioners have never heard of Said, Spivak, Bhaba, Fanon and other post-colonial thinkers? Why are these texts on Eurocentrism so rarely found on reading lists on Development Studies courses in Development Studies departments and institutions?
>
> (Kothari, 1996: 13)

> Cosmopolitan jet-setters in São Paulo live one kind of development while women in sub-Saharan Africa walking for hours to collect water experience a completely different kind of development. How do we listen to this difference?
>
> (Radcliffe, 1999: 84)

Postcolonialism is a difficult and contested term not least because it is far from clear that colonialism has been relegated to the past. Its meaning is not limited to 'after-colonialism' or 'after-independence' (Ashcroft *et al.*, 1995: 2) but seeks to highlight the complexity and ambivalence of the imperial encounter in crucial ways. In one sense, the word filled a gap left by the abandonment of the term 'Third World', particularly for those influenced by postmodern or post-structural thought. Typically, the postcolonial has been defined in relation to a period of time which is marked by the power of the colonising process (Slater, 1998). It raises important questions about how the 'core' and the 'periphery' mutually make and constitute each other by focusing on relations between the colonisers and colonised. In some ways postcolonialism is also related to postmodern critiques of development in that they share a concern with agency, resistance, hybridity and discourse, and a common desire to explore the legacies of Enlightenment thinking and to challenge dominant 'western' histories of progress. It is thus helpful not just to think of postcolonialism as something coming literally *after colonialism* and therefore signalling its demise, but to think about this more flexibly 'as the contestation of colonial domination and the legacies of colonialism' (Loomba, 1998).

Postcolonialism thus refers to ways of criticising the material and discursive legacies of colonialism (Radcliffe, 1999: 94) and represents a number of perspectives which can broadly be termed 'anti-colonial' (McEwan, 2001: 127). The associated term 'postcoloniality' on the other hand refers to 'the historical experience of living in a time after colonialism' (McDowell and Sharp, 1999: 208) and points to a way of criticising the legacies of colonialism in the South which are both material (found in the state or society) and to do with ideas (how people think about themselves and their relationship to other worlds) (Radcliffe, 1999). The terms 'postcolonial', 'postcoloniality' and 'postcolonialism' evoke responses in *both* the metropole and the periphery. There is much debate and discontent, however, about the manner in which these terms have entered the lexicon of development discourse – ranging from whether the terms should be hyphenated to their very legitimacy and their growing currency within academic circles. Many argue that postcolonialism neglects important issues of poverty, resource distribution and development (Rajan,

1997; Sylvester, 1999; Pieterse, 2001a) and object to the way in which postcolonial approaches make a virtue out of being free-floating and open-ended (Sylvester, 1999).

In some ways it seems that the critics of postcolonial approaches view their contribution as one of ornamentation rather than actually advancing our understanding or analysis of development, suggesting that at bottom it is possibly just a kind of liberal multiculturalism (Kelsall, 2002). Its theoretically abstract nature seems somewhat at odds with the need to adequately connect to the specific, concrete and local conditions of everyday life (Jacobs, 1996: 158). Thus how do we combine the important cultural focus of postcolonial studies with the economic focus of the study of development (Schech and Haggis, 2000)? This is not an easy task since, in many ways, postcolonial studies are premised on a critique of western and Eurocentric models such as those generated in the name of 'development'. In some ways then, postcolonial studies and development studies are rarely seen as interconnected fields: 'Two giant islands of analysis and enterprise stake out a large part of the world and operate within it – or with respect to it – as if the other had a bad smell' (Sylvester, 1999: 704).

In addition, postcolonial writing and the kind of critique of development it offers have generally been informed by postmodernism and post-structuralism which is seen by some critics as of limited relevance to interpretations of non-western cultures and societies because they are 'infused with French social theory' (Paolini, 1997: 84). Postcolonial theories are therefore also often inspired by many of the same streams of social thought as the post-development school, sharing common interests and concerns with diversity and difference. One important feature of debates about postcolonialism is the recognition that European imperialism took various forms in different times and places but that none the less the prestigiousness and power of the imperial culture usually took centre-stage. Indigenous cultures did not simply passively accept the cultural practices of colonialism however. Imperial culture was (re)appropriated by the colonised in projects of resistance to colonialism that sought to contest and undermine imperialism and the power of its knowledge. Postcolonialism is also seen to be born out of important streams of literary and cultural criticism and as such is: 'a celebration of the particular and the marginal which envisages peoples of the Third World carving out independent identities in a de-Europeanized and hybrid space of recovery and autonomy' (Paolini, 1997: 84).

Thus much of this work is seen to have an emphasis on the marginal and on the 'carving out' of identities in postcolonial times. It involves giving 'voice' in western academe to 'Third World peoples' (and this term is still used as if it is unproblematic in many of these debates). Highlighting the Eurocentrism of certain kinds of scholarship about the 'developing world', the 'Third World' or 'the Tropics' is therefore important here. The work of Edward Said (1978) in his book *Orientalism* gave particular impetus to these debates – Said had looked at how colonial discourses represented the 'Orient' and sought to manage and dominate it. In his later work *Culture and Imperialism*, Said (1993) argues that postcolonial identities are intertwined, intermixed and complex, a point that continues to be relevant in the context of globalisation and the resultant cultural hybridities that characterise the 'global' era (see Chapter 7). This chapter seeks to examine how postcolonial literatures are also largely the result of this interaction between imperial culture and the complex of indigenous cultural practices. These literatures have emphasised the agency of the colonised in transforming their own societies and subverting or remaking colonial power relations. Postcolonial theory builds on this by forming a critique of the way in which the West has 'made' knowledge about the South and as such can help us to seek new ways of learning about and understanding development.

The extent to which it is necessary or even possible to de-centre the West in our representations of global development today is a key question here. As McEwan (2001: 127) argues, postcolonialism is also at core about a critique of the spatial metaphors deployed by western institutions as well as a challenging of the 'experiences of speaking and writing by which dominant discourses come into being'. Are there other kinds of representations and knowledge in non-western regions which may also be important? Whatever postcolonialism is about it is based fundamentally in the 'historical fact' of European colonialism. Postcolonialism has also tended to be very much a concern in countries that were former imperial powers (e.g. France, Britain and Portugal). Here it serves as an umbrella term for the experiences of decolonisation in the metropole. We can define the hyphenated form of the word 'post-colonial' to mean both the material effects of colonisation and the huge diversity of everyday and sometimes hidden responses to it (Crush, 1995). Assumptions are, however, made about the commonality of imperial rule or the universality of certain colonial experiences which as we have seen varied in a number of ways in different times and spaces. There are clearly numerous problems with the process of classifying countries and with the terminology used to understand a country and its characteristics and history. Thus for some observers there has tended to be a preoccupation with the West and with western forms of knowledge/power (Sylvester, 1999), while the concerns of postcolonialism cannot be translated easily into 'on the ground' practice. There have also been suggestions that the oppositional stance of postcolonial studies will not actually make much impact on global inequalities and power imbalances and that, as a whole, it looks too much to the past.

One important point to remember here, however, is that postcolonialism has developed unevenly at the global level (McClintock, 1992: 87). Argentina is not 'postcolonial' in the same way as Angola, nor is Brazil 'postcolonial' in the same way as Botswana. In setting up a 'postcolonial condition', it is possible that such approaches 'inevitably compress the Third World into a single dimension' (Paolini, 1997: 87). Postcolonialism is not some sort of singular condition therefore and, if anything, postcolonial writings seek to avoid or at least be critical of 'framing' the conditions of others. The term 'postcolonial' is extremely ambiguous and complex, relating to a whole variety of different cultural, political and economic experiences. A principal objection for many critics is that a concern for the postcolonial obscures the 'actual' or material political economy and sociology of development, making it difficult to see where 'inequality' fits into the equation. Global political and economic inequalities will continue to remain obscured if these issues are not engaged as an important point of departure. This notion of the 'actual' political economy of development being downplayed and marginalised in the silence of postcolonial writings is thus important to acknowledge here. A concern for the relationships between postcolonialism and global capitalism remain absent from many contemporary debates. None the less, postcolonialism is not somehow entirely separate from the material world as many critics seem to suggest. In part, postcolonial literatures are important precisely because they provide reflections on a whole variety of themes relating to social and political inequality and the unevenness of the material world. Critical attention to culture, difference and the spatial imagination of progress and development does not necessarily imply that we have to dispense with a concern for the material geographies of colonialism and their legacies (Nash, 2002: 222).

Arguably, postcolonialism is not totally opposed to a concern for material realities but also allows us to deconstruct the mythical material futures of development that were predicted for many countries. It calls into question, for example, how countries can be presented with a 'future in a rearview mirror' scenario, as Nyamnjoh (2001) puts it in referring to James Ferguson's reading of contemporary Zambian development (see Box 4.1, p. 88). This important reading of Zambian development discourse focused in part on the prosperous future prescribed and predicted for Zambia, exploring the shattering of these illusions in postcolonial times. Ferguson shows that in the late colonial period high hopes and expectations were held for Zambia which failed to materialise in postcolonial times. In some ways, therefore, many development concepts and categories were incapable of accounting for or theorising the subsequent reversal and retrogression of Zambia in the 'postcolonial' period, where there had been a decline in several sectors of the national economy.

Issues and themes of exile and Diaspora are not, however, universally relevant everywhere and could be seen as the product of a 'fixation on margins created directly by certain parts of the West' (Sylvester, 1999: 714). In this sense, the term 'postcolonial' has more to do with a particular generation of scholarship:

> 'All along I wonder whether there is such a thing as postcolonial theory. To me it seems more the intellectual wave-length of a generation than a theory – the generation of decolonisation (Fanon, Cabral, Abdel Malek), and postcolonial diaspora (Said, Hall, Spivak, Appiah, Rushdie), interconnected in a loose patchwork of themes and approaches. These themes relate to the South and to the South in the North . . . On all scores this now lies behind us.
>
> (Pieterse, 2001c: 92)

The term 'Diaspora' literally refers to the scattering of populations and dates from the Roman conquest of Palestine; it remains a key theme of postcolonial debates (Afzal-Khan and Seshadri-Crooks,

2000). Thus in one sense the postcolonial refers partly to a generation of intellectuals and to the coherence of an epoch rather than to a particular 'theory' (Pieterse, 2001a). This is none the less a particularly important generation of scholars whose work does have a direct relevance to the theory and practice of development today – more on these authors below. Postcolonialism is also a term very much bound up with the migration forced by the slave trade, involving transnational connections which criss-cross national boundaries and lead to new forms of communities and identities. A very geographical notion, the focus on diasporas emphasises the routes that different migrations involve and the new transnational connections of peoples that result from this. Massey (1994) refers to these as 'stretched out' geographies of flows and connections which produce particular forms of culture and identity. In some ways, postcolonialism is partly about thinking through the implications of stretched-out geographies, making connections and understanding the important flows and movements between North and South. None the less, we must remember that the location and development of geography as a discipline is 'inescapably marked' by its beginnings as a western-colonial science (Sidaway, 2000: 593). This would suggest that in seeking to deal with and understand these stretched-out geographies, the discipline of geography is not itself untainted by colonial legacies and power relations.

All postcolonial societies in Africa are still subject in one way or another to overt or subtle forms of neocolonial domination which independence has not solved. Occasionally the associated term 'neocolonialism' is deployed to describe these relations of continuing domination and exploitation which were once characteristic of colonialism. In some ways, the terms 'neocolonialism' and 'postcolonialism' are not particularly incompatible but depend on a writer's particular conceptual and thematic persuasions. Thus neocolonialism has often been associated with Marxist writings and also with the structuralist tradition of dependency writings. The term is problematic, however, in that it overplays the power of imperial centres, enframing the Third World as passive and 'continually captured' (Slater, 1998: 654) without saying much about the impacts of colonialism on western societies. Indeed, for some commentators the term 'postcolonial' is a replacement for the idea of the 'Third World', but the main difference between the two is that 'Thirdworldism' saw a common project of linked resistance to neocolonialism, whereas a sense of commonality is occasionally lost in postcolonial studies with its characteristic focus on difference.

These debates are not reserved exclusively for countries that have achieved independence in the past fifty or sixty years but have also been extended to countries that were never formally colonised or which became independent more than a century ago. Thus postcolonialism is in some ways preferable in that it is more nuanced and can serve as *a kind of counter-discourse which seeks to disrupt the cultural hegemony of the modern West*, with all its imperial structures of feeling and of knowledge, wherever they have been manifest and in whatever capacity this hegemony exists. The term 'postcolonial' does need to be fully interrogated and contextualised however, historically, culturally and of course geopolitically (Pieterse and Parekh, 1995). Is this counter-discourse relevant to other regions of the South beyond Africa however, particularly those that have not experienced colonial rule as recently as many African peoples and those that were never colonised at all? In order to answer this question it is important to understand the relevance of postmodernism in development thinking today, since in many ways postcolonial and postmodern thinking are very closely related and intertwined (Simon, 2002). This has important implications for the discipline of development studies since, as Kothari (1996: 3) has argued:

> Development studies is a relatively new discipline which has its origins in a colonial past. . . . Development Studies is a neo-colonial discipline in which particular gendered and racial formations constructed through colonial processes are re-presented and re-articulated.

Thus we can bring postcolonial critiques to bear in understanding the ways in which development studies (and other disciplines) have assumed the power to name and represent other cultures and peoples today given that they were initially created and produced by colonialism. Do these disciplines reproduce unequal relations of power as Kothari (1996) has argued? This chapter seeks to outline the key themes and considerations of a postcolonial geography of development. In so doing, it does adopt something of an Afro-centric focus in places and as a result it is important to remember that not all postcolonialisms and not all conditions of postcoloniality can be understood through the lens of postcolonial debates about Africa. There are important differences and parallels that can be drawn with development theory and practice in a variety of other contexts (such as the Middle East) which need to be brought out here. Indeed, this sensitivity to cultural, political and economic differences between countries is one of the fundamental issues of debate in postcolonial studies.

This chapter asks where geography as a discipline can make a contribution to a redrawing of the imagined geographies of contemporary 'postcolonial' development. Is it possible that thinking about postcolonial discourses of development helps us to understand the emancipatory possibilities of new configurations of social relations across spaces, places and scales? The first section focuses on themes of nationhood, national identity and citizenship that have arisen in postcolonial debates. We then move on to look at related issues of how language and cultural identity have been explored in postcolonial contexts and literatures, before shifting direction to explore the legacies of colonial notions of trusteeship and their importance in shaping development thinking about First World/Third World 'partnerships' today.

POSTCOLONIALISM AND NATIONAL BELONGING

Postcolonial theory involves discussion about experiences of various kinds: migration, slavery, suppression, resistance, representation, difference, race, gender and place (among others). It also involves a response to the claims of imperial histories which wrote colonised 'others' out of the script or denied them a place in history altogether. In many ways these debates are characterised by the notion that the 'Empire writes back' (Ashcroft *et al.*, 1995), in order to comprehend and reinterpret the colonial experiences and identities of 'Third World' peoples today (Paolini, 1997). Postcolonial critiques stress the need to destabilise the dominant discourses of development (with their ethnocentricity and origins in imperial Europe). In this sense they provide a much needed 'corrective' to the Eurocentrism of development discourses which prioritise western histories and paths to 'industrialisation' or mass consumption (McEwan, 2001). Some of these writings are also very instructive in that they deconstruct and destabilise the histories and geographies of terms such as 'Third World' and 'Third World woman', acknowledging the diversity of perspectives, approaches and paths to development that are masked behind these labels. Just as we have set out to explore the 'other side of the story' about the development project or 'the view from below, the views of women, the view from the South' (Munck, 1999: 200), so postcolonialism can help us to understand 'history from below' (Dirlik, 2002). In this way such literature can provide useful insights into debates about the making of colonial and postcolonial nationhood and identity, by revealing stories and histories that had been suppressed in earlier accounts, by challenging the claims of some development 'experts' and by challenging the claims of other narrations of the past. Whereas modernisation consigned claims on history to the past, seeing them as backward reflections of tradition, these claims have in a way reappeared in the contemporary context of globalisation where social and political relationships have been reconfigured. Reading postcolonial literature can give us an idea of how these sorts of transformation are experienced in the South and how people negotiate the complex boundaries of social and cultural power and identity in their respective societies.

One important source of 'postcolonial' writing in the South has been the work of literary figures such as Gabriel Garcia Marquez, Ngugi Wa Thiongo, J. M. Coetzee, V. S. Naipaul and Chinua Achebe. What is particularly interesting and important about some of these writings is the way in which they often reflect on the nation, on national communities and on questions of national belonging. In some cases these writers also point up some of the continuity (and discontinuity) between colonial relationships and structures of power in the past and their resurfacing in the present. Some particularly interesting questions have been raised by African writers, notably about the hybrid nature of identity. In postcolonial studies, hybridity is theorised as an intrinsic dimension of the colonial encounter with hybrid subjects (Slater, 1998). On the other hand, 'postcolonialism' is about mental revolutions, about decolonising the mind, about the realisation that 'we too might have a story to tell':

> The nationalist movement in British West Africa after the Second World War brought about a mental revolution which began to reconcile us to ourselves. It suddenly seemed that we too might have a story to tell. *Rule Britannia* to which we

Figure 6.1 Twenty-five years constructing the future of Mozambique

Source: Frelimo Forum Mulher

had marched so unselfconsciously on Empire day now stuck in our throat.

(Achebe, 1987: 12)

Following on and inspired by the work of a group of Indian historians known as the 'Subaltern Studies Group', it has been argued that the project of decolonisation involves the recovery of the lost historical voices of the marginalised, the oppressed and the dominated (Guha and Spivak, 1988). One founding member of this group, Ranajit Guha, had argued that the elitist bias of colonial histories of India had denied peasants recognition as subjects of history, as makers of rebellions (for example) or as people with consciousness. The word 'subaltern' here meant subordinate in terms of class, race or gender, for example. It is important to remember however, as Chakrabarty (1998: 460) has argued, that subaltern studies can be seen in part as a 'mutant of area studies'. In this way these important studies emerged out of Indian national historiography and became a way of writing history 'from the bottom up' by focusing on dominated groups and by exploring the dynamics of domination and oppression in colonial and postcolonial India (McDowell and Sharp, 1999). By 'recovering' and giving voice to the silences, this kind of scholarship participates in this process of decolonisation by seeking to subvert and contest the elite and ruling versions of the past. These kinds of histories also focused on the agency and creativity of the subaltern.

Akhil Gupta (1998) argues that in India, the dominant discourse that enshrined national 'development' as the key to progress after independence in 1947 is a central condition of the 'postcolonial condition' today in that country. Gupta's work suggests that the interface between postcolonialism and development is a crucial and fruitful one to explore. Interestingly, Gupta (1998: 39–40) argues that the state-guided development project in postcolonial India which sought to address 'underdevelopment' was not simply a 'structural location in the global community of nations' but an important 'form of identity in the postcolonial world'. Thus underdevelopment figured in these state-guided discourses not just as a material relation of social and geographical position but also as a state of mind, a form of identity, a notion of citizenship. What is particularly interesting about some of Gupta's work is the way in which he maintains a focus on the uneven materiality of the lives of people affected by British imperialism in India. Some writers have warned, however, that postcolonial history can never simply make the marginalised, the oppressed, the subaltern, the subject of their own history. Who gives academics and these writers the right to speak for the marginalised?

In a study of the legacies of late colonialism, Mamdani (1996) argues that colonisation was fundamentally about differentiating between peoples of the colony at the national level where administrators and settlers were given the status of citizens and the indigenous peoples were seen as subjects with no rights. In this way the institutional segregation of the colonial state was, according to Mamdani, later reproduced in postcolonial times, through complex formulations of identity which constituted African peoples simultaneously as both citizens/subjects. Mamdani argues that we need to break away from the entrapment of Africa within 'history by analogy' however, which either exoticises Africa or represents the continent's history as part

Figure 6.2 The UN and indigenous peoples, 1992 (Spanish version), commemorating the International Year of Indigenous Peoples, 1993

Source: UN Department of Public Information, design by Jan Arnesen

of larger European stories of the past or metanarratives. The African state itself takes hybrid, 'transculturated' forms – with European roots and beginnings but with a variety of African 'inflections' and complexions (Ahluwalia, 2001). We can examine the variety of ways in which postcolonialism informs our understanding and critique of development discourse and helps to decolonise established ways of apprehending North–South relations (McEwan, 2001). These debates can help us to understand the ways in which national states and administrations may themselves be understood as legacies of the past with important implications for the creation of citizens and subjects in the present. These writings can also extend the boundaries of conventional thinking about development and difference in important ways.

Concerns with the language of development are not esoteric but rather are central to the process by which worlds and geographical difference are imagined and enframed. In this way postcolonialism as a kind of counter-discourse articulates and interlinks with associated critiques of national and international development discourses such as those formulated by feminist writers. Postcolonial feminisms, for example, allow for and accommodate competing and disparate voices (Radcliffe, 1999) and try to deconstruct the myth of the homogeneous and stereotypical 'Third World Woman' (Mohanty, 1988). This myth constructs a powerful form of 'Third World Difference' where all 'Third World Women' are viewed as powerless, burdened and oppressed. These women tend to be seen then as passive victims, objectified in time and space. Thus postcolonial approaches to feminism force a move away from a singular national or international feminism (which was based upon the vantage point of white, middle-class, western feminists), thereby highlighting the failure to acknowledge the important differences of ethnicity and class between women (McEwan, 2001: 129). By citing some well-known Third World women, some geographers thus seek the right to speak about Third World women, abdicating responsibility for engaging critically and personally with the global power relations that classify 'Third World' others (Radcliffe, 1994). Post-structuralist critiques of the (feminist) subject have, however, made it quite difficult to define feminism's ambitions and subject (McDowell and Sharp, 1999). Postcolonial perspectives on gender see women as active agents of change and as highly diverse and heterogeneous in their concerns and priorities rather than assuming that all women in the 'South' are universally poor. Rather, attention has focused instead on social constructions of masculinity and femininity in each context, seeking to understand the different forms of identity and subjectivity that are produced through different (historical) discourses of development. In an important sense, these debates have raised crucial questions about the role of women in defining national identity and belonging in postcolonial times.

One of the most recent examples of feminist writers confronting this particular dilemma of the boundaries of national citizenship in postcolonial times comes from the work of novelists and campaigners such as Medha Patkar and Arundhati Roy, both of whom have been prominently involved in campaigning against development and dam projects in India. Medha Patkar is a founder, principal spokesperson and key organiser of the Narmada Bachao Andolan (NBA), the popular movement which has opposed plans for the construction of a series of dams on the Narmada River in Northwest India (see Chapter 9). The NBA has questioned the whole process of development planning in India and (under Medha's leadership) has campaigned for knowledge and information concerning plans that affect people's lives to be made more public and earlier on in the planning process. As a result of her writings and campaigns, issues concerning the socio-cultural implications of the dam projects have been placed firmly on the agenda, while the World Bank and the Indian government have been forced to rethink their plans for the river and its diverse communities. Medha Patkar has also helped to establish the National Alliance of People's Movements, a network of activists that works throughout India and tries to change the relations between people and the state and to formulate alternative visions of development. The Booker prize-winning novelist Arundhati Roy has also criticised these dam projects proposed for Narmada. Roy has been part of several very high-profile demonstrations in India for which she was 'symbolically imprisoned' for one day in March 2002 (*Guardian*, 7 March 2002). Women have thus been at the forefront of many contemporary struggles against 'development' projects and programmes, and have been central to the radical environmental and social justice movements. As we shall see, the struggles of women across the South have been crucial to redefining and rethinking development in many ways (Bunting, 2002) and at various spatial scales.

LANGUAGE, CULTURE AND IDENTITY

One of the main reasons that postcolonialism *is* relevant to the study of development is that it can allow us to understand the material and cultural geographies of colonialism and the politics of identity and belonging as it varies across spaces. A common focal point of many debates about postcolonialism and development revolves around the notion of subjectivity, which refers to the range of subject positions or identities that an individual human being as agent or subject actually mobilises or embodies. In general, postcolonial approaches are particularly sceptical of the notion of a unitary, self-authoring subject posited in the Enlightenment era and tend to incorporate a critique of the 'unreflexive projection of subjectivities as universals' (Werbner, 2002: 2). Thus we might also dispute the universalisation of a single image or definition of poverty and poor people, for example. Subjectivities are shaped and defined morally and politically in postcolonial times or even economically, through consumer identities *vis-à-vis* the global market. Africans are thus 'both citizens and subjects ... sometimes they are more citizen than subject and other times they are more subject than citizen' (Nyamnjoh, 2001). It is important when exploring development to consider the extent to which subjectivities are determined by discourses, political economy, state structures and personal dispositions (among others). What we can begin to do, with the aid of postcolonial theory, is to open up the notion of agency, to deepen our understanding of subjectivity by looking at its multiple forms, influences and meanings and opening up the spaces where development's subjects are constructed.

It is often assumed that marginalisation, dispossession and exploitation form a singular common ground for the making of subjectivities in 'neocolonial' times but we can widen the debate about where and how this takes place. It is therefore important to understand the plural arenas in which economic, cultural and political identities are made in postcolonial times (Werbner, 2002). Subjection to the discourses of development is not just about relations of power and domination but also resistance and reconstruction. In this sense postcolonialism focuses on the need to understand the formation of a collective, social memory of shared histories and cultures. In some ways, postcolonialism seeks to 'replace' memories of colonisation and calls into question the distinction between oppressors and oppressed in order to rewrite history from a more multicultural perspective in a way that seeks to write back into history those left out of it or condemned to backwardness (Dirlik, 2002). Other approaches to and ways of thinking about history are crucial here (e.g. in film or fictional literature).

Postcolonial literature re-centres development processes within the lived experiences and consciousness of those subjected to development at all levels: local, national and international (Perry and Schenck, 2001). They can allow us to understand concepts of 'time as lived' (Mbembe, 2001) and to focus more on the terms and lived experiences of ordinary people. Questions of literature are important here since postcolonial studies as a field itself

BOX 6.1

Postcolonial cinema and 'third spaces' of representation

Different ways of reflecting on the history and geography of colonialism come through quite clearly in the study of 'postcolonial' and 'Third World' cinema. A range of 'postcolonial' subjectivities are constructed through film.

> In relation to cinema, the term 'Third World' is empowering in that it calls attention to the collectively vast cinematic productions of Asia, Africa and Latin America and of minoritarian cinema in the First World ... Just as peoples of color form the global majority, so the cinemas of people of color form the majority cinema ... it is only the notion of Hollywood as the only 'real' cinema that obscures this fact.
> (Shohat and Stam, 1994: 27)

The term 'postcolonial cinema' usually includes a whole variety of different types of films which are aggregated together, including 'Thirdworldist' films produced by and for Third World peoples, films which adhere to the principles of 'Third cinema', cinematic productions of Third World peoples, films made by First and Second World people in support of Third World peoples and recent diasporic hybrid films, for example, those of Mona Hatoum or Hanif Kureshi, which both build on and interrogate the conventions of 'Third cinema'. For some commentators, both 'Third World cinema' and 'Third cinema' retain important tactical and polemical uses in terms of cultural practices and political transformations (Shohat and Stam, 1994). The term 'First cinema' describes commercially structured film industries (e.g. Hollywood), 'Second cinema' refers to avant-garde, personal, art or *auteur* films and 'Third cinema' offers resistance to imperialism and to oppression (Solanas and Gettino, 1976). Reflecting upon the *Cinema Novo* of Brazil, Perón's 'Third Way' in Argentina and the cinema of the Cuban revolution, Solanas and Gettino (1976) defined Third cinema as that which recognises in anti-imperial struggles a process of 'decolonising culture'. The purpose of 'Third cinema' according to the originators of the term is to create a 'liberated space' for emancipation (Armes, 1987; Gabriel, 1989). The important contribution made by Solanas and Gettino (1976) was their assertion that, following Fanon's injunctions about national culture, in order for an alternative cinema to emerge in the peripheral 'liberated spaces' colonial aesthetics had to be dissolved:

> [we must] insert the work as an original fact in the process of liberation, place it first at the service of liberation, place it first at the service of life itself, ahead of art; *dissolve aesthetics in the life of society*; only in this way, as Fanon said, can decolonisation become possible and culture, cinema and beauty – at least, what is of greatest importance to us – become our culture, our films, and our sense of beauty.
> (Solanas and Gettino, 1976: 60–61)

In 'postcolonial' Mozambique, cinema came to reinterpret histories of colonialism and to represent the ideals of planned socialist development (as well as those of its subjects). There were also a variety of important Cuban films such as *Memories of Underdevelopment* (Chanan, 1990). Directed by Tomás Gutiérrez Alea, this was one of a number of films produced by the Cuban Film Institute (ICAIC). The film's main protagonist is Sergio, a man caught between the old and the new of Cuba after the socialist revolution in the late 1950s and early 1960s. The film provides a fascinating insight into Cuban life at this time and some of the identity crises that the revolution had produced. The film's title suggests that, now the revolution is over, colonisation has been replaced by (inconsolable) memories of underdevelopment.

Contemporary films such as *Life and Debt* (Stephanie Black, 2001, New Yorker Films) are also an important part of this debate about postcolonial cinema and representations of the past. This was a documentary that explores the implications of the free trade agenda for Jamaica, providing important insights into the nature of postcolonial geographies of Jamaican development. The film looks at the

Figure 6.3 Cuban Film Institute with film posters
Source: Alix Wood

BOX 6.1 (contd.)

destruction of Jamaican industry and agriculture, and exploitative relations between the Jamaican economy and global capatialism. The country is seen to become simply a market for North American goods and a source of underpaid labour. Imported products from the USA (e.g. powdered milk) are seen to destroy local production and create dependency. Del Monte seeks to dominate banana production in the country, controlling all bananas produced nationally while farmers remain incredibly poorly paid. At one point a Jamaican minister (Michael Manley) remarks: 'You ask, whose interest is the IMF serving'; 'Ask who set it up' comes the answer. Local competition with imported goods and services is penalised, as the IMF is seen to charge extortionate interest rates and be dominated by a handful of countries, chief among them the United States (http://www.ru.org/).

emerged from the literature of 'Commonwealth criticism' which began to develop in the 1960s, giving expression to a variety of nationalist modes of resistance. In a way these writings help us to read the 'local' and 'global' in new ways and to begin to understand alternative perspectives, finding other ways of interpreting development practice. As a result it is clear that 'literature, as a crucial record of subject formation, should take its place in an interdisciplinary, international development debate' (Perry and Schenck, 2001: 200). In such work we can find important records of the process of subject formation in a variety of places and spaces undergoing 'development'. These works can tell us much about the dialogues, voices, languages and social constructions of postcolonial states. Forms of resistance are also often evident in such writings which can help us understand how resistance is and can become transnational, linking men and women in different apparently 'local' situations across the world.

The multiplicity of different stories of development that have been forthcoming confirm further that development cannot be seen as emanating from a monolithic, singular ideological text or doctrine. Thompson and Thompson (2000) use the metaphor of two trees – the baobab and the mango – invoking literary sources from postcolonial writings, to suggest that these are important metaphors through which we can understand the 'lessons' of development in Ghana and Thailand, two very different countries. The baobab, as a symbol of the aridity of the African savannah, represents the top-down domination of a landscape, a predatory stifling of growth and a swallowing up of its surroundings after taking root. The mango on the other hand is a reassuring symbol of abundance that can be used modestly without damage if given due care and attention. Simon (2002) argues, however, that the baobab is also a symbol of endurance and survival against the odds, providing resources and sustenance for different communities. The point here is not to argue over which tree provides the best metaphor but to open up a space for thinking about alternative ways of 'reading' and interpreting the geography of development in such varied locations.

Figure 6.4 Still from the Cuban film *Memories of Underdevelopment*
Source: M. Chanan, courtesy of ICAIC

Another important question is whether fiction – its making and its consumption – is a luxury when there are so many more pressing and urgent needs. When there are wells to be dug, demonstrations to be attended, injustices in the world to be fought, why should the inventing of fiction have any sort of importance? With respect to India, Ghosh (2001) has argued, for example, that popular fictions 'represent the corruption of the era as well as the desirability of its modernity', creating interesting and important paradoxes. Thus Indian fictions tell the story of the 'deferral' of western modernity in the 'imaginary of the postcolonial nation'. This writing provides alternative insights and interpretations and offers us a different approach to development issues. Important questions need to be raised, however, about the privileged positions from which such texts are written and consumed. As Ghosh (2001: 955) puts it:

> In postcolonial India, the cultural and gendered politics of Indian nationalism can be read through the texts of popular novels where constructions of the 'modern' women in the service of the Nehruvian national development project and its successors are presented for the consumption of the literate middle class.

Ghosh's account discusses how these fictions present a picture of the liberated Indian woman characterised by processes of westernisation, where women westernise themselves to the extent that they discard their spiritual or traditional roles in favour of the 'development' of the Indian nation. In some ways therefore, such fictions offer essentially flawed representations of the modern Indian nation and only 'false' promises that all citizens could be provided with the necessities of life and the desirable symbols of modernity. Thus in reading postcolonial literature we can combine new types of 'data' such as imaginative literature to advance our understanding of development and how it is received, experienced and articulated in countries such as India, as a way of understanding struggles over the meaning and practice of development.

The work of Aimé Césaire, in particular his *Discourse on Colonialism*, is useful here (Césaire, 2000) in advancing an understanding of the literary critique of colonialism. Born in Martinique in 1913, Césaire (along with Senegalese poet Léopold Sédar Senghor) gave substance to the concept of *Négritude*, a key part of the process of decolonisation in French West Africa. For Moore-Gilbert (1997), Césaire's writings anticipated not only the work of Frantz Fanon but also other key postcolonial texts such as Edward Said's *Orientalism* (Said, 1978). *Discourse* is particularly important because it 'offers new insights into the consequences of colonialism and a model for dreaming a way out of our postcolonial predicament' (Kelley, 1999: 13). The book was first published in 1950, in the age of decolonisation and revolt in Africa, Asia and Latin America and just five years before representatives of the Non-Aligned Movement (NAM) met in Bandung (Indonesia) to discuss the freedom and future of the African continent.

Discourse, although short on proposals for future change, offers a powerful 'poetics of anti-colonialism' which provides both a reflection on the material and spiritual havoc created by colonialism as well as a critique of colonialism and its need to reinvent the other, to turn 'the other' into a barbarian, which he called 'thingification' – a process which had ultimately degraded Europe. The book does not set up a global power struggle between capitalism and socialism but rather focuses on the need to overhaul a racist system in order to enable other imaginations of the world. Thus Africa is seen to have been stripped of its history by an entire generation of 'enlightened' scholars, casting Africans as 'little more than beasts of burden or brutish heathens' (Kelley, 1999: 10). Although his writings were never intended to offer a road map or a blueprint for revolution, Césaire warned those waging anti-colonial struggles not to replicate the Manichaean world of backward/forward dualities in representations of the non-western and western, and to avoid the danger of trying to return to some mythical pre-colonial harmony. Instead, Césaire argued that these struggles needed to avoid following European footsteps and carve out altogether new directions.

Another key question in debates about postcolonialism and decolonisation has been the issue of how language is used and understood. Chinua Achebe's writings in particular seem to express a sense of unease with whether the language of the coloniser, English, will be able to 'carry the weight' of his 'African experiences', arguing that this would need to be a new English 'altered to suit new African surroundings' (Achebe, 1975: 62). Using the 'mother-tongue' (which is invariably not the language of the former colonial power) produces phrases such as 'dreadful betrayal' and 'guilty feeling' but using foreign languages is something which is positively embraced. Achebe revised this opinion, however, in some of his later interventions (Achebe, 1988). Can English ever 'carry the weight' of African experiences of postcolonialism and development? Can there be an African English (just as there are Canadian, Australian, American versions)? For Kenyan writer Ngũgĩ wa Thiong'o, language was the very means of the spiritual subjugation of African people. For Ngũgĩ the actual domination of a people was very much a process about culture, language and identity, suggesting that the imposition of the languages of the colonising nations was crucial to the domination of the 'mental universe of the colonised' (Ngũgĩ, 1986: 16). This then is an important part of understanding the implications of postcolonialism, by examining the ways in which colonial ideologies dominated the 'mental universe' of the colonised or continue so to do.

Again, the language and culture of the colonising power (in this case Britain) took centre-stage as Kenyans, some of whom spoke Gikuyu as their mother-tongue, had to bow before this in deference. Ngũgĩ's point is that this took Kenyans further and further from themselves to 'other selves' or 'from our world to other worlds'. Thus wherever European languages and cultures have been imposed on people of colour there have been psychological ramifications. Algerian writer Frantz Fanon termed this process 'alienation', looking at the psychological alienation that results from racial and class domination in colonial situations (Fanon, 1967, 1968). By alienation he was referring to the many conditions ranging from the superiority complex of the coloniser to the inferiority complex of the colonised. Colonial language policies varied according to the identity of the coloniser in each context (Mazrui, 1993). For Fanon, to speak means above all to assume a culture, to support the 'weight of a civilization' (Fanon, 1967). Thus with the acquisition of a language (native or colonial) comes an entire set of cultural underpinnings. Fanon argued that colonial education and the Christian missionary enterprise had produced a whole series of racist images of the 'native' which were perpetuated through colonial ideologies of cultural difference. Through colonial education and missionary activity there was an attempt to elevate the culture of the coloniser and debase the culture of the colonised. The process led Africans to identify with the European explorer, the missionary, the 'bringer of civilisation'. The overall effect on the mind of this educational and religious war is alienation according to Fanon.

Postcolonialism and postcolonial theories are partly about thinking through the consequences (after independence) of this alienation, raising important questions about how these stereotypes were internalised by the colonised. In many ways the coloniser's language was seen as elevated, 'scientific' almost, which impacted upon postcolonial politics in important ways. In addition, how have language and culture been creolised or hybridised in postcolonial times? Postcoloniality is partly about negotiating the identities and power relations that histories of colonialism have bequeathed, leading to a whole variety of hybrid cultural and political forms. Globalisation in particular is having important impacts on the nature of language and cultural identity in Africa – through the growing importance of US and European news, politics and culture, for example (Ake, 1995). Thus Fanon's ideas (though clearly quite specific to the African context) continue to be relevant in the contemporary era. For Ali Mazrui (1993: 358), writing about the postcolonial appearance of alienation:

> [Fanon] takes language as a totality, as a macro system and looks at its psychological impact on the colonised in light of the particular social connotations of inferiority and superiority, for example, which it has come to acquire as a direct result of colonial and racial relations of domination. This cycle did not come to an end upon the attainment of political independence in Africa. The neo-colonial Africa that followed . . . has essentially continued to promote relations of dependency and domination in favour of alienation, albeit through the mediating role of a local bourgeoisie.

Thus for Mazrui, the term 'neocolonial' is preferable here, allowing him to characterise the relations of dependency and domination that have continued after former colonial rule ended. Postcolonialism is relevant to African conceptions of identity and development in that it begins with people's own perceptions and considers these as something to be discovered rather than something that can be asserted *a priori.* None the less, it is important to understand the limits of the focus on hybridity in the context of the continuing asymmetrical relations of power between North and South and the continuing dominance of a (not very heterogeneous) neoliberal doctrine of development (see Chapter 7).

PARTNERSHIP OR TRUSTEESHIP?

The distinguishing feature of the late twentieth-century question of development is that trusteeship, the integral of the nineteenth century

Figure 6.5 Statue commemorating first Mozambican President Samora Machel
Source: The author

doctrine of development, has been renounced as the source of action toward development. From the immediate historical experience of formal imperialism and its end, it is easy to see why trusteeship should be renounced when the goals of development for post-colonial Third World states are explicated by people who are neither of the state or the Third World.

(Cowen and Shenton, 1996: 446)

In understanding the continued relevance of the term 'neocolonial' we can critique the ways in which colonial humanitarianism has been reinvented after the formal end of colonial and imperial rule. In some ways a theory of 'trusteeship' was built into the construction of development from the very beginning of the Enlightenment period. This saw science and state direction coming together to secure the basis of social harmony through a process of national development (Cowen and Shenton, 1996). There were even notions of socialist trusteeship (although the language was very different) in some of Marx's writings. One of the ways in which the idea of development has roots in the eighteenth- and nineteenth-century concerns of the Enlightenment stems from the reinvention of notions of 'trusteeship' after 1945 as a strategy for supervising the territories of the 'Third World' or 'postcolonial world'. Trusteeship in colonial administration was all about the mission to civilise others, to strengthen the weak, to give experience to the 'childlike' colonial peoples who required supervision. When development took over where colonialism left off one of the important ways in which decolonisation was imagined was through the idea of development administration and debates about trusteeship and 'partnership'. Provided for under Chapters 12 and 13 of the UN charter, the trusteeship system was intended to promote the welfare of the 'natives' and to 'advance' them towards self-government. This was supervised by the UN Trusteeship Council which primarily involved permanent members of the Security Council. The administering authority was required to produce reports on the economic, political and social advancement of the territories' peoples (UN, 2002). The power of the administering state included full legislative and juridical authority and even the right to treat the territory as if it were part of the administering state.

With the independence of the Pacific Island of Palau in 1994 (administered by the USA as part of the Pacific Islands Trust Territory since 1947) the Council ceased to function and no longer meets annually. Petitions were considered from the inhabitants of trust territories and the Council would visit the territories periodically to make inspections. In the case of the Italian Trusteeship Administration of Somalia in the 1950s, the centrepiece of initiatives was a series of seven-year development programmes introduced in 1954 which drew on blueprints provided by the US Agency for International Cooperation (later USAID) and the UNDP. The British colonial administration was also involved and did not terminate its rule of Somaliland until 1960. Trusteeship was often rejected after 1945 because of its colonial connotations, but the idea 'implicitly reappears' many times in postwar conceptions of international development (Cowen and Shenton, 1996: 446).

After the Second World War, many former colonial administrators went on to take posts with NGOs such as Oxfam or turned their hand to teaching undergraduate or postgraduate courses on development administration and management. Cooke (2001) points out that some ex-colonial officers were even teaching as full-time staff members in the late 1990s at the University of Manchester and its Institute for Development Policy and Management (IDPM). In 'postcolonial' times, development management (which is now based around the logic of foreign aid and continues to have an elitist bias) blurred the core principles of trusteeship. Colonial ideas about First World interventions in the operation of Third World states and societies have therefore continued beyond the formal end of colonialism and imperialism. The contemporary focus on 'good governance' in countries such as Britain therefore has to be understood in this context, as a legacy of the history of colonial administration and its management of colonial development. This is not to view colonial administration as a homogeneous set of practices and ideas but rather to seek to understand the continuities to 'postcolonial' times.

Despite all the talk of 'empowerment', 'development' is still very much something that is defined and enunciated by the 'First World'. Just as in colonial times, the frameworks and strategies of development are authored outside the country concerned and are grounded in foreign (neoliberal) ideologies. Just as colonial states sought to govern their territories and administer their peoples, so the IFI's agenda of 'good governance' today seeks to discipline the realm of local politics, to prescribe the kinds of political change that are possible in the South. In this sense Cooke (2001) argues that the practice of empowerment and participation 'always takes place within first world boundaries', defined, measured and conceived by the administrators rather than the administered. Development administration was also bound up very much with Cold War geopolitics, a key element of US counter-insurgency techniques in Vietnam, for example, waging an 'unarmed managerial struggle against communism in the underdeveloped nations by engineering the transformation to capitalist modernity' (Cooke, 2001: 12). Development assistance, as we have seen, has a long and close connection to Cold War geopolitical discourses. In the contemporary era, participation and partnerships partly become a way for development agencies to counter accusations of neocolonialism (Cooke and Kothari, 2001). The seductive language of partnership is thus regularly used in postcolonial times but has an important colonial history where outside agency was prioritised:

> Terms central to participatory development which came into vogue through colonial anthropology, like 'community', 'village', 'local people' and so on are all elements in colonial and postcolonial discourses which depict the world in terms of a distinction between 'them' and 'us'.
> (Cooke, 2001: 14)

'Partnership' is a classic example of one of the 'plastic' words of development (which we will be examining further in Chapter 8) and represents a term that is often over-used in the development world (Brehm, 2001). The idea of North–South partnerships especially became popular after the publication of the 1969 Pearson Report on *Partners in Development* (Jolly, 1999: 40). More recently, the OECD report *Shaping the Twenty-first Century* included an annex on 'development partnerships in the new global context', ideas which have been echoed in Britain since 1997 by New Labour's development proposals (DFID, 1997, 2000a,b).

At present, the theme of partnership serves an ideological role in the new neoliberal policy framework in that it conveniently 'papers over contradictions and the rollback of government' (Pieterse, 2001a: 17). It also often involves quite naive assumptions about the power and power relations that really exist between postcolonial partners. As we shall see in Chapter 8, the contemporary managerial language of 'ownership' used by various development agencies today (to show how development is a local, place-specific product of autonomy) still involves creditors telling recipient nations how their development should proceed and requires outside approval and endorsement despite the fancy window dressing. In one section of the 1997 White Paper on international development produced by Clare Short and her Department entitled *Working with British Business*, the UK government argued that 'the pursuit of short-term commercial objectives, such as the previous government's support for the Pergau [dam] project or Westland helicopters should be avoided' (DFID, 1997: 41). The prioritisation of business and commercial involvement in development partnerships is crucial to New Labour's vision however, and is a central strand of their particular variation of Third Way politics (Slater and Bell, 2002). Maxwell and Riddell (1998: 264) note that potential 'partners' may thus interpret the UK's perspective on partnership in the following terms:

> we know how best to achieve development . . . we know how you should alleviate poverty . . . either you accept the approaches we think are right for you or you will not qualify for a long-term partnership with us . . . if you do not accept our view of development then we will not provide you with aid.

Thus the 1997 report on international development does not really move away from the long history of unequal relations of power and representation that characterise conditionality debates.

BOX 6.2

Tony Blair's neocolonialist vision for Africa

> Blair stood in parliament unashamedly to say the British government should stay ready to recognise and support the victory of MDC and should not stay ready to recognise the victory of ZANU-PF. . . . But of course we say: Go to hell. Our people have decided and that is what matters to us. It's not the right or responsibility of the British to decide on our elections. We don't decide on their own [elections] and why should they poke their pink noses in our business?
>
> (Robert Mugabe, cited in the *Independent*, 2 March 2002: 5)

One way to consider the ways in which postcolonialism in Africa and other regions is relevant today is to examine briefly the strategy of British Prime Minister Tony Blair which set out Britain's agenda for 'future imperialist intervention in the continent' (Talbot, 2002). The Prime Minster apparently arrived in and moved around West Africa in a style that 'increasingly resembles that of a colonial missionary' (Talbot, 2002: 3). On several occasions Tony Blair has referred to Africa as a 'scar on the conscience of the world', pointing out that an African child dies every three seconds. According to Blair, 'no responsible leader can turn their back on Africa'. The PM visited Nigeria, Ghana, Sierra Leone and Senegal in February 2002, where Blair said he would 'take up the issue of Africa' with other G8 powers and through the next round of WTO talks. The visit to these countries (four of which are former British colonies) also linked to the New Partnership for Africa's Development (NEPAD). Blair claimed he could bring peace and economic growth to Africa in what was a neat diversion from his domestic policies and an attempt to reassert British power and influence in Africa. During the visit the PM insisted that British and western support were needed to help prevent the 'collapsed' and 'failed' states seen in the past which he regarded as a breeding ground for global terrorism since September 2001. During the visit to Nigeria, for example, Blair said 'there is no leafy suburb that is so far from the reach of bad things and bad people' (cited in Talbot, 2002: 1).

Under Blair the sale of arms to Africa has increased, with a quadrupling since 1999 (with Morocco, Egypt, Kenya, Gambia and Zimbabwe chief among the recipients) (Talbot, 2002). One of the most interesting moments of Blair's heavily stage-managed visit to Africa came during his visit to a village in Northern Ghana, where the PM began to sound off about unfair trading restrictions imposed by the West as if he (and other conscientious leaders like him) had no part to play in the making of such a scenario. Clare Short, UK Secretary of State for DFID, went further still by referring at the time to 'a conspiracy from France and the EU to lock Africa into poverty', pointing out that at the same time 'Europe preaches free trade' (Short, cited in Talbot, 2002). Neither Blair nor Short is leading the campaign for less protectionism on behalf of poor Ghanaian farmers however. So why construct themselves and Britain as expert advisers about economic management and effective trade? One issue in particular that Blair has 'championed' at previous G8 meetings is that of international debt relief, offering only relatively small reductions in debt repayments and only ever as a condition of the usual package of neoliberal economic reforms, ensuring that, through debt relief, 'a clear track record of economic reform is established'. Why and how is Tony Blair qualified, however, to say what is a 'clear track record' and why is only a very specific kind of economic reform being welcomed here? A staggering $250 million *a day* is transferred from African countries to western banks in debt repayments according to the World Bank's own figures (World Bank, 2002). Jubilee Research and Drop the Debt campaigns calculated that of the $300 billion in debt owed by the world's fifty-two poorest countries, only a pitiful $18 billion has so far been cancelled. For the first twenty-two countries that qualified for the HIPC scheme, annual debt repayments are being reduced by only a quarter, despite their status as heavily indebted (which usually means spending more on debt repayments than national healthcare).

Tony Blair has boasted that Britain has doubled its global aid contribution since 1997, but international debt relief has not been forthcoming on anything like the scale that is needed (or that is claimed publicly by British politicians). Jubilee South argues that only a trebling of international aid and the total eradication of debt will lead to the MDG of halving world poverty by 2015 being successful. Blair speaks of offering countries a 'hand-up' rather than a 'hand-out' but the biggest way to offer such a 'hand-up' would be to bring about radical plans for debt relief and the abolition of protectionism in global trade markets. Thirty-four of the HIPCs are African, all of which, according to

BOX 6.2 (contd.)

Jubilee Plus and the New Economics Foundation, need to have a complete and immediate write-off of their debts. Thus the British Department for International Development (DFID) has committed itself and Britain to a leading role in global poverty reduction initiatives, while failing to address adequately the major and growing issue of indebtedness. The Jubilee Debt Campaign (JDC) consists of a coalition of almost fifty national organisations and over sixty local and regional groups, and has lobbied British MPs about this issue. For many commentators, the DFID's claim to be offering a hand-up and to be leading global poverty eradication seems quite shallow and rhetorical in postcolonial Africa. Arguably, Blair's style of 'Third Way' politics is of limited relevance to Africa, characterised as it is by a 'rhetorical gloss over a new style of right-wing politics' (Petras, 2000: 8). Aid provision to Africa is conditioned increasingly on the compliance with this 'neo-mercantilist agenda' (Petras and Veltmeyer, 2002). We also need to historicise the emergence of a 'metropolitan concern' in Britain for African terrains and the ways in which it gave rise to a specific kind of humanitarian gaze (Lester, 2002) and a certain model of 'humanitarianism'. Arguably the policies of Blair and his New Labour Party in Britain are far from being free of a 'postcolonial paternalism' in their approach to development. The British Prime Minister has failed, for example, in his relations with postcolonial Zimbabwe and in his attempts to oust the regime led by Robert Mugabe through the Commonwealth Secretariat. For a man who has offered himself up repeatedly as the saviour of Africa 'it must be galling . . . to be so publicly rebuffed by Africa's leaders' (Milne, 2002: 20). Britain's relationship with Zimbabwe has been severely strained in recent years as the Labour administration has singled out Zimbabwe for international denunciation because of political violence and intimidation and the restriction of democratic freedoms (these are not unique to Zimbabwe however). For both New Labour and Mugabe's ZANU-PF, colonialism and its legacies have a crucial bearing on the question of Zimbabwean democracy and development today.

Figure 6.6 A reminder that western countries should 'Act like it's a Globe, not an Empire'
Source: © Protest Graphics

Are partnership and conditionality not, in some ways, contradictory objectives and as a result are these partnerships not generally 'intrinsically one-sided'? (Slater and Bell, 2002: 346). It is not always particularly clear what this term means however, particularly when used by donors; nor is it often specified clearly how 'partnership' can systematically be put into practice given the varied legacies of colonial power relations. If there has been a traditional donor/recipient relationship between 'partners', how can policy discussion ever be considered an equal process? North–South linkages, as we have seen, are important in a number of senses, but the supposedly 'beneficial' nature of these relationships for southern organisations can be questioned. Southern NGOs often have an incredibly important local knowledge, understanding and presence in the countries where they operate. In connection with a range of resources provided by northern NGOs, these partnerships can potentially be very fruitful.

One recent study (Brehm, 2001) examined the 'partnerships' forged by ten European NGOs with southern partners and found that many of the ten organisations tended to focus on the concepts rather than on the purpose and practice of partnership. The study also revealed that 'systematic principles' of practice among the European NGOs were rare and that funding issues put in place important power relations between the 'partners'. The guise of 'dialogue' between worlds is therefore often a shallow one. In this way it is also important to ask about the process of 'capacity building' for and by NGOs: for what and for whom is this done? 'Civil society' is one important concept which has increasingly come to shape northern NGO perceptions of the southern organisations with which they work. In situations of political instability and conflict do NGOs understand the nature of national military and security organisations (Koonings and Kruijt, 2001) and are the latter engaged as 'partners' for reform? According to the website of DANIDA, the Danish government development agency, a 'Partnership 2000' scheme has been chosen as a new strategy for working with southern organisations in recognition that:

> Denmark by itself cannot create development for the poorest sections of the world population . . . this must necessarily take place in partnership with the developing countries, their governments and civil society – not least the poor, the groups and structures that represent them – and with other partners in development work.
>
> (DANIDA, 2001, http://www.um.dk/publikationer/)

Different conceptions of development and the presence of local hierarchies and social relations can often complicate this sense of co-operation. Progressive intentions do not always translate easily into development practice. It is possible that the radical edge of participatory methodologies and practices has been blunted as the agendas of NGOs are co-opted by the institutional mainstream such as the World Bank and USAID, for example (Long, 2001). Some geographers also have 'mixed' experiences of working on research projects where collaborative arrangements were insisted upon by the funding agencies, such that establishing North–South partnerships in the process of conducting research often becomes 'fraught with difficulties in each case and ultimately unsuccessful' (Simon, 2001: 15). According to David Simon:

> [t]here is a real dilemma over the basis of North–South and indeed, also South–South partnerships. Unequal power relations in the production of knowledge – and even in terms of defining what constitutes knowledge – remain profound and fundamental.

Thus even the process of forming South–South partnerships is not uncomplicated in itself. Alliances of northern and southern NGOs, as in the Fair-Trade Network, are often able to help individual communities of producers change their lives while engaging simultaneously in global trade campaigns and progressive policy advocacy (Simon, 2001). International NGOs and NGO networks such as the *Jubilee 2000* campaign for the alleviation of HIPC debt offer, at least in part, the hope of a way forward (see Chapter 9). The UK-based World Development Movement (WDM) is another example of an NGO that has forged international alliances and 'partnerships' around trade issues and negotiations. These are important examples of what has been called a 'locally grounded but globally informed praxis' (Mohan and Stokke, 2000). This is a key issue for development geography: to what extent can 'grounded' local interventions be informed by a global 'praxis'? Here it is also important to think about the interdependence of the local and the global. As Simon (2001: 23) puts it, 'the interdependence of social, cultural, economic and political groups at different scales and constituted on different bases is now *in*creasing rather than *de*creasing' (emphasis in original).

The New Economic Partnership for African Development (NEPAD) is a recent initiative formulated by African states that seeks to build on these interdependencies by acknowledging the important role of partnership and co-operation in international development. It has requested annual commitments of some US$64 billion in aid, loans and investments, and seeks an agenda that is set by African peoples 'through their own initiatives and of their own volition', while focusing on the way in which African peoples must 'shape their own destiny' (Bond, 2002: 1). There is an argument, however, (as we shall see in Chapter 8) that this is far from being a locally authored strategy that reflects and arises from Africa's rich traditions of political struggles. Many African intellectuals and social movements are united in opposition to the proposed partnership, however, because it surrenders too much to international institutions and submits to established power relationships that many feel it should be doing more to overhaul. Libyan leader

Muammar Qaddafi criticised NEPAD for its obeisance to 'former colonizers and racists' (quoted in Bond, 2002: 1). In response, South African President Thabo Mbeki (a strong backer and advocate of the partnership) insisted that 'We do not want the old partnership of a rider and horse' (quoted in Bond, 2002: 1). Closely shaped by US and UK politicians and endorsed by the IFIs and the interests of international capital:

> [NEPAD] empowers transnational corporations, Northern donor agency technocrats, Washington financial agencies, Geneva trade bureaucrats, Machiavellian Pretoria geopoliticians and Johannesburg capitalists, in a coy mix of imperialism and South African subimperialism.
>
> (Bond, 2002: 2)

There have already been protests around NEPAD which is seen by some critics to be a plan and partnership that was authored by technocrats and elites in places such as Geneva and Washington rather than by the social movements of the continent most directly concerned. It supports existing poverty reduction strategies such as the HIPC and PRSP processes and talks about vague public–private partnerships rather than reflecting on or at least acknowledging the wave of anti-colonial popular struggles that united peoples and places across and beyond the continent. Again NEPAD repeats long-established notions of 'bridging the gap' with rich countries, while simultaneously *reproducing that gap* by backing existing development strategies and accepting international political and economic (in)equalities as they currently stand. Arguably then it has also adopted a problematic postcolonial strategy for African development which reflects long-standing colonial ideas about development management and who should administer it.

CONCLUSIONS: POSTCOLONIALISM AND THE VOICES OF THE POOR: CAN ANYONE HEAR US?

> Postcolonial studies has the potential to be a new and different location of human development thinking. . . . Not infected (as much) by the know-all history of development studies, but just as embroiled in thinking about the West, it is freer to criticize colonialism and creeds of progress.
>
> (Sylvester, 1999: 717)

A good example of the value of a postcolonial critique of development comes from a reading of *The Challenge to the South*, a report prepared in 1990 by the South Commission. The Commission had been chaired by former Tanzanian president and advocate of self-reliant development Julius Nyerere. Although the Commission held no public hearings, it did bring together some very interesting members of the South establishment who had attended many of the 'big' development events and conferences of recent decades (Rist, 1997: 199). As a result of the dominance of this particular kind of 'establishment' very few new ideas and ways forward were articulated and few core principles run consistently through the report. The final document also reaffirmed a 'development imperative' that ought to be realised through 'sustained economic growth' (quoted in Rist, 1997: 201). What is particularly interesting about the report is the way in which it seeks to speak on behalf of the whole of the South and is based on the idea that the South speaks with one voice. Further, the report also failed to view the wealth of the North as a key part of its conception of poverty in the South. It did, however, talk about enabling people to utilise their potential in its definition of development, and in his Preface, Chairman Nyerere argued that 'the responsibility of developing the South lies in the hands of the people of the South'.

Postcolonial critiques can also allow us to deconstruct the emergence of what Ahluwalia (2001) refers to as 'Afro-pessimism', where Africa's future is seen as bleak, meaningless, already a tragedy – assuming that Africa has neither the political will nor the capacity to deal with its own problems. It is absolutely crucial to contest this image of Africa's future as meaningless. We must remember that as Mudimbe (1988) points out, 'Africa' is an invention, a western signifier. This means of signification is being currently reconstructed by western institutions that exert their power through 'knowing' what is best for the continent – namely development and modernity: 'It is through their amassing of statistics and surveillance that an underdeveloped, primordial, traditional and war-ravaged Africa is (re)produced' (Ahluwalia, 2001, 133).

Development is about statistics and surveillance and the image or picture of underdevelopment that is (re)produced. Postcolonial critiques of development thus often seek to explore how specific ideological formations and persistent normative assumptions and expectations have 'flowed from a colonial discourse into a development discourse' (Kothari, 1996: 5). Indeed, for Schuurman (2001) this normative preoccupation with inequality, with the poor, the exploited and the marginalised is the 'proper' focus of development studies, rather than a concern with diversity and difference (Schuurman, 2001: 9). In this way it is argued that it is inequality 'which should constitute the main focus within the explanandum of development studies'. Development geographers, however, may not see these concerns as mutually exclusive and should also seek to deconstruct the idea of an explanandum, an all-encompassing and explanatory interpretive schema.

Although the formal political management of colonised territories came to an end with the collapse of European empires in the first three decades following the Second World War, the process of decolonisation is not complete. Normative assumptions about progress are an essential component and enduring feature of development, as we have seen, prescribing solutions for other regions and normalising this process around an intellectual discipline or institution. Colonial constructions and representations of 'Third World others' therefore persist to a degree and themselves require decolonisation. People and identities cannot be seen as singular composites, simplistic images (such as that of the 'Third World woman') which efface and erase difference and diversity in development. Whereas 'Thirdworldism' flattened the differences and heterogeneities between states, so postcolonialism seeks to bring out and celebrate this difference and diversity. This does not mean that we should oppose a positive decolonisation to a negative of neocolonisation; that is far too simplistic. A key part of this process involves recognising that postcolonial citizens and subjects mobilise not just a single identity but several fluid identities which are constantly revised, meaning that it is necessary to think about the flexible nature of citizenship and the dynamics of belonging and identity (Ong, 1999). In an important way then, African subjects reimagine themselves by confirming the 'very porous borders of Africa as a discourse of geography, history, culture, nation and identity' (Ashcroft, 1997: 137).

Postcolonialism is fundamentally about porous borders and fluid identities, shaped in part by a variety of historical and cultural discourses of 'nation'. If globalisation processes are transcending nation-state boundaries, new forms of identity are beginning to emerge and it is thus necessary to consider the meanings of terms such as 'local/global' and to think again about the porous borders between and within continents. In some ways, postcolonial literature helps us to think in new ways about the interface between the local and the global and to focus on the lived experiences and consciousness of ordinary people by providing a crucial and different record of subject formation which combines a new conception of data and sources for studying development. It is necessary here to question the extent to which decolonisation is incomplete and how migrations and displacement force us to revaluate notions of close connections between nations and states in an era of global cultural flows (Appadurai, 1996).

Edward Said (1993) has pointed out that imperialism bound disparate societies, peoples and cultures together but in a way which was at root profoundly unjust. We can thus extend this concern by asking: How does international development bind disparate societies, peoples and cultures together in postcolonial times and how might this be unjust (for some)? Thus we can also examine the geopolitical trajectory of societies that have been subjected to varying forms of both colonial and imperial domination (Slater, 1998). There is a real sense, however, that these debates need to be more grounded in the local and everyday realities of particular peoples and places (Paolini, 1997). If it is uprooted from specific locations, then the term becomes harder to investigate meaningfully, obscuring the relations of domination that it seeks to uncover (Loomba, 1998). It is therefore necessary to view the role of former imperial powers such as Britain and France in postcolonial Africa in a historical context, examining the emergence of metropolitan senses of moral responsibility for distant others during the slave trade, colonial settlement and imperial conquest (Lester, 2002). The roots of the Rwandan genocide, for example, can be linked directly to Africa's colonial past but also to its postcolonial present, and it is this interaction between past and present that postcolonialism is fundamentally about. This is relevant because it recognises that Africa 'has to deal with its past in order to understand its present and confront its future' (Ahluwalia, 2001: 133). Debates about partnership and ownership of development need to be understood in this context, examining the extent to which concepts and notions of 'trusteeship' are truly a thing of the past and the possibility that they may have been reinvented in 'postcolonial' times.

Figure 6.7 The United Nations and the campaign for slavery reparations

Source: Cartoon by B. Farrington

Figure 6.8 The legacies of colonial history continue to have a contemporary relevance in countries such as Sierra Leone

Source: © Dave Brown, *Independent* (2001)

As we have seen, the cures for the ills of recipient countries in need of humanitarian assistance were first forged through experiences of colonisation and settlement during imperialism – which has proved to be of enduring significance. These forms of humanitarianism, born in the nineteenth century, have been dismissed by many postcolonial critics as little more than a legitimating screen for imperialism despite the fact that they often sought to challenge official and settler practices (Lester, 2002). Gandhi (1998: 27) thus argues that it is important to distinguish the different groups that produced humanitarianism from those that sought to oppose it during imperial times. This moves us away from the idea of drawing simplistic lines of exclusion and inclusion when mapping the power of development as an idea in imperial times. What is important here is that this humanitarian enquiry 'created a sense of property in the objects of compassion' (Lacquer, 1989: 179). This meant that peoples seen as the object of humanitarian concerns were appropriated to the 'consciousness of would-be-benefactors', enabling humanitarians to 'speak more authoritatively for the sufferings of the wronged than those who suffer can speak for themselves' (Lacquer, 1989: 180). This also encouraged the formulation of 'prescriptive principles' aimed at relieving suffering and improving the sufferers. A sense of responsibility for distant strangers may thus never be able to escape the declaration of 'epistemological sovereignty over [their] bodies and minds' (Lacquer, 1989: 188). It is precisely the establishment of this sovereignty over the minds and bodies of distant others that can and must be challenged through postcolonial geographies of development. Jonathon Crush (1994: 303) defines the aims of a postcolonial geography of Africa as about:

> the unveiling of geographical complicity in colonial dominion over space; the character of geographical representation in colonial discourse; the delinking of local geographic enterprise from metropolitan theory and its totalizing systems of representation; and the recovery of those hidden spaces occupied and invested with their own meaning, by the colonial underclasses.

Thus where the 'colonial underclasses' had been written out of colonial histories of development, ignored, exploited and overlooked, postcolonial histories and geographies seek to 'recover' these kinds of hidden spaces. In addition, these regions must also no longer be denied an autonomous space in debates about 'global' development today. The ambivalence of African relationships to modernity is not always adequately recognised and interconnections between western and non-western are sometimes neglected. African cultures and peoples are not somehow outside 'the West' or 'western modernity' but a central part of its make-up and origins. One of the major contributions of postcolonial scholarship, however, is the concentration on themes of resistance and hence the field of the possible for the marginalised and subaltern peoples that are the subjects of postcolonial development (Paolini, 1997). This is also one way in which it is possible to focus on particular, specific contexts in thinking about postcolonial identity today.

For many critics, globalisation is simply a restatement of imperial ambitions where the continuing legacies of imperialism may be seen to be relevant to many globalisation discourses (this time led by the United States) (Ahluwalia, 2001: 123). As we shall in Chapter 7, unequal relations between cultures (which separate and divide worlds) are being 'productively reproduced under globalisation' (Ahluwalia, 2001: 123–124). For others, globalisation also involves continuity with colonial discourses in those representations of 'emerging markets' promoted by western-led financial institutions which can be read in part as a contemporary reformulation of colonial idioms (Sidaway and Pryke, 2000). Africa has barely featured in many of these discourses. Postcolonialism and globalisation come together for Ashcroft *et al.* (1998) around themes of local/global interaction and of cultural transformation. Where development discourses of the 1950s and 1960s sought to force 'Africa' and other world regions into a map of 'globally normative patterns', postcolonialism emerges to disrupt conventional geographies of 'contact zones', of North–South 'interaction', and to focus instead on the postcolonial hybridity of cultures and identities, which forces us in turn to rethink the very spatiality of development and discourse, to reconsider the meaning and theorisation of peoples and places.

In this sense we must engage with surveys like the World Bank's 'voices of the poor' analysis of 60,000 poor women and men which involved conducting studies in the 1990s in fifty countries around the world (see Chapter 8). In total eighty-one Participatory Poverty Assessments (PPAS) were conducted which the Bank argues were based on 'discussions' with poor people. Although there are many problems with the way the Bank has sought to draw upon these 'voices of the poor' to legitimate its discourses about opportunities and markets (Narayan, 2000), the study did engage in an ambitious attempt to understand people's own perspectives and experiences of poverty and was itself conducted in many 'postcolonial' contexts. We thus need to think about how issues of subjectivity and definitions of knowledge and poverty are articulated in this and similar surveys and the extent to which history and culture are seen as important to understanding poverty and its spatial variation. Fieldwork and conducting social research in developing countries proved difficult for the Bank and emotionally challenging for many of the team employed to conduct this study, some of whom later required counselling for the traumatisations they had experienced as a consequence of what they had seen and heard (Narayan, 2000). At the same time fieldwork can allow us to explore issues of locally grounded but globally informed praxis, to transcend and reinterpret the relationships between researcher and researched and the gulfs and separations between worlds of global development. Studies such as the World Bank's 'voices of the poor' survey, which claim to use 'participatory' approaches and methodologies, assume that they too have created a space in which the marginal can speak, but their very representation as 'poor people' (very few of whom are actually even named) often robs them of their voices.

The representation of Africa as the dark continent continues unabated, with the continent viewed as a kind of repository of disease, war and pestilence. Africa has also been seen as the 'hopeless continent' (*The Economist*, May 2000) or as a 'basket case' (*African Business*, January 1999), surrounded by negative stereotypes and metaphors which obscure more nuanced (and decolonised) interpretations. Contemporary African, Asian, Middle-Eastern or Latin American cities are, however, increasingly part of global networks linked by new communication technologies (which have themselves become a key arena of struggle for many people). Hybridity is a feature in many of these spaces and places, where modernity is defined and consumed in different ways, gaining meaning through everyday practices and realities. Abdou Maliq Simone (2001) has explored the legacies of colonialism for African cities and the social spaces of urban areas, arguing that they were not designed with African rhythms and sensibilities in mind but have been transformed as physical and cultural spaces through the process of decolonisation. Similar transformations can be made to the ideological spaces of development if postcolonial theories and debates are engaged in more widely by those interested in development studies of various kinds.

There has also been the persistent charge that 'recent post-colonial theory fails to engage with the "real" world' (Ahluwalia, 2001: 134), an important weakness if connections are to be made to global development debates today. As Robinson (2002: 533) has argued, there is 'still considerable work to be done to produce a cosmopolitan, post-colonial urban studies', one which is free from presumptions about global hierarchies, problematic categories

Figure 6.9 Consensus and the conference on racism
Source: Cartoon by D. Cagle

such as 'Third World cities' and the developmentalist myths of urban poverty. A decolonisation of the presumptions in urban studies about 'city-ness' is required in order to explore urban futures in the postcolonial world (Robinson, 2002). By adopting a concern with the postcolonial, it is possible to critique earlier categorisations of cities and countries into 'western' and 'Third World' varieties, a categorisation which emphasised difference and deviation from the norm.

Despite the closeness of this interface between postcolonialism and development, many critics continue to ask how this 'theory' can be translated into practice on the ground, since knowledge and power are so intimately linked. While useful in encouraging us to think on the edge or along the margins (Slater, 1998), the term and the debates that have surrounded it also have their problems, ambiguities and dissonances:

> There is the persistent idea that western academia has the answers for the rest of the world and only by gaining access to education in the West and adopting Western theories can people from the 'Third World' understand their own histories and economies.
>
> (Kothari, 1996: 10–11)

It is therefore worth thinking about the extent to which these debates represent the interests and concerns of western-based, urban intellectual elites. Clearly it is not possible to homogenise the lives of the subaltern, the marginal and the poor. Other ways of understanding and knowing the people and challenges of development are possible which do not rely on 'neocolonial' disciplines and access to western 'experts'. Postcolonial critiques of development discourses have been very important but there is room for a much further and deeper engagement between geographical debates and exchanges about postcolonialism and development. Language and literature should figure prominently in these debates, as might questions of fieldwork and the actual process of conducting research in the 'developing world' or the global 'South' where global inequalities sometimes become dramatically obvious and hard to negotiate. Another of the key potential points of engagement must be issues of economic relations and effects which have so far tended to 'elude representation in much of postcolonial studies' (McEwan, 2001: 130). Berger (2001a), for example, is very sceptical of the potential of postcolonial theory, arguing that it does not fit well with the contemporary circumstances of Latin America (whose colonial histories are less recent). Thus, as Sidaway (2000) has argued, 'mapping' the 'postcolonial' is a problematic and contradictory process. Both Sidaway and Berger also share some important reservations about the relevance of these debates to Southern and Eastern Europe and the post-Soviet and post-Yugoslav republics, arguing that postcolonialism fails to capture the timing and character of their quite specific historical experiences. According to Berger (2001a: 220):

> More generally, the notion of the postcolonial is problematic when it is used explicitly or implicitly to encompass Latin America, the Middle East, or Asia and even Africa . . . can such diverse paths as those taken in India, Brazil, South Africa, Hong Kong, Fiji, Egypt, East Timor, South Korea, Vietnam and Thailand . . . be framed by the postcolonial condition?

If we look at colonial Southeast Asia, for example, between 1800 and 1941, Thailand did not fall under European control at all and there were British, French and Dutch variations of colonisation. Varieties of economic and political transformations were thus unfolding across Asia which were not easy to summarise and had contradictory implications for those involved (Barwise and White, 2002). Postcolonialism thus has diverse origins which include contributions from Latin American authors. The problem that Berger and many other critics of the relevance of postcolonial scholarship partly have is that these ideas are seen to be 'of less relevance when grappling with the new post-Cold War order' (Berger, 2001a: 221). None the less, the Middle East is very important to debates about postcolonialism and the making of international development discourses and theories more generally. There are, however, problems of framing the region in this way, given the continuing importance of 'internal colonialism', where Israel (for example) continues to have a presence in Gaza, the West Bank and southern Lebanon (Sidaway, 2000). None the less, we need to recognise that colonialism had a major impact on politics in the Middle East. Dabashi (1993: 509), for example, talks of how '[t]he demons of colonialism and neo-colonialism had so deeply penetrated Muslim politics, they are not to be exorcised so easily'. It is important then to theorise the complex and varied forms that imperialism and neocolonialism take in the contemporary world. A variety of 'national' struggles in the Middle East were necessarily international in their implications but one of the principal limitations of postcolonial studies has been that they often fail to adequately recognise and comprehend these kinds of horizontal (as well as vertical) cross-border alliances:

> Had postcolonial studies taken on more fully the import of such cross-border and multi-axial alliances, this would surely have generated a less

BOX 6.3

Neocolonialism and identity in the Middle East

Debates about neocolonialism and identity have also been important in a number of countries of the Middle East. In the specific case of Iran, Dabashi (1993) refers to an 'Other-centred' culture of defining everything in relation to 'the West' which has denied successive generations of Muslims the chance of coming to terms with their colonial past and postcolonial present. They are, he argues, 'entrapped' by 'postcolonial rhetoric' (Dabashi, 1993: 511), a figment of their own imagination, in an act of their own creation. This imagination has survived the end of the revolution in Iran. For some postcolonial Muslims, a turning point can occur only if a different kind of 'other' is gradually constructed, an 'other' quintessentially different from 'the West' (Dabashi, 1993: 513). In this way some Muslims have looked to 'the East' for a substitute 'other', a process that has been futile and irrelevant since, *vis-à-vis* 'the West', Iran and all other parts of the non-western world are already seen as belonging to 'The East':

> In the imaginative geography thus created, whereby on the Rubik's Cube of world divisions Japan, Israel and South Africa are constructed as part of 'The West', India and China could not register agitated memories or mesmerizing illusions in any significant way.
>
> (Dabashi, 1993: 514)

The Cube can be moved in various directions however. We need to move away from the illusion of an omnipotent West manipulating the destiny of a whole people; there is also a need to dethrone these constructs and find new 'others'. Thus themes of postcoloniality are very relevant in the Middle-East. As Halliday (2000) suggests, the greatest challenge presented to Islamism is often seen as the non-Islamic (and especially) western world, but Muslim people themselves need to look within the region for 'the ability to find and implement a viable economic development strategy' in the contemporary era (Halliday, 2000: 150–151). This also necessarily involves a certain renegotiation and definition of citizenship, identity and national belonging. As Newman (2002: 305, emphasis in original) argues, with respect to Israel:

> the changing nature of collective identity in Israel and amongst the Jewish Diaspora has meant that alternative and diverse visions of Israel's geopolitical position are emerging, relating to . . . notions of citizenship and identity and the question of *who is an Israeli*.

There is also some suggestion that new gender identities have emerged which challenge the dominantly held male views of what it means to be an Israeli (Fenster, 1997). During periods of decolonisation in the Middle East revolutionaries and activists from a range of countries and contexts also sought to transport their wars of liberation across national borders, understanding (for example, through Marxism) that their 'national' struggles were necessarily international in their implications.

> rigid 'centre–periphery' conceptual mapping than that which has characteristically organized the field.
>
> (Boehmer and Moore-Gilbert, 2002: 11)

These rigid and enduring 'centre–periphery' conceptual mappings, as we have seen, are certainly not unique to the field of postcolonial studies. Postcolonial geographies can help us to work through the tensions between global and local, core and periphery, between the global-level 'grand narrative of colonialism' (Nash, 2002: 228) and the varied and localised projects of resistance to colonial values and power relations. These debates do not offer an all-embracing theory and explanation of development (and nor should we expect them to) but they do offer an opportunity to rethink relations across the spaces of development, focusing on the important mutually constitutive relations between West and non-West, centre and periphery, raising questions about new diasporas and border-zones. For some observers however, the approach does not really move beyond binaries and is unable to generate new insights into the failures of modernisation and dependency approaches (Kelsall, 2002). Postcolonialism does, however, enable us to re-emphasise relations with other cultures as an integral part of studying development and allows us to challenge the authority of development discourse and the way it constructs the world in postcolonial times. This means going beyond the rhetoric and tyranny of 'participation' and 'participatory' studies of development and poverty to deconstruct development, to question the construction of development knowledge, the language of development theorists and practitioners and also the power relations that exist between these theorists and practitioners and the people and places they seek to study and represent. This also involves decentring the West and incorporates a certain rejection of the 'truth claims' of one part of the world on behalf of the rest.

BOX 6.4

Chapter-related websites

http://www.africanbookscollective.com/
African Books Collective, a wide range of titles from African publishers, with catalogues available.

http://www.pen.org.mk/68th/index.html/
World Pen organisation website.

http://www.interlinkbooks.com/
Interlink - Emerging Voices series of Third World fiction, with a special emphasis on women writers.

http://www.heinemann.co.uk/
Heinemann with their African Writers Series and Caribbean Writers Series.

http://www.indexoncensorship.org/
See also *Index on Censorship*, 'Word Power', 1999.

http://www.magabala.com/
Magabala Books publishes work by Australian Aboriginal and Torres Strait Islanders. Their fiction list can be ordered through their website.

http://www.akuk.com/ e-mailak@akedin.demon.co.uk
AK Press and Distribution Independant distributors also have an extensive international list in the UK.

http://www.gateway.library.uiuc.edu/lat/lit.html/
The University of Illinois hosts links to Latin American literary resources and online bookstores.

http://www.umiacs.umd.edu/users/sawweb/sawnet/
Website dedicated to South Asian female novelists, giving a brief description of authors and links to articles and reviews. Also links to children's books.

http://www.is.rice.edu.~riddle/play/sasialit/findbooks.html/
Useful website for finding Asian books outside of Asia, with links to numerous outlets around the world.

http://www.ipcs.org/au/links.html/
The Institute of Postcolonial Studies – contains links to various sites connected to the study of postcolonialism, journals, reading lists and study

7 Globalisation, Government and Power

INTRODUCTION: GLOBALISATION AND THE NON-WESTERN WORLD

> Most people are put off by the world of economics. Too boring. Too obscure. Too technical. . . . After all, it's what economists themselves would have us believe. They are the priesthood and we are the novices, waiting for the mysteries of the faith to be revealed. Despite their fancy forecasting tools and econometric equations, orthodox economists have been completely hopeless at predicting financial disasters or managing the international economy. They don't know any better than the rest of us.
>
> (Ellwood, 2000: 1)

> What we are doing to the poor in the name of globalization is brutal and unforgiveable.
>
> (Shiva, 2001: 59)

For the first time in its thirty-one-year history, the World Economic Forum (WEF) met outside Davos (Switzerland) for its 2002 annual get-together. Protesters were met with a violent response from New York Police, however, who secured the Waldorf Hotel conference centre (where the meeting was held) with military-style precision. In recent years, the annual meetings of the WEF have become important sites of protest and contestation in themselves in debates about the world economy and how it should be managed. The WEF represents a group of representatives, usually drawn together from around 1000 international companies, which claims to engage business and society to 'improve the world', but their meetings have become massively contested affairs, with widening protest and demonstrations. Some argue that WEF gatherings are a kind of 'Business Olympics', 'a big cocktail party for the global corporate elite' (SFGJ, 2000). Attempts to publicly debunk the myth of free trade have sought to question the deep and numerous holes in the conceptualisation of 'trade' and the expectations of what might flow from it. At the protests surrounding the 2000 WEF meetings, a group called Students for Global justice (SFGJ) argued that their objective was to open up the 'ideological space' of development in search of alternatives, protesting because they felt it was necessary to debate alternative guiding principles for the global economy, putting on the agenda the possibility of transformation away from the current 'pro-corporate malaise' (SFGJ, 2000: 3). This chapter raises the question of how contemporary debates about 'globalisation' are connected to pro-corporate institutions such as the WEF and seeks to examine how this promotion of capitalist discipline by the agents of neoliberalism can be questioned.

The World Social Forum (WSF) on the other hand is a newly established coalition (proposed by a number of Brazilian civil society organisations) which meets (around the same time as the WEF) in the city of Porto Alegre, Brazil for its annual conferences. The slogan of the first forum in 2001 was 'another world is possible', a theme that has since been widely adopted by a variety of social movements and campaign groups, many of which use the Internet to highlight 'globalisation and its effects seen by those who resist' (as declared the Porto Alegre–WSF website). At the 2001 WSF meeting more than 700 social movements from 122 different countries were represented. Resistance was still seen as primarily 'local' (Houtart, 2001) and one of the key concerns discussed was the fragmented and piecemeal nature of resistance, isolated into different sectors, which potentially militated against the formation of common grounds for global resistance (see Chapter 9). Up to 60,000 people amassed in Southern Brazil (among them trade unionists, parliamentarians, community, environment, poverty and human rights groups along with representatives from landless and peasants movements).

The forum thus brought together a diversity of movements, uniting fisher folk from India with farmers from East Africa and trade unionists from Thailand, for example, offering a successful counterpoint to the WEF. The WSF II meeting even included a people's tribunal on debt (highly appropriate in a country drowning in debt to the tune of an estimated US$140 billion). The social forum's key themes included the production of wealth, access to wealth and sustainability, civil society and the public realm, power, politics and ethics. WSF meetings were thus beginning to explore the possibility of many alternatives in terms of efforts in communications, practical economics and new forms of political and social organisation – all with a view to creating a supposedly 'post-capitalist' world (Houtart, 2001: vii). This is clearly a long-term process but it is beginning to raise questions about the possibility of a land reform which favours peasants, about mechanisms to ensure democratic control over finance capital, about the redistribution of wealth through social security and about public reappropriation of collective public resources, such as water, knowledge, seeds and medicines (see Chapter 9).

As we have seen, many global development agencies give the impression that global poverty is falling as a direct result of their particular poverty reduction strategies, declaring that the world is unquestionably 'on the right track' in this regard. World income inequalities, according to some of these institutions (e.g. the WEF and the World Bank), are falling *as a direct consequence of world trade and transnational enterprise*. These agencies share a common assumption that 'globalisation' is a process that is inherently good for poor and 'thirsty' debtor countries, which must seek to avoid 'lagging behind'. To globalise is, therefore, to 'catch-up'. If, as is the case, the answers to these questions depend on how income and distribution are defined and measured, how real is the scenario that world trade is leading the attack against world poverty? Many 'western' politicians are beginning to wake up only gradually to the realities of these inequalities. Many see these concerns as those of distant geographies, worlded beyond the universe of immediate moral concern:

> Elite western policy-makers seem to regard the growing income equality gap as they do global warming. Its effects are diffuse and long term and fears of political instability, unchecked migration flows and social disruption are regarded as alarmist.
>
> (Wade, 2001: 80)

For how much longer can the effects of world poverty be regarded as diffuse? If another world is possible, how much will poverty be a feature of this new world? The point here is that for an organisation with incredibly well-resourced research capacities, the self-styled 'knowledge bank', the World Bank, knows very little about world income distribution, its causes and consequences. Globalisation is first and foremost about the power of these organisations to shape global development agendas, to globalise their ideologies and to set the terms of debates, choosing to research only those forms of economics that support their own arguments.

This chapter focuses on the theme of international government and power in development and navigates a number of recent debates about the globalisation of the world economy, examining the implications for the 'Global South'. If the world is in the throes of 'globalisation', can every nation expect to do equally well and what are the specific challenges facing geographers who seek to examine the interface between development and globalisation? This is complex because there has often been a careless use of both terms in much of the

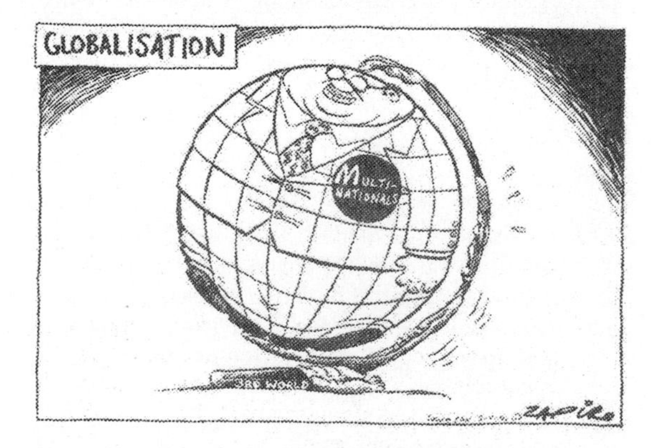

Figure 7.1 Globalisation and time/space compression (the inflated role of multinationals)

Source: Jonathan Shapiro for *The Sowetan*, 3 May 2001

literature which makes it difficult to analyse what is happening on the ground and which actors and agents are actually involved. It has been 'fetishised' to the extent that as an idea it is now seen as having 'an existence independent of the will of human beings, inevitable and irresistible' (Marcuse, 2000: 1). This chapter argues that it is necessary to retain a sense of how globalisation actually exists and impacts on peoples and places every day, on the ground, exploring how these processes *are* related to the will of human beings. It has to be recognised that this is partly about an expansion of capitalist relationships that began centuries ago rather than just in the past few decades.

A distinction also needs to be made between technological globalisation and the globalization of power. What we need to focus on in rethinking development geographies is the globalisation of capitalist crises (e.g. from Thailand to Argentina) on a world scale. In this sense neoliberalism is the ideology of those in power and something that is being universalised as a consequence of that power. It is necessary therefore to examine the extent to which a network of power and government has emerged at the global level, whereby these organisations shape the agendas of nation-states and localities in a number of ways. Neoliberalism is seen here as an 'ongoing ideological project' (Peck and Tickell, 2002: 401) that takes different forms at different geographical scales. This chapter, however, also argues that there is a need for a certain kind of 'deliberalisation' of development and the spatial relations that this involves. Neoliberalism provides a kind of 'ideological software' for 'competitive globalisation', a set of prescriptions all countries must follow at global, national and local levels. Thus it could be argued that the neoliberal 'offensive' needs to be mapped carefully by development geographers, charting how its 'local' institutional forms are linked to more general ideological characteristics operating in other parts of the world. How is the neoliberalism adopted in Brazil similar or different to that of Benin or Britain, for example? The terms 'neoliberalism' and 'neoliberalisation' refer to different processes, as we shall see. Escobar's (2002) recent work, for example, seeks to map the neoliberalisation of Latin American spaces of development.

BOX 7.1

Globalisation: some key trends and features

Key interrelating political and economic trends together summarise what we usually take the term 'globalisation' to mean:

- The increasing importance of the financial structure and the global creation of credit, leading to the dominance of finance over production.
- The growing importance of the 'knowledge structure', knowledge as a significant factor of production.
- The transnationalisation of technology and the increasing rapidity with which technologies become redundant increase the emphasis on 'knowledge industries'.
- The rise of global oligopolies in the form of TNCs: corporations must 'go global', acting simultaneously in a number of different contexts.
- The globalisation of production, knowledge and finance, leading to a decline in the regulative power of nation-states.
- The new 'freedom' of capital from national regulative control and democratic accountability, which is said to have led to increasing poverty, environmental destruction and social fragmentation.

***Source:* Adapted from Rikowski, 2001a**

These questions of macroeconomic governance, the role of global institutions and their impacts at different spatial scales are absolutely central to a reworked and critical geography of development. 'Governance' is a term that usually refers to the act or process of governing and is therefore sometimes seen as being synonymous with government. In terms of development debates however, the term focuses on the wider range of governmental and non-governmental institutions and actors that shape policy outcomes. The term also refers to the relationships between different actors (ranging from government institutions to non-government organisations or NGOs and social movements or private companies). We shall be examining the possibility that there has been a spatial restructuring of governance as new networks of power are being composed at different supra-national or supra-territorial spatial scales.

The Bangkok-based organisation Focus on the Global South has focused attention on the impacts of organisations such as the World Bank, IMF and WTO. They argue that the WTO in particular 'is

Figure 7.2 The World Bank: mixing the 'development soup' and its ingredients by formulaic recipe
Source: Bretton Woods Project

founded and maintained on many myths'. In November 2001 the organisation took part in demonstrations outside the Bangkok World Trade Centre along with 1500 farmers, jasmine rice producers, trade unionists and HIV/AIDS activists – calling for the WTO to 'get out of agriculture' and an 'end to the patenting of life and drugs'. How have transnational corporations sought to increase their influence in the South and how do international patterns of trade and investment impact upon levels of world poverty?

Africa's share of world trade has decreased from 5 per cent to just 2 per cent in the past fifty years (White, 2001), but why is this continent becoming increasingly marginal in terms of trade in a globalising world? An important theme of this chapter is how globalisation has been theorised in different ways across different disciplines and in different cultures. Culture and globalisation are linked in these discourses of development and we need to examine the extent to which there is a relationship between globalisation and postcolonialism. What is particularly interesting about the huge volume of literature about globalisation and development is the way in which non-western regions are depicted and constructed as having sought to gain from these processes. Few of these sources seem to say anything about the role of Africa in globalisation despite the fact that, as Ahluwalia (2001) puts it, the continent helped to 'inaugurate' the forces of modernity itself:

> Africa, which was at the centre of the economic forces which inaugurated modernity itself, as part of the transatlantic slave trade, sadly is now marginalised, with scant attention paid to it by those who proclaim a new globalised world.
>
> (Ahluwalia, 2001: 113)

Similarly, Amin (2001b) criticises what he calls the 'incense bearers of globalisation' for ignoring the fact that, for Africa at least, the ratio of extra-regional trade to GDP did not change significantly through the twentieth century. Globalisation is thus by no means a new or novel phenomenon but rather is closely related to the making of European modernities and the global expansion of European empires in regions such as Africa. Neither in this sense is resistance to the process of incorporation into a globalising economy altogether 'new' when peoples around the from world have long resisted this if their material circumstances or cultural values are threatened (Wallerstein, 1998). What is so astonishing about some of these debates is that they assume that this is in some way a novel process for Africa, delivered courtesy of the greater dissemination of western progress and enlightenment and the growing (and inevitable) scale and reach of global capitalism. Using these terms (global/capitalism) together in this way is not unproblematic however, since this would suggest a singular, homogeneous capitalism which holds sway (Pieterse, 2001a). It must be remembered that there are many capitalisms which 'both co-operate and compete' (Pieterse, 2001a: 91). Globalisation is sometimes also seen as an entity in itself, distinct from capitalism and somehow now the underlying force transforming the world. We can thus share Wallerstein's reservations about the use of the term 'globalisation' in contemporary contexts:

> As used by most persons in the last ten years, 'globalization' refers to some assertedly new, chronologically recent, process in which states are said to be no longer primary units of decision-making, but are now, only now, finding themselves located in a structure in which something called the 'world market', a somehow mystical and surely reified entity, dictates the rules.
>
> (Wallerstein, 1998: 107)

BOX 7.2

The World Trade Organization (WTO)

Headquarters: Geneva
Director-General: Supachai Panitchpakdi (Thailand)

The WTO replaced the General Agreement on Tariffs and Trade (GATT) in 1995. GATT was one of the original Bretton Woods initiatives which established a set of rules to govern world trade. Its aim was to reduce national trade barriers and to stop the 'beggar-thy-neighbour' actions that had bedevilled trade in the pre-Second World War period. Seven rounds of tariff reductions were negotiated under the GATT treaty, the 'Uruguay Round' beginning in 1986. The WTO vastly expands the GATT's mandate in new directions. It includes the GATT agreements, which focus largely on trade in goods. But it folds in the new General Agreement on Trade in Services (GATS), which covers areas like telecommunications, banking and transport. There are also agreements covering trade-related intellectual property rights (TRIPs) and trade-related investment measures (TRIMs). These new treaties have far-reaching implications for environmental standards, public health, cultural diversity, food safety and many other areas.

The old GATT had no legal teeth to enforce rules but the WTO can impose tough trade sanctions. According to the Director-General of the WTO, Dr Supachai Panitchpakdi, the role of director general of the WTO is different to the CEO persona of the IMF and World Bank leaders in that it is 'mostly a strategist, coordinator and inter-national diplomat who tries to placate all negotiating partners and keep them on board' (*Geobusiness*, October 2001: 31). Panitchpakdi goes on to say that 'The WTO is just like a bicycle, and to keep it in motion we need some mobility, we need more dynamism' (*Geobusiness*, October 2001: 31). Why is only free trade and not also fair trade the WTO objective and why has its trade policy review mechanism (which ensures that all countries are abiding by its regulations and implementing agreements) been so weakened by diplomacy? Bizarre distinctions operate between trade and non-trade issues (labour relations, human rights and the environment) which are kept apart. There were 135 member countries present at the Seattle Third Ministerial Conference in 1999 and a further thirty-five nations had observer status there (Bakan, 2000; Rikowski, 2001a). By 2001 the WTO had 142 member nations (Rikowski, 2001a). What distinguishes the WTO from other international agreements is its Dispute Resolution Panel – even the USA and European countries are supposed to obey the organisation's rules. The WTO is supposed to operate on a system of consensus but in practice it is driven by what Bakan (2000) refers to as 'the Quad': the EU, USA, Japan and Canada, whose representatives meet daily in Geneva to discuss non-trade issues and are lobbied extensively by TNCs. The outlook underpinning the WTO is regulation, giving transnational capital more freedom to do what it wants, where and when it wants. The idea is that trade liberalisation has led to more competition, greater market efficiency and so, necessarily, to a higher standard of living. Trade is of course corporate-managed, as set out in some of the 700-plus pages of WTO rules:

> The neoliberal ideological underpinning of corporate-managed trade is presented as TINA – There is no Alternative – an inevitable outcome rather than the culmination of a long-term effort to write and put in place rules designed to benefit corporations and investors, rather than communities, workers and the environment.
>
> (Rikowski, 2001a: 6)

Seattle has 'cast a long shadow over the WTO' according to *The Economist* (17 November 2001). Doha illustrated yet further the WTO's desire to bathe itself in conciliatory talk regarding the interests of poorer nations (Monbiot, 2001). The organisation is particularly well placed to promote the interests of poor countries, by challenging protectionism and promoting fair trade, but quite often that challenge is weak and incoherent. Exploring and contesting the myth of free trade is a key objective in rethinking development today. Oxfam, for example, estimates that northern protectionism costs developing countries some US$100 billion a year, twice what they receive in aid (Oxfam, 2002). Uganda, for example, could export peanuts to the USA if only Ugandan producers did not face tariffs of up to 164 per cent, or rice to Japan if only 93 per cent of the Japanese market was not reserved for Japanese growers. Africa's share of world trade is just 2 per cent but the rules of the trade game are clearly loaded against the continent, and so it is unlikely that this situation will change in the near future.

The 'world market' is often portrayed as this kind of 'mystical and surely reified entity' in much contemporary global development discourse. None the less, it is not now and only now that some countries find themselves a part of world markets, to which they have been connected for centuries in many cases by colonial experiences. Non-western cultures and economies were thus intertwined with these processes of globalisation *throughout* the twentieth century rather than just at its culmination. Nevertheless, new information, communication, transport and manufacturing technologies have allowed production, commerce and finance to be planned and organised on a global scale (see Box 7.1). Transnational corporations or (TNCs) are seen to traverse national boundaries and borders in a way that eclipses national state institutions and their capacity to 'broker' development within national territory. We must also reject, however, the myth of the powerless 'Third World state' viewed as a monolith, where the state is seen to have abdicated its power to the world marketplace in this Orwellian language of globalisation (Marcuse, 2000). Not all TNCs are domiciled in 'the West' or the 'First World'; they can come from 'developing countries' as well (Yeung, 1999; Sklair and Robbins, 2002).

Thus more attention is being paid to the diversity of meanings attached to the term 'global', which may surface in debates about the formation of flows where an accelerated movement of money, images, information, migrants, drugs or new technologies is considered (Slater, 1998). These flows are increasingly seen to transcend the territorial confines of nation-states. Many participants in the globalisation debate seem to agree on the decreasing importance, culturally, economically and politically, of nation-states which are seen as being hollowed out from above (by global structures and organisations) as well as from below (by popular resistance in particular localities) (Schuurman, 2001). What we will be seeking to examine is how the central role of nation-states in development is de-emphasised in favour of global governance. In Chapter 9 we will be also be looking at how 'civil society' and local resistance have also de-emphasised the role and authority of nation-states. This does not mean that the nation-state is no longer relevant since in some ways 'it is naïve to write off nation-states as important players in the globalization game' (Schuurman, 2001: 12). In debates about globalisation and development we can call this notion of the global and of global change into question, raising concerns about the way in which spatial scale and interconnectedness are theorised in these accounts, many of which are ripe for deconstruction. Insights from postcolonial perspectives can also be employed here to help us rethink issues of globalisation (Slater, 1998). None the less, globalisation *is* leading to supra-territorial or 'trans-border' relations and to changing human geographies where local, national and global actors are affected differentially:

> This sense of intensifying global interdependence and interconnectedness, which stretches across a variety of spatial scales (global, national, regional, local, communal), and which is characterised by a persistent growth in the spatial density of connections, communications, networks and circuits, is consistently present as a defining feature of many analyses of globalization.
>
> (Slater, 1998: 648)

The global may also be seen to be linked to the strategy of a TNC (as an image that will sell a commodity), or can even serve as the basis for calls for mobilisation against environmental degradation. A key argument of this chapter is that there are multiple tensions and contradictions in theories and practices of globalisation, but we will focus on the shifting spatial scales and networks of governance that affect 'poor countries'. A number of recent debates about how we think about and theorise governance at different spatial scales are important here and are central to our reworking of development geographies.

The first section of the chapter explores the concern with 'actually existing' globalisation and the importance of thinking spatially about the implications for territories, nation-states and places. The chapter moves on to discuss the relationship between globalisation and debates about 'free trade', focusing on the ideologies that lie behind its promotion in the global development agenda and looking briefly at the example of Indian agriculture. The next section discusses the notion of a 'post-Washington consensus' in development thinking today, where some have argued that the IFIs have moved forward by adopting progressive-sounding terms such as 'social capital'. The final, concluding section asks about the possibility of a different kind of multilateralism in the context of a supposed 'new' world order in international relations.

NEOLIBERALISM AND 'ACTUALLY EXISTING' GLOBALISATION

Despite the massive volume of literature for and against 'globalisation', most commentators seem to agree that this is something that is actually happening (Wallerstein, 1990; Appadurai, 1996; Hoogevelt, 1997). As with the study of development more generally, there is no 'standardized way to fathom the influence of globalization' (Schuurman, 2001: 13). More work is needed on the theme of 'actually existing' globalisation and neoliberalism, focusing on the outcomes on the ground. How do the politics of neoliberalism that exist in particular localities come into contact and conflict with global discourses of neoliberalism? This helps us to understand the more variegated nature of development and how it inhabits institutions and places but also the 'spaces in between' (Peck and Tickell, 2002: 387). There is consensus none the less that globalisation is a key idea by which we ought to understand the transition of human society into the third millennium. Globalisation is seen to have linked diverse economies which seek to supply the insatiable demands of the West's consumption needs (which are in turn often mimicked locally).

Globalisation has often been discussed in relation to concepts such as Coca-Cola, cars or clothing sweatshops which bind complex geographies and link producers and consumers across spatial scales. These and other consumer products such as footwear, flowers, bananas, chocolate and coffee have also received increasing attention among western consumers as fair trade movements are organised around them which focus on the prices paid by consumers and the prices paid to producers. Some form of international consumer standardisation and a narrowing of consumption differences is often envisaged by many when the term 'globalisation' is used. As Gibson-Graham (1996) argues, mainstream and left-wing theorists alike share a view which sees capitalism as hegemonic, as the only present form of economy which will therefore continue its dominance into the future. There is thus a cavalier assumption 'that the West carries all before it' (Paolini, 1997: 103). Everything is centred on or by capitalism, even resistance. The incoherence and contradictions of capitalism are often overlooked, as are important 'non-capitalist' social sites. Capitalism has been endowed with such an overpowering dominance that it has become impossible to think social reality in another way or to imagine another world. Gibson-Graham (1996) demolishes the 'straw-man' of what was termed 'capitalocentrism', clearing a path for the consideration of the multiplicity of reality (in this case capitalism) and the various struggles around it:

> In the globalization script . . . only capitalism has the ability to spread and invade. Capitalism is presented as inherently spatial and as naturally stronger than the forms of non-capitalist economy (traditional economies, 'Third World' economies, socialist economies, communal experiments) because of its presumed capacity to universalize the market for capital commodities.
>
> (Gibson-Graham, 1996: 125, 130)

This increasingly familiar 'script' or story of globalisation clearly lacks nuance and assumes that all other forms of economy will fall away as a result of its universal and overwhelming force in the (largely passive) space of the 'Third World'. This region, according to the CIA report *Global Trends 2015*, will see a compressed process of globalisation, seen as 'rocky' and 'volatile' for the frail or ailing economies of the Third World (CIA, 2000: 7). Here capitalism's ability to 'spread' and 'invade' is a key focal point of the discussion. This might leads us to ask how we can challenge the similar representation of globalisation as 'capable of "taking" the life from non-capitalist sites, particularly the "Third World" ' (Gibson-Graham, 1996: 125). It is thus important here to call into question the 'naturalness of capitalist identity as the template of all economic activity' (Gibson-Graham, 1996: 146). What is called 'non-capitalism' becomes damaged, 'violated, fallen, subordinated to capitalism' (Gibson-Graham, 1996: 130). In some ways then, many writers have assumed the inevitability of capitalist 'penetration' in the South in much of the literature on globalisation. Non-capitalist ways of organising have been numerous and widespread in the South (e.g. through co-operative and communal organisations). These ways of organising also deserve further attention and could lead to new ways of thinking and speaking about 'class' and other social relations.

This moves us away from the tendency to ask 'Is globalisation "good" or "bad" for the poor?' (as has often been posed by many agencies) and instead to pose more important questions about the meanings of global and local (Massey, 1992, 1994) or about the nature of 'global' communication and how it can be used to build resistance to development theories and practices. Places and identities are never completely reshaped by capitalism and development; there are numerous points of resistance and contestation which lead to their remaking. This also opens up a space for thinking about alternative economic forms which moves us away from the simplistic 'template' approach of 'capitalocentrism'. These are key issues for the future of post-development thinking (Escobar, 2002: 202). It is also a useful reminder of the blinding power that central concepts of 'modernity', 'development', 'capitalism' and so on can have in organising and to a degree restricting our view of the world and our sense of the possibility of alternatives.

Clearly linked into globalisation processes, however, is a tier of unaccountable international governance institutions (IMF, WB, WTO, NAFTA, OECD) which have a major impact on the political and economic affairs of all nations. Rowbotham (2000: 3) argues that these agencies are drifting 'towards a cumbersome global government' and are 'over-centralised, unwieldy and inherently incompetent'. The crucial question here is: can and how should 'unprecedented' global trade transfers of capital, products and services enhance the global campaign to eradicate poverty? As Rowbotham (2000: 5) has put it: '[v]irtually the entire planet now follows a single economic ideology, founded on the concept that foreign investment is inherently and unquestionably desirable.' The globalisation of trade and the debt crisis are also closely linked in that the neoliberal model has always been based upon a notion of trade that is 'avowedly corporate-friendly' (Rowbotham, 2000: 5). The important point here is that foreign trade and investment are represented as inherently and unquestionably positive forces at the forefront of the global campaign to eradicate poverty.

It is also worth remembering then that international debt and poverty stem directly from the monetary system set up by the USA after the Second World War. More specifically, the origins of the debt crisis lie in the petrodollar recycling mechanism used by the Organisation of Petroleum Exporting Countries (OPEC) which emerged from the recirculation of OPEC's surplus capital funds in the early 1970s (Saddy, 1982). Today, debates about the movement of capital are central to many discourses on globalisation, as many critics of globalisation have focused on the rootlessness and speculative nature of these capital flows, high-

Figure 7.3 The IMF–World Bank–WTO troika represented in a series of protests in Washington involving people from around the world

Source: NBA

lighting the decisive impacts this can have on 'wobbly economies' (*The Economist*, 8 September 2001). An American economist, James Tobin, has suggested that a small levy on capital movements would discourage this, an idea which solicited approval from French politician Lionel Jospin in the run-up to the presidential elections of 2002. Jospin signalled that his government was keen for a debate on globalisation involving all countries, NGOs and the IFIs, starting in 2001, suggesting that this would be good for the countries' reputation: '[w]hat better way for post-colonial France to reassert its moral and intellectual authority?' (*The Economist*, 8 September 2001: 45).

The number of 'developing countries' now subject to the orthodoxies of neoliberalism is quite staggering. This is difficult to appreciate however, given that 'writing about globalisation has been mostly and unashamedly from a First World, Western perspective' (Paolini, 1997: 88). Thus a great deal is written about globalisation in the 'Third World' which does not build upon or acknowledge

'Third World' scholarship on these themes. In addition, much of the literature on globalisation and world cities consigns whole areas of the South to structural irrelevance (Robinson, 2002), discussing only a few groups of cities and their interactions with the world economy, as if all other cities were not relevant. Yet the global economy 'is of enormous significance in shaping the futures and fortunes of cities around the world' (Robinson, 2002, 538). All cities of the world (and not only western/world cities) have important interfaces with this globalising world economy and each has its own social, cultural and historical legacies which it carries into the era of globalisation (Shatkin, 1998). Globalisation affects all cities (albeit in different ways) and not just those at the top of the global hierarchy (Marcuse and Kempen, 2000). Views from off this partial map of urban studies are thus crucial (Robinson, 2002).

Globalisation is often seen to have important spatial dimensions, to represent a *stretching of space* where what happens in one region of the world has ramifications in many others, but also a *deepening of space* and a reconsideration of the terms 'local' and 'global'. Each geographical 'place' in the world is adjusting to new global realities, leading to new social relations and roles (Massey, 1994). Here we need to understand the ways in which the speed, intensity and volume of economic and political resources into and out of places and spaces creates new geographies of inclusion or exclusion in a globalising world. In many ways these 'new' geographies reproduce many of the uneven geographies that characterised older, colonial relationships between North and South. Due to the distinctly 'un-global' nature of much that is written about 'globalisation' it is not always clear what impact it has had in specific localities, for example, in Havana, Harare and Hyderabad. Furthermore, very little has been written about how 'Third World people see themselves and their society, its past, present and future' (Paolini, 1997: 91).

Thus the 'agency' of *people themselves* as *agents of development* is denied, since little attention is paid to the embeddedness of globalisation (or responses to it) in particular places. It is also as if it is being assumed that this question of agency does not matter, given the overwhelming dominance and supposed inevitability of the triumph of global capitalism. As Paolini (1997: 99) puts it, 'the impression we are left with is of puppets playing out prearranged roles'. Contrary to the popular image of globalisation leading to homogenisation of peoples and cultures there has actually been more *differentiation* in terms of inequality and the growing income gaps between the haves and 'have nots' of this newly globalised world economy (Ahmad, 1992). It cannot be automatically assumed that the globalisation of lifestyles will lead directly to an erosion of cultural distinctiveness. One persistent theme in the literature on globalisation is the idea of a convergence around a notion of a single 'global human condition' (Held and McGrew, 1993: 262–263), but it is not always clear how this relates to the particularities of place and individuality.

It is important to examine the question of why western discourses on development have not accorded Africa an autonomous space in these debates or have preferred to see the continent as *outside* the forces of globalisation and modernity. The particularities of Africa and other regions have been misread in these debates as boundaries become blurred and assumptions are made about African intersections with the forces of global modernity. What is needed here is a 'progressive concept of space' (Massey, 1994) where places are not seen as static entities or in a simplistic state of 'non-modernity' but as related to processes, relations and interconnections which link the global and the local such that we might insist on a 'global sense of place'. This is an idea developed further by Escobar (2002: 203) who argues that 'place is central to issues of development' and calls for a focus on 'place based practices'. The interconnectedness that globalisation brings about is often cited but rather less frequently understood and accurately theorised. As we have seen, different regions of the world were intertwined in the making of European modernities and the same is true of the making of globalisation today. There is no single response to modernity nor is it always inherently negative (Dussel, 1998). The making of global development geographies today involves a whole variety of regions, nations, places and continents and attention to the making of their various forms of modernity and struggles to become 'modern' and 'developed'. Globalisation then is a multidimensional process which has major implications for states, nations, places and communities.

In discourses on globalisation much is often made of the limitless, utopian freedoms promised by a borderless world. In this sense territory is often understood in isolation from the complex of state power, identity and geography of which it is a part

(Ó Tuathail, 1999). Sovereignty and democracy are now defined at a variety of spatial scales: local, regional and global. Territorially, political communities are being reconfigured with important implications for the theory and practice of development. Debates about 'globalisation' and development (which have often been abstract and lacking in clarity) have focused on the 'annihilation of space by time' and the increased ease of travel and communication which characterise the contemporary world (Holloway and Hubbard, 2001). Some commentators refer to the 'end of geography', envisioning a time when all places will have similar social and cultural characteristics as global corporations and media agencies. Geography is thus seen to matter less as global agencies spread similar kinds of products and images across the globe. This process has been termed *deterritorialisation* and refers to the changing significance of territory, suggesting an 'emptying out' of space and time which many see as related directly to the changing geographies of global capitalism. As Ó Tuathail (1999: 140) argues, to speak of deterritorialisation is to suggest that the complex of geography, power and identity is being dismantled in some way and to speak of the 'transgression of inherited borders, the transcendence of assumed divides, and the advent of a more global world' (Ó Tuathail, 1999: 140). In this way we might challenge the idea of deterritorialisation as the consequence of 'unstoppable globalisation' and reject the assumption that the world was previously made up of discrete borders and boundaries, calling into question the claim that globalisation will today lead to a unifying 'liberation' for the repressed and poor peoples of the world (Ó Tuathail, 1999). Just as some boundaries are therefore supposedly coming down, new ones are being erected all the time, composing new spaces of exclusion from international development.

In the specific case of the North American Free Trade Agreement (NAFTA), US trade policy statements call for the deterritorialisation of North American national spaces by the flow of capital (Slater, 1998). A common theme in the literature on globalisation then is precisely this perception of the increased marginalisation of states in a globalising capitalist world economy. State sovereignty in particular is now seen to be driven by new processes of flexible accumulation or by transnational corporations. Global financial markets are seen to prevent states from regulating their own currencies (which are also often defined by global development institutions such as the IMF or the World Bank). Media agencies and crime syndicates can also challenge the authority of the state, disseminating information across borders in ways that elude state regulation and control. In the drive for liberalisation, restrictions are removed on the movement of capital across national boundaries which supposedly help the poor and promote growth. The evidence is shaky, however (Cobham, 2001), while the impact of FDI is quite often ambiguous (see Boxes 7.3 and 7.4).

The macroeconomic instability of Asia at the end of the twentieth century is a strong example of the damaging effects that this volatility can have on poor people's livelihoods. Recent events in Argentina also illustrate how the poor will take to the streets to illustrate the extent to which this instability has come to disrupt their communities. The borderless world of freedoms that NAFTA proposes adopts quite unproblematised assumptions about deterritorialisation, offering 'sweepingly superficial representations' of development geographies in the process (Ó Tuathail, 1999: 142). What is being signalled here is the ideological uses of the term 'globalisation' which pedal neoliberal assumptions by focusing on a borderless world where there is unfettered movement of capital. In this sense we can ask 'For whom is the world borderless?' (Ó Tuathail, 1999: 149). All too often neoliberal discourses represent this presumed borderlessness as leading to a more integrated and connected world of development, implying that unevenness will fall away in its wake. That *new* spaces of deprivation and disconnection are emerging is papered over in this way and is rarely acknowledged or problematised.

The notions of national culture and of national development are clearly in need of reconsideration as they are redefined by changes in the nature of global capitalism and the way this is (re)constituted globally. As we have seen, a variety of debates have centred on the legitimacy of the state over a given territory which is supposedly being called into question by globalising processes (Ahluwalia, 2001). In Africa, for example:

> [C]apitalism constitutes and reconstitutes itself in a variety of forms in order to be able to penetrate different areas of the world. If globalisation manifests itself in such a manner, what room is there for the role of the state and civil societies?
>
> (Ahluwalia, 2001: 116)

BOX 7.3

The free trade 'rip-off' and the 'Asian crisis'

The rules governing international trade have been based increasingly on 'free trade' or 'regulated' trade ideologies. When it is remembered that the borrow/invest/export/repay model relies absolutely upon successful exporting by debtor nations, it may be seen that 'trade conditions are clearly of critical importance' (Rowbotham, 2000: 69). In June 1999 a delegation of Indian farmers toured Europe with the slogan 'Fair trade, not free trade' (taking in the WTO, NATO and the biotech company HQ of Monsanto and others (Rowbotham, 2000)). Free trade is unfair; the very word 'free', however, implies that the alternative is 'unfree', in some sense oppressive and restrictive. Free trade supposedly leads to wealth creation and global diversity in trade specialisations. The 'free' flow of capital is crucial here too – from developed to developing. Then there is the expectation that free trade will lead to a decrease in the inequalities between people and between nations, through 'a kind of convergence' of wealth distribution. Is the Asian crisis an example of the failings of the free trade myth? This is the global gamble which underwrites the World Bank's (Faustian) bid for world dominance (Gowan, 1999). It is sometimes called the 'Asian malaise, the Asian contagion, the Asian debacle and the Asian crash' (Rigg, 2001).

This is usually dated from 2 July 1997 when the central Bank of Thailand, after spending some US$9 billion, gave up its attempts to protect the baht and its link to the US dollar. The newly floating baht collapsed, soon rippling out to other countries in the region (Indonesia, Malaysia, the Philippines and South Korea, Singapore, Hong Kong and Taiwan, Vietnam, Lao PDR). Economic growth in some of the most acute cases stood at 5.5 per cent but soon dropped to negative growth in 1998. The countries of South Asia were largely immune, with the great bulk of Asia only marginally affected (Rigg, 2001). The Asian crisis was thus a misnomer. Initially mainstream commentators linked the crisis to the nature of Asian capitalism. They stressed the degree of 'crony capitalism' and corruption in the region (Masina, 2001). It was also a political crisis, undermining the legitimacy of states seen to have achieved this vaunted growth in the first place, leading to changes of government in Thailand, South Korea and Indonesia and also to President Suharto's resignation in May 1998 which impacted on East Timor. Some people have viewed the crisis as an opportunity to move away from the destructive policies associated with fast-track industrialisation (Bullard *et al.*, 1998).

Figure 7.4 The inexorable rise of globalising forces
Source: CWS and Ammer

This question of what role there is for the state in relation to capitalism and development is often asked. This varies in each place and locality but there are a number of generic features of neoliberalism, articulated (for example) in the World Bank's annual scripting of global development issues, the WDR. The 2002 WDR purports to champion the cause of the poor and speaks volumes about how the Bank views development (BWP, 2001a). The market is again prioritised as the primary mechanism for growth and poverty reduction and so, it is argued, institutions *need to serve markets* more effectively. Building institutions to serve markets enhances the opportunities available to poor people, according to the Bank. The WDR again fails to acknowledge alternative perspectives however, as it would seem institutions work for markets, not people. Corporate monopolies of the market and corporate governance failures are also (rather conveniently) not discussed. With 2002 as a year of scandal in corporate America this seems difficult to accept.

Pieterse and Parekh (1995: 45–46) argue that the focus on modernity and capitalism in studies of globalisation has meant that much of the debate has been defined by Eurocentric terms and has been plagued by 'westernisation', much as the modernisation discourses of the 1950s had been. Modernity is assumed to automatically equate with homogeneity but instead Pieterse argues that globalisation processes need to be viewed as 'hybridisation which gives rise to a global mélange' (ibid. 45). In this way it is necessary to recognise the new global cultural complexities that are being produced through globalisation processes. As we have seen, the notion of a 'Third World' has been severely challenged but in some ways has been replaced by the possibly more problematic notion of a 'postcolonial world'. Postcolonialism is useful here however, because it groups together all formerly colonial societies (despite differences in their relation to the global capitalist system), while at the same time 'offering a point of entry for the study of those differences' (Hoogevelt, 1997: xv).

This 'point of entry' then is crucial here. How (in the aftermath of colonial relations) are different societies changing as a result of the present transformation of the global political economy? From where and how does this point of entry come? Africa has thus been partly included and partly excluded from these processes. Indebtedness for many commentators allows western capitalist economies to 'manage' the periphery and to continue to extract a kind of surplus (for example, in interest payments), which maintains relations of dependency. The World Bank has a list, for example, of forty-two HIPCs and manages the criteria for relieving indebtedness in ways which confer far too much influence on the organisation given the actual sums involved. As a consequence of this:

> Structural adjustment has helped to tie the physical economic resources of the African region more tightly into servicing the global system, while at the same time oiling the financial machinery by which wealth can be transported out of Africa and into the global system.
>
> (Hoogevelt, 1997: 171)

Globalisation and localisation occur simultaneously, rather than globalisation automatically leading to the simple effacement of local identities. Globalisation is taking place through rather than in spite of the multiple identities that people have (for example with allegiances to kin, group or nation). 'Glocalisation' is a word used to illustrate the extent to which the local is constructed through a global/local nexus rather than only within local spaces (Massey and Jess, 1995). These glocalised spaces (particularly in Africa) are still profoundly asymmetric in terms of the power relations that produce them. As with other binaries used in development thinking, it is necessary to look at the intersections between local and global at all points and in all spaces in a way which does not assume that the former is always eroded by the latter. The political language of a global development community is worth exploring further in this regard – examining how the key ideals of freedom, democracy and social justice emerge through these languages and discourses of development. One problem with these and many other discourses of globalisation is that they have tended to homogenise the world in their representation of non-western societies, for example denying African specificities (Mazrui, 1999) and assuming that globalisation has been uniform across these spaces.

Africa does not figure prominently in theorisations of globalisation today. As a result of the increasing marginality of Africa in the processes of globalisation it is not clear what position it occupies in the globalising world economy. What is needed therefore is a more nuanced approach to the local/global nexus (Ahluwalia, 2001). The reason is that identities can be multiple or, as Edward Said puts it:

> No one today is purely one thing. Labels like Indian, or woman, or Muslim or American are no more than starting-points, which if followed in to actual experience for only a moment are quickly left behind. Imperialism consolidated the mixture of cultures and identities on a global scale.
>
> (Said, 1993: 407)

It is this mixture of cultures which is important here, illustrating further how the politics of labelling in development discourse is becoming increasingly problematic. As a result of the 'overworlding' of regions such as Africa in debates about globalisation and development the continent occupies a space from which we are 'able to view the blindspots and indiscriminate presumptions of the academic discourses under review' (Paolini, 1997: 93). In other words, because Africa has so often been seen as a *non-space of development* it is possible to expose limitations of these presumptions about progress in other worlds. 'Globalisation' has introduced the illusion of a trans-territorial world of multicultural dialogues, with flows moving in all directions, but in reality flows and connections occur inside and around established centres of power (Slater, 1998) rather than evenly across the periphery in a web-like pattern of connections.

GLOBALISATION AND THE MYTH OF FREE TRADE

> We may end up with cheap bananas or a bargain-basement microwave. But at what price for society as a whole?
>
> (Ellwood, 2000: 12)

Debates about trade were an important part of the Enlightenment as writers like Adam Smith and David Ricardo tried to explain why some nations prospered and experienced growth. This was the period of classical economics and was embedded in broader debates about political economy and philosophy. There isn't space to go into these debates in detail here but it would be misleading to suggest that discussions about trade and progress are new or just a twentieth-century phenomenon. At the end of the nineteenth century there were further important debates in what has been called 'neoclassical economics' where the central theme of economics changed from a focus on the growth of national wealth to questions about the efficient allocation of resources (Peet with Hartwick, 1999). In addition, Marxist perspectives on political economy also began to emerge towards the end of the nineteenth century that focused on the contradictions within capitalism, where wealthy states were seen to force trade at disadvantageous terms and imperialism was seen to accelerate the accumulation of capital (see Peet with Hartwick, 1999). As we have seen, theories of dependency picked up on these themes about the monopolistic nature of capitalism and its capacity (through imperialism) to put in place, around trade, exploitative relations between core and periphery.

Nearly a hundred and fifty years later, the writings of people such as Adam Smith and Karl Marx continue to have an important bearing on the way in which free trade is conceptualised today. In November 2001, at the fourth ministerial session of the WTO in Doha, Qatar, a range of delegates from NGOs around the world highlighted the imperialist nature of organisations such as the WTO and sought to contest some of these long-standing ideologies and the unrestricted nature of trade flows. Standing at the entrance of the summit building, they held up signs with the words 'No voice at the WTO' and had tape over their mouths. Doha, a small city of 600,000 people, was turned into a fortress with massive security preparations. Inside, the G-77 group of countries from the South pushed for the implementation and resolution of issues discussed at the Uruguay round of talks (the final round of which was held between 1986 and 1994!), while the USA and EU tried to introduce 'new issues' of trade, government, competition and investment, playing down contentious issues such as labour standards which had erupted in Seattle in September 1999. New negotiations for further trade and economic liberalisation then began in Doha, opening negotiations on existing agreements and initiating new tariff discussions. In addition, new, non-trade areas have been brought within WTO jurisdiction – the so-called 'Singapore issues' of investment, competition policy, government procurement and trade facilitation (Bello, 2002). Outstanding issues for countries of the South remained from the previous Uruguay round however, of which there were some 104 which were incomplete according to the G-77 countries. Once again the excessive protectionism that restricts the reach of 'developing country' markets was also not addressed and some critics have suggested that this will even intensify in view of the global economic downturn in 2002 and early 2003.

Each year countries in the South lose about US$700 billion as a result of trade barriers in wealthy countries, meaning that for every US$1 given in aid and debt relief by rich countries, poor countries lose US$14 because of trade barriers (Sogge, 2002: 35). Some progress was made on the issue of trade-related intellectual property rights (TRIPs) which restrict the capacity of G-77 countries to protect their public health systems, but as ever only vague declarations and 'commitments' were offered. These are only political declarations however, and are not legally binding to any particular pharmaceutical TNC. The united front that was presented by the G-77 countries in Doha was encouraging, however, and will prove crucial in the years ahead. In some ways the EU and US representatives were particularly astute at finding new ways to split these emerging coalitions, while in a wider sense the monarchy of Qatar could control and limit the number of NGO activists and other dissonant voices entering the country and from making their voices heard. Economic and military aid packages were also offered to some countries such as Pakistan and Nigeria to solicit their support and silence their potential opposition. The infamous 'Green Room' was reinstituted in Doha when twenty select and handpicked countries were isolated from all others and invited to devise a final declaration. A delegate from Uganda, trying to infiltrate these exclusive meetings, was forcibly rebuffed (Bello, 2002). There are few records of these discussions, illustrating the lack of transparency and accountability which characterises them.

When WTO head Mike Moore visited India in 2000, he joked (rather uncomfortably) that in no other place on Earth had so many effigies of him been burned (Ainger, 2001). At the Doha meeting of the WTO in November 2001 the organisation pledged to co-operate more closely with the World Bank and IMF for more coherence in 'global economic policy-making'. The World Bank has repeatedly reiterated its commitment to work with the WTO to encourage trade barrier reductions, with trade seen as a tool of poverty reduction and development. For some observers, convergence between the three (ideologically) signals a kind of 'death of development' – a process begun in 1994 but one which really took off at the Seattle ministerial meeting in 1999 where the three issued a joint declaration indicating a shared commitment to trade liberalisation as the mechanism for global economic growth and stability. Organisations such as the WTO have since become prime targets for the criticism of 'anti-globalisation' movements, for advocating an uneven and unjust model of change which needs to be contested.

In documents packaged with relatively uncontroversial titles such as 'Growth is good for the poor', the World Bank argues that the incomes of the poor rise through overall growth only if 'fiscal discipline' is pursued, a government spends less and inflation stabilises. Greater participation in world trade, according to this view, will directly increase the income of the poorest one-fifth of the world's population. The impoverishment of African peoples and economies will cease, so this logic follows, if African countries trade further with other countries and regions and begin to accelerate economic integration, leading to convergence and therefore to the disappearance of income inequalities. Trade liberalisation has many negative impacts on the agricultural policies of southern countries however, shifting the priorities of the government from their own citizens to the needs of a volatile international market. Agriculture accounts for less than 10 per cent of world trade in total, some 70 per cent of which comes from 'developing countries' (Panos, 2001); yet the global agro-food system is skewed against the interests of poor farmers of the South. In 1974 developing countries accounted for 28 per cent of total food imports; by 1997 that figure had risen to 37 per cent. At the same time developing countries have not been able to break into protected western markets and consequently their share of total world food exports rose only slightly between 1974 and 1997, from 30 per cent to 34 per cent (White, 2001). Those who have broken into these markets have not always seen incomes rise, since prices have actually fallen for a number of agricultural commodities (e.g. bananas and coffee), which has not always led to 'pro-poor growth'. How can development be sustainable in such circumstances? As a recent Panos Institute report puts it: '[Liberalisation] militates against the kind of individual or local support that is implied by the livelihoods approach to poverty reduction' (Panos, 2001: 35).

The question of the 'livelihoods' approach to poverty reduction is explored in Chapter 8, but this issue of how liberalisation 'militates' against individuals and localities is an important one. For this reason many countries of the South (e.g. India) have called for a 'development box', a system of allowing countries with large numbers of poor farmers to protect and invest in their farms and communities rather than just concentrate on the removal of trade barriers (*Guardian*, 3 September 2001). The crisis and despair experienced by many farmers in the Punjab as a result of the globalisation of Indian food and agriculture comes across strongly in the work of Vandana Shiva (2001), who writes about the 'brutal' and 'unforgiveable' consequences of liberalisation processes for the Indian poor. Highlighting the diversity of farmers and small farm systems, Shiva explodes the myth that industrial monocultures using GM seed are most productive. Women's agricultural contributions are rendered invisible by global free trade discourses which see them as non-productive or even economically inactive:

> When growth increases poverty, when real production becomes a negative economy, and speculators are defined as 'wealth creators', something has gone wrong with the concepts and categories of wealth and wealth creation.
>
> (Shiva, 2001: 61)

The marginalisation of women is about more than texts and discourses but it is also based very much around actual material processes such as those of global patriarchy. As geographers we can extend this concern for discourses and the textuality of development by also paying close attention to the power relations (e.g. in male-dominated or patriarchal systems) that operate at each geographical scale, highlighting the material implications in terms of poverty and social exclusion, for example. Shiva argues that local economies and food cultures are being destroyed across India *because of the globalisation of the food system.* Here, the knowledge of the poor farmer is converted into the property of global biotechnology corporations where the poor pay for the seeds and medicines *they themselves* developed and cultivated. Shiva also refers to 'market totalitarianism' whereby the global free trade economy is a threat to sustainability and the survival of the poor and communities – all in the name of market 'competitiveness'.

This constitutes a different kind of violence on poor people and poverty in that globalisation, for Shiva, has 'become a war against nature and the poor' (Shiva, 2001: 65). Depicting a world of monocultures, monopolies, appropriation and dispossession, Shiva (an ecologist and physicist) writes of how women are disempowered, overlooked and their knowledge quite literally 'stolen'. Corporate-controlled agriculture is seen as the productive future for countries such as India and their agro-food systems which are becoming increasingly globalised. Shiva also highlights the links between WTO rule-making procedures and their 'on-the-ground' implications for farmers in India who are dispossessed by these rules and structures and their anti-democratic procedures. In an interesting article entitled 'The poor can buy Barbie dolls', Shiva (2002) recounts her experience while appearing on a recent Indian TV chat show. The panel of which she was a part were assembled to discuss the impact of removing import restrictions (Quantitative Restrictions (QRs)) in India. This has allowed a free flow of foreign goods into Indian markets, celebrated as India's 'consumer bonanza'. Shiva (2001) bemoans the lack of attention to India's poor in these debates as 'imported goods drive out domestic production and livelihoods'. When the issue of poverty was raised on the show, one panellist (an economist) claimed that free trade was good because it meant that 'the poor can buy Barbie dolls': 'It is the mindset of elite India – blind to the growing hunger and destitution of the people of this country but enthralled by the junk that can now be imported' (Shiva, 2001: 2).

Figure 7.5 Farm subsidies and the USA
Source: Cartoon by M. Lane

Shiva's argument is not with Barbie dolls but with the idea that opening up Indian markets for the sake of consumer choice is inherently good for the livelihoods of India's poorest farmers and producers. What is particularly interesting about the account Shiva provides is that debates about trade, liberalisation and investment are seen to make certain kinds of assumptions about consumption, assuming that social improvement means having better access to the consumption of imported goods and services.

GLOBALISATION AND THE POST-WASHINGTON APOTHECARY

> [T]oday's developers are like the alchemists of old who vainly tried to transmute lead into gold, in the firm belief that they would then have the key to wealth.
>
> (Rist, 1997: 46)

For most of the 1980s and into the early 1990s, the Washington consensus dominated development theory and policy – the term denoted a series of measures that would produce wealth and freedom from poverty (fiscal and monetary austerity, the elimination of government subsidies, moderate taxation, free interest rates, lower exchange rates, privatisation and encouragement of FDI and the liberalisation of foreign trade). FDI inflows to the 'developing world' reached US$1.3 billion in 2000, but a predicted downturn in the world economy is likely to lead to more intense competition for less foreign aid and investment. As a result proactive investment promotion measures are proliferating in the South as countries seek to establish investment promotion agencies to capture a slice of these flows and construct a 'favourable investment climate'. UNCTAD has thus created the World Association of Investment Promotion Agencies (WAIPA) within the UN structure to aid poorer countries in pooling their knowledge of successful investment promotion practices. There are now at least 164 national IPAs worldwide and another 250 subnational ones (UNCTAD, 2001a). On average, annual IPA budgets worldwide amounted to US$1.1 million in 1999. In some cases regional investment agencies have been established, such as the Inter-Arab Investment Guarantee Corporation set up in 1975 with membership of nearly all Arab countries seeking to foster and enhance inter-Arab investments.

According to UNCTAD, agencies in OECD countries 'apply the most focused approach to investment promotion with investor targeting and after care as prime functions' (UNCTAD, 2001a, 1–2). Thus, IPAs from the South must learn from their compatriots in the advanced capitalist world. One result of these sorts of expenditure and the existence of agencies such as WAIPA (which works with the World Bank's Multilateral Investment Guarantee Agency (MIGA)) is that investment promotion is becoming increasingly sophisticated as many countries scramble to pander to the needs of wealthy corporations, arguably taking expenditure away from other more critical social concerns. One investment guide to Mozambique includes a section of text written by UNCTAD entitled *Of Risks and Returns: Investing in LDCs* (see Box 7.4). This begins by asking why anyone would want to invest in a 'least developed country' when the risks 'are sky-high and profits precarious' (UNCTAD, 2001a: 9). It warns of casually dismissing a quarter of the world's countries as locations for investment, while positioning Mozambique as one of twenty-two African 'LDCs' (of which there are forty-nine worldwide). It also comments on the country's image as an LDC and rails (ironically) against the perception that all these countries are essentially the same:

> One problem with the association of high risk with LDC's is that it treats 49 countries as though they were all clones of a single national type. In truth there is much variation. Some LDC's are riven by civil war and some destabilised by coups and counter-coups. There are others however that have established a track-record of political stability and sustained growth (Uganda and Mozambique) or shown great resilience in the face of national calamities (Bangladesh) . . . Is there a moral here? Yes, one that can be summed up in a single maxim: *Differentiate*. Investors need to differentiate among the 49 LDCs. Some will confirm their prejudices; yet others will shake them.
>
> (UNCTAD-ICC, 2001: 9, emphasis in original)

It seems far more likely that in the context of a global apothecary of development, free-market ideologies will *confirm* many prejudices about least-developed others in their 'cloning' of states with 'good government'. Many critics of these ideologies of neoliberalism have noted that they have often had a strong US flavour and that many of the world's leading transnational corporations are US-based, while most FDI is recorded in US dollars. Washington is thus often seen as the 'undisputed political, economic and ideological centre of the world' (Fine *et al*., 2001: x). The intellectual superiority of western-trained economists is an issue here as economics has become the 'naked emperor of the social sciences' (Keen, 2001). They are the 'priesthood' and we are the 'novices' here. As one commentator has put it, the World Bank 'is to development theology what the papacy is to Catholicism, complete with yearly encyclicals [the WDRs]' (Holland, 1998: 5). Although once an unpopular sect with no influence, neoliberalism can be seen as a world religion, with its dogmas, its doctrines, its priesthood, its lawgiving institutions and 'its hell for heathens and sinners who dare to contest the revealed truth' (George, 2001: 9). To support this priesthood there is now a huge international network of research centres, institutions, publications, scholars and writers who seek to package

BOX 7.4

Mozambique's 'turnaround'

Mozambique is one example of how structural adjustment (and more recently the Poverty Reduction Strategy Papers (PRSPs) actually operates in practice. In 2001 an *Investment Guide to Mozambique* was published as part of the UNCTAD-International Chamber of Commerce (ICC) joint series on investment. UNCTAD has 190 members and is based in Geneva, while the ICC is a world business organisation representing 130 countries. The secretary-generals of both organisations (Rubens Ricupero and Maria-Livanos Cattaui respectively) write in the Preface to the guide that they seek to foster a dialogue between investors and governments through this series and to help bring together 'firms that seek new locations and countries that seek new investors' (UNCTAD-ICC, 2001: vi). The average bilateral and multilateral donor assistance to Mozambique runs into hundreds of millions of US dollars. Very little attention is paid to the country's complex historical background in such investment guides but we do learn that Mozambique has gone through an amazing 'turnaround' since the mid-1990s after the end of the civil war between Frelimo and Renamo, experiencing a rate of growth in the economy of up to 10 per cent per annum. The reason for this change in fortunes is seen to be the transformation of Mozambique during the 1990s into a multi-party democracy with a market economy. The liberalisation of the country's economic and political environs, through such measures as currency devaluation, privatisation, the removal of state subsidies and restrictions on foreign investors, are all seen as universally successful for all Mozambicans in the investment guide. The government is praised throughout, while many sympathetic investors from Shell and other corporations confirm this support for the government's neoliberal reforms. We are left with the impression that there is thus no alternative except to remove fetters on the free flow of foreign capital and to deregulate financial markets. UNCTAD even inserts an 'improvement index' to measure 'Africa's competitiveness' and improvements towards that end (+0.3 per cent for Mozambique as opposed to −0.5 per cent for neighbouring Zimbabwe) (*Africa Competitiveness Report*, 2000–2001). Four brief paragraphs are given to the country's history and government (from Vasco da Gama to Samora Machel).

Special economic zones, investment, land and financial laws are detailed alongside a picture of Mozambique's investment 'potential'. Mozambican human resources become a tiny part of this quest for profit margins through investment, while in practice this document does anything but differentiate Mozambique from other 'least developed countries' and aligns the country with other neoliberal orthodoxies on trade and investment which are not at all context-specific but rather are structured and arranged by a whole array of foreign agencies. UNCTAD advocates flows of private capital because they are seen to transfer technology and with that expertise and knowledge, helping to improve access to international markets and greater capital inflows. The Investment Marketing Services Department of MIGA (part of the World Bank group) recently produced an 'investment promotion toolkit' which is available to Investment Promotion Agencies (IPAs) in countries such as Mozambique for the princely sum of $600 (payable of course to the Bank). In one of the nine module guides, entitled 'Servicing investors', the toolkit spells out how MIGA and the Bank feel that investors ought to be treated if they plan a visit to a potential investment location (let us say Maputo):

> Clients should be met at the airport and transported to a quality hotel located close to the IPA office. Make sure the check-in goes smoothly and tell the executives what time a car will pick them up for their first meeting. . . . Giving clients your home telephone number (as well as cell phone and beeper number if applicable) is also a personal touch that shows that they are a top priority.
>
> (MIGA, 2001: 7)

Most FDI in 'developing', 'emerging' or 'transition' economies is resource-seeking, aiming to 'exploit a country's comparative advantage' (MIGA, 2001: 3), for example in oil or minerals or labour costs. There is also, however, 'market-seeking' FDI, which is investment that seeks to build local markets and manufactured products for consumption (from soft drinks to washing powder and tobacco). In order to increase the efficiency of their global operations, TNCs have broken their production processes into discrete functions and source their inputs on an increasingly global basis, assembling products which are exported worldwide. These corporations often account for a significant proportion of a host country's exports, since most FDI is export-related

BOX 7.4 (contd.)

and geared to export away from the producing country rather than for local consumption. Evidence suggests, however, that not all countries benefit from FDI in terms of the impact on their national growth trajectories (UNCTAD, 2001a), particularly the lower-income countries such as Mozambique which do not have the resources to compete with IPAs from other countries. In this sense UN agencies such as UNCTAD and WAIPA do little to question the relevance of these sorts of 'development activities' for some of the weaker and poorer states. Conferences resemble beauty contests where all the main agencies (including the World Bank group agencies, UNCTAD and the OECD) struggle to create the image of success and progress, blazing a trail to prosperity for and on behalf of the poor. According to Joe Hanlon, multilateral donors actually promote corruption in Mozambique, while the World Bank 'sees what it wants to see' in the country because it needs the 'myth of the Mozambican success story' and the mirage of 'good performance' even when local movements have called for anti-corruption measures (Hanlon, 1996; BWP, 2002b).

For more on donors and corruption in Mozambique http://www.mol.co.mz/analise/

and push the core ideas of this doctrine relentlessly (George, 2001). In Chapter 8 we shall see how this world religion is diffused and disseminated and how it relates to 'development theology' more generally, in all its mystique and promise.

The 1990s have seen growing levels of discontent with the policies and perspectives of these institutions, as poverty reduction has made limited headway and neoliberalism has also failed in parts of the former Eastern bloc countries. East and Southeast Asia remains intriguing, however, in that between the 1970s and 1990s, 'growth rates' (the favoured indicators of the Bank) have grown and have been represented as success stories tiger economies that had responded to the prescriptions of the IFIs (in that these countries were seen to create their own comparative advantage in the world market). Growth is a far more complex process than these agencies recognise, rooted as it is in the specific dynamics of place and varying considerably over time. Historically the economic theory that has informed these institutions has 'remained characteristically blind to the fact that the market is itself an institution' and that non-market and institutional factors generally might be important (Fine *et al.*, 2001: xii). The social and political conditions (e.g. class) that make possible various market exchanges (about which the Bank speaks so often and so favourably) are frequently overlooked by the mainstream development community. The emergent 'post-Washington consensus' that some writers have referred to is based on the belated recognition that market 'failures' and institutions are important conditioning factors in the outcomes of development. It seems premature to speak of a 'post-Washington consensus' however, when free trade and privatisation are still vitally important, and fiscal and monetary issues remain very conservative domains in these institutions. The World Bank's approach to privatisation was once described by Joe Stiglitz, former World Bank economist and Nobel Laureate, who referred to the process as more closely akin to 'bribarisation' – for which he was fired in 1999 by World Bank President James Wolfensohn for outspoken criticism of the IFIs.

Unlike the dogmatism of the Washington consensus, the post-Washington consensus is seen by contrast to represent certain degrees of recognition that 'development' is a complex social process, that institutions work differently in each society and that social relations with institutions and markets are

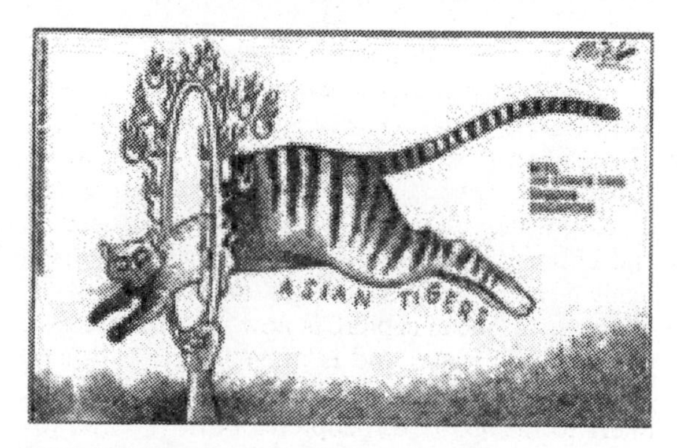

Figure 7.6 Recession claws back the Asian Tigers
Source: CWS and Miel

Figure 7.7 Singapore skyline
Source: Lily Kong

important questions. The whole notion of '*social capital*' adopted by these organisations can be critiqued for the extent to which attention is shifted away from inequalities of power (Fine, 2001). For geographers this concept is important, since it raises questions about spatial patterns (Mohan and Mohan, 2002). It also illustrates further the 'colonisation' of the social sciences by neoliberalism (and economistic ways of thinking about development and society) at the expense of alternatives (Fine, 1999; Mohan and Mohan, 2002). If relations between social groups are relevant in the making of development, so too are relations between spaces and between people and places, yet none of this is really touched upon in such debates.

The term 'social capital' is now used to discuss a vast range of issues and is often linked to discussions of civil society (see Chapter 8). This idea refers to the ability of people to work together for common purposes in groups and organisations and has been bandied around by the Bank as a testament to its close proximity to contemporary research. Proponents of this idea at the World Bank argue that it will lead to new forms of practice, facilitating conversations across disciplines, but the reality is that 'the room to manoeuvre opened up by notions of social capital is extremely small' (Hart, 2001: 654). It was, however, the beginnings of a recognition that there are many different types and varieties of capital which are not always embodied in land, factories and buildings but in human knowledge and skills (Thomas, 2000: 37).

The IFIs seem not to understand that development is about the reproduction of society on a 'different level and in a "modern" way, bringing about quite sometimes radical changes to social relations' (Fine *et al.*, 2001: xiv). The principal problem is that the Bank still has a rather primitive notion of social differentiation and social change. Women are often central to the forms of social capital that development agencies seem keen to mobilise in poverty relief programmes but the terms of women's insertion into these programmes is rarely problematised by the Bank (Molyneux, 2002). Although the emergence of this concept is widely seen as signalling a welcome (if belated) engagement with the 'social', it has not been allied to and integrated with the work of the Bank's Gender Unit. Once again, this 'deliberately fuzzy concept' (Harriss and de Renzio, 1997: 921) suffered nearly thirty years of criticism about the 'top-down', 'growth-first' policies of the Bank before it was finally taken on board. Throughout this time, networks, community ties, experiences of participation and new kinship relations were being forged across the South, particularly in Latin America (Molyneux, 2002). Social inequalities thus remain unchallenged either in theory or in policy while women (through their roles in the reproductive economy) continue to serve as the unrecognised 'shock-absorbers' of economic crisis (Elson, 1991). This concern for social capital thus partly represents a 'sticking plaster' approach to development (Molyneux, 2002), which masks structural issues (notably the uneven distribution of income and assets) and cannot substitute for effective economic policy and redistributive measures. Today the concept is in some ways 'destined to collapse under the weight of its own floppiness' (Hart, 2001: 653). How and why is social capital supposed to serve people, who benefits and why? These are all questions that the Bank seems unable and unwilling to answer. It is thus also important to remember that:

> The concept of social class is also necessary, as are those of generation of economic surplus, its division among classes and its utilisation for consumption and investment. More broadly, the concept of social reproduction is vital in analysis of the social relations to be found at the workplace, in agriculture, in the schools, within the family, in state organisations and between state institutions and enterprises
>
> (Fine *et al.*, 2001: xiv)

The IFIs seem to have a quite limited grasp of and interest in the historical, geopolitical and cultural spaces which impact upon and condition its work. Although the 'new' consensus attempts to broaden the theoretical agenda of development economics, it does so within 'the same narrow, reductionist framework as its neo-liberal predecessor' (Fine *et al.*, 2001). What is missing therefore from the IFIs' agenda is a concern for the political economy of development, with a focus on social class, social reproduction and state–society relations. Perhaps 'consensus' is not the right word here since neoliberal orthodoxies have always been contested and violent arenas of struggle (Bond, 2000a) and have seen a whole variety of forms of resistance: local, regional and global (see Chapter 9). The Washington institutions still continue to adhere to the idea that there is a 'textbook' notion of what constitutes the 'natural path' of development. This agenda (particularly in situations of endemic corruption as in India or Angola) may actually aggravate and exacerbate the problem, depending on the distribution of political and class power in each context.

In the final analysis, key assumptions remain off-limits for the IFIs, protected domains on to which new concerns about social capital or empowerment can be grafted, leaving the underlying premises relatively unmoved or unchanged. The Bank is thus constructed as dynamic and progressive when in reality the core foundations of its ideology remain unmoved. As Pincus (2001) puts it, 'change is easier to effect in rural development through the virtual world of development rhetoric rather than in the concrete realities of lending operations'. Thus the Bank and many other global development institutions continue to operate in this 'virtual world of development rhetoric'. Social and economic phenomena ranging from savings behaviour to fertility patterns are explained through these limited economic lenses but social and economic issues have never been and will never be central to the technocrats who staff these institutions and determine their changing priorities. The resignations of Ravi Kanbur (author of the 2001 WDR) and Joseph Stiglitz (Chief Economist of the World Bank) also suggest that there has not always been 'consensus' within the Bank, but often dissonance and uncertainty, characterised by a paranoia about the Bank's public image, its priorities and dreams but also the perception that it is not responsive, that it is 'out of touch' or that it is top-down and bureaucratic. All of these perceptions seem increasingly appropriate. These institutions are not, however, simplistic condensates but varied and complex, and therefore it is necessary to know more about how they operate in each place and locality, to examine the extent to which the Bank's neoliberalisation of the spaces of development is an incomplete and unfinished project, contested in each place.

CONCLUSIONS: GOVERNANCE, GEOPOLITICS AND DEVELOPMENT: TOWARDS A NEW KIND OF MULTILATERALISM?

> Do not the poor of the world look at you and say 'well, here are those guys lecturing us about free markets and telling us of the values of free trade but what can they do to prevent the United States raising trade barriers and pouring untold sums of money into subsidies for American farmers?'
>
> (Jim Cousins (MP) to Horst Koehler (IMF), quoted in BWP, August 2002: 7)

In 1974, the General Assembly, reeling in the wake of world oil price rises in the previous year, passed a formal resolution which called for the restructuring of the world economy and the creation of a New International Economic Order (what became known as the NIEO). The NIEO called for more aid, expanded trade and (of course) economic growth. By the early 1980s it had become clear that this UN 'initiative' (like many of its predecessors) had largely failed but the pace of internationalisation and globalisation in the world economy was accelerating, with major political and economic implications for the South. In some ways the framework and vision of the NIEO confirmed the South's dependence on the North rather than combating it (Rist, 1997). Debates about modernisation were central to calls for an NIEO and in a way they reinscribed modernisation geographies by calling for aid flows and an expansion of trade diffusing from North to South. In a related sense, contemporary discourses on globalisation may be seen as a reinvention of the modernisation discourses (Smith, 1997). Neoliberalism adds a new twist to the age-old liberal faiths and theologies of free trade of goods and services – an unflinching, dogmatic belief in the rights of private capital. It is a system with a western (and particularly a US) worldview which puts economic values and goals first and social goals a distant second.

Former World Bank Chief Economist Joseph Stiglitz (2002) argued recently that the Bank's developing country assistance strategies are generally constructed around limited country-specific inspections (usually of five-star hotels) and pointed out that each minister is handed exactly the same four-step strategy, irrespective of context or country:

> *1.* Privatisation (aka 'bribarisation'). *2.* Capital market liberalisation. *3.* Market-based pricing (countries may experience some 'social unrest' as the Bank puts it or riots to you and I). *4.* Poverty Reduction Strategy Papers (PRSPs).

The PRS papers (as we shall see in Chapter 8) that were developed for many countries have simply not worked and are highly anti-democratic in their construction. Every time their reforms have failed the IFIs have often simply demanded *more* free-market policies (Palast, 2001) or have sought to blame the poor for their lack of 'entrepreneurial spirit' and market orientation (De Soto, 2000). The IMF, the WTO and the World Bank are the institutions which manage and enforce the laws of neoliberalism. The result is that national governments, both North and South, have become subordinate to the dictates of supra-national organisations with apparently limitless power. These agencies can define what they think constitutes 'competent' governance for many states. Globalisation, it has been suggested, is 'a new name for the Empire of Northern capital' (Appadurai, 1999: 229) or perhaps more specifically, the 'Empire of the Dollar' (Rowbotham, 2000). It is also necessary to think about the interpretation of local and distant influences, to understand further what Amin (1997b: 129) has called the 'out there–in here connectivity' and the simultaneous processes of globalisation and localisation. Places have continued to be important in the global era in that the construction of local resistance involves people in particular places declaring their opposition to national and global elites. Globalisation is unequally and differently experienced by people, but global and local are so thoroughly intertwined that in some ways 'speaking of "local" resistance to the "global" is an overly simplistic representation' (Kelly, 2000: 158).

Figure 7.8 Globalisation – everyone for themselves
Source: CWS and Bado

We can argue that deterritorialisation is not a 'new' process and that it cannot be divorced from wider considerations about the complex interconnections of geography, power and identity (Ó Tuathail, 1999). Postcolonialism is also important here in allowing us to re-theorise identity and to adopt a concern for the grounded, actually existing nature of these processes and changes and their effects. The unstoppable juggernaut of globalisation is often misrepresented as something which is new to previously discrete national spaces, but we have seen that such representations are allied closely to the project of cementing an ideological discourse around the alleged necessity of free-moving capital. For all the talk of worldwide communications and a 'global village' there is a sense in which new forms of apartheid are emerging:

> While transformations in markets and telecommunications are creating a global village, this village is characterised by a functional global apartheid that separates and segregates certain affluent and wired neighbourhoods from other deprived and disconnected zones and neighbourhoods.
>
> (Ó Tuathail, 1999: 143)

TABLE 7.1 THE BURDEN OF CONDITIONALITY (*STRICTLY* DEFINED)

Region	*Total conditionalities (average)*	*Of which governance-related conditionalities*	*As a percentage of total conditionalities*
Africa	23	9	39
Asia	17	4	24
Central Asia and Europe	36	24	67
Latin America	33	23	39

TABLE 7.2 THE BURDEN OF CONDITIONALITY (*LOOSELY* DEFINED)

Region	*Total conditionalities (average)*	*Of which governance-related conditionalities*	*As a percentage of total conditionalities*
Africa	114	82	72
Asia	84	49	58
Central Asia and Europe	93	55	59
Latin America	78	41	53

***Note*: Data based on IMF Letters of Intent and Policy Framework Papers (PFPs) between 1997 and 1999 for the sample of twenty-three countries that had a programme with the IMF in 1999: Africa: Djibouti, Gambia, Ghana, Guinea, Madagascar, Mali, Mozambique, Rwanda, Senegal, Uganda, Tanzania, Zambia; Asia: Cambodia, Indonesia, Republic of Korea, Thailand; Central Asia and Eastern Europe: Kazakhstan, Kyrgyzstan, Latvia, Romania; Latin America: Bolivia, Brazil, Nicaragua.**

***Source*: Adapted from Santiso (2002: 22).**

TABLE 7.3 EXAMPLES OF THE BURDEN OF CONDITIONALITY (*STRICTLY* DEFINED)

Region	*Countries*	*Total*	*Governance-related conditionalities*	*As a percentage*
Africa	Mali	26	13	50
	Mozambique	22	12	55
	Senegal	27	9	33
	Zambia	18	6	33
Asia	Cambodia	30	9	30
	Indonesia	18	8	44
	Rep. of Korea	10	4	40
	Kazakhstan	27	17	63
Eastern Europe	Albania	43	33	77
	Latvia	28	20	71
	Romania	43	25	58
Latin America	Bolivia	32	21	66
	Brazil	38	21	55
	Nicaragua	29	18	62

***Note*: Data based on IMF Letters of Intent and Policy Framework Papers (PEPs) between 1997 and 1999.**

***Source*: Adapted from Santiso (2002: 22).**

This seems to strike at the core of the issues under discussion in this chapter. Despite all the talk of development in a borderless world, one very particular doctrine continues its hegemony, while spaces of exclusion and disconnection have not magically disappeared. Thus globalisation may be accompanied by a process of re-nationalisation and reterritorialisation. Globalisation is not about

TABLE 7.4 EXAMPLES OF THE BURDEN OF CONDITIONALITY (*NARROWLY DEFINED*)

Region	*Countries*	*Total*	*Governance-related conditionalities*	*As a percentage*
Africa	Mali	105	67	64
	Mozambique	74	58	78
	Senegal	165	99	60
	Zambia	87	59	68
Asia	Cambodia	83	65	78
	Indonesia	81	48	59
	Rep. of Korea	114	44	39
	Kazakhstan	114	69	61
Eastern Europe	Albania	72	47	65
	Latvia	65	28	43
	Romania	82	34	41
Latin America	Bolivia	95	44	46
	Brazil	89	45	51
	Nicaragua	50	34	68

eliminating nations but is rather a process that is 'complicating the construction of collective identities' (Schuurman, 2001: 12). A very select group of transnational capitalists and corporations benefit from these removals of borders and national restrictions to international flows which are themselves promoted by a particular cult of doctrinaires and dogmatists. This chapter has focused on the emergence of dominant neoliberalising agencies since the role of these agencies in macroeconomic governance and in the setting of political agendas must be central themes in a reworked geography of development. What we are seeing is the 'hollowing out' of states from above by the increasing significance of international political and economic organisations which interfere in the conceptions of development formulated by particular states. The multidimensional spatiality of boundaries and territories produced by states (both rich and poor) does not simply disappear in the face of the 'global', swept aside by an all-powerful 'global capitalism' more generally.

Much has been made of the existence of a post-Washington consensus since the catastrophic 1997 economic meltdown in Asia and the subsequent crises in Russia, Brazil and parts of Latin America. World Bank economist William Easterly has alleged that aid financing, including US$1 trillion (£685 million) in World Bank and IMF loans, have 'failed to achieve their targets of poverty alleviation' (Easterly, 2001). The 'knowledge bank' responded by beginning disciplinary proceedings against him for not respecting media clearance procedures where all interviews/articles are pre-screened by the Bank (BWR, October/November 2001: 7). In this way the Bank has its own 'thought police' who seek to intercept those people and documents that are not 'on message'. There is none the less now a growing popular movement to redesign the Bretton Woods architecture 'from the ground up'. And that means more than just tinkering with the wiring (Monbiot, 2001). Instead, we need a radical rethink that will put humans in control, at the heart of the global economy rather than at its periphery (Ellwood, 2000). According to an editorial in the *New Internationalist* magazine in August 2001, a common sense of frustration is allied to a desire to plan for an alternative world:

> [W]e keep waiting patiently for a miracle of progress which never seems to appear. The truth is the global economic system has broken down. Tired plans to dust it off and prop it up aren't going to do the trick.

The World Bank has sought to respond to some of these criticisms and launched a briefing in April 2000 entitled *Assessing Globalisation*, which examined the implications of international trade for poverty, inequality and the environment (World Bank, 2000a). The document set out to challenge a recent declaration of the International Forum on Globalization (IFG) that had argued that rather than leading to economic benefits for all, economic globalisation has brought the planet to the brink of environmental catastrophe, unprecedented social unrest, and an increase in poverty, hunger, landlessness, migration and social dislocation. This is a view shared by the Third World Network (TWN), which has argued that the economies of most countries are in 'shambles' and that these neoliberal experiments 'may now be called a failure' (Third World Network, 2001). The Bank document concludes that the poorest least developed countries are not being impoverished by globalisation but rather are in danger of being excluded from it. The simplistic logic of neoliberalism has it that just being a part of export markets will automatically lead to better technology. Yet even the Bank's own research has found that there is no clear association between growth and inequality.

The World Bank actually began as an institution for postwar reconstruction and development. Since 11th September 11 2001 the US has abandoned its 'isolationist approach' in favour of 'the pleasure of multilateralism' (*Time Magazine*, 15 October 2001). US subscriptions to the UN have (finally) been paid, increased aid finance is promised and international cooperation has become a new challenge for 21st-century America. Development assistance is thus once again being explicitly linked to post-Cold War geopolitical concerns, with 'help' dependent on ideology and political affiliation to this or that particular coalition. The US government readily removed aid sanctions on Pakistan, for example, and helped facilitate favourable debt treatment and speedy new IMF financing for the country in return for its co-operation with the US-led coalition. The 50 Years is Enough! Campaign was quick to comment on this problematic practice of using the IFIs as instruments of the US (geo)-political agenda. Andrew Rogerson, a Bank representative in Brussels, denied this, claiming that 'the Bank is not facing pressure from member countries to take decisions based on geopolitics' (Rogerson, citied in BWP, October/November 2001: 1). This and other major global development institutions have however, as we shall in Chapters 8 and 9, always and only ever made decisions based on geopolitics. This is an amazing contention for the Bank to make when we consider that a key aspect of the globalisation project has been 'the reconfiguration of nation-states into neoliberal states in the context of the Cold War and its aftermath' (Berger, 2001: 1079). The political 'theory' that informs the vision of the perfect neoliberalising state gives the appearance of being a universal, yet this has been resolutely and stubbornly Eurocentric (Lummis, 2002). For some commentators, the explicitly 'western-centric' character of contemporary political theory in development thinking is now 'too obvious to require a lengthy demonstration' (Lummis, 2002: 64).

As with all bodies of (development) knowledge, we need to view such models and theories of the state as inextricably 'rooted' in particular locations (e.g. Washington) and particular histories, cultures and traditions. This model of good governance and adequate statehood is thus universalising but not universal. It is none the less predominant and increasingly influential. To suggest that the IFIs operate in a parallel universe, outside the messy and conflictual realms of the political, is quite frankly ludicrous and profoundly contradictory. Central to the IFIs' dominant and highly influential vision is the idea of an:

> inexorable march towards global democracy and universal free market prosperity which defines democracy in minimalist terms (elections, universal suffrage and relative press freedom) and tends to downplay or ignore the connection between unequal capitalist development, social inequality and political instability.
>
> (Berger, 2001: 1079)

The models of neoliberal 'success stories' favoured by these organisations – countries such as Thailand, South Korea and (until recently) Argentina – were heavily based around state-centred notions of how to create development, yet still an approach is advocated where the state is heavily rolled back before the market. As we have seen in earlier chapters, the dominant market-oriented liberal conception of national citizenship has changed in many countries (e.g. India and Mexico) which had experienced long periods of state-guided development histories based around single-party politics. For states such as Angola or Colombia, states in deep crisis, the 'good governance' agenda may not have much to contribute and may even accelerate the possibility and likelihood of political and economic disintegration.

Figure 7.9 Cartoon of financial crisis
Source: CWS and Wonsoo

Figure 7.10 Cartoon of Asian markets
Source: CWS and Gable

'Good governance' is a term that is typical of the 'plastic words' of the international aid and development community (Sogge, 2002: 131), a popular catch-phrase that seems morally beyond reproach while appearing clear and positive.

These plastic buzz-words and languages can be deconstructed, as we shall see in Chapter 8. What is most interesting is that pathologies which seemed unique to Africa and other 'backward' regions according to IFI discourses are now 'starting to appear with increasing frequency in the nation-states of Northern America and Europe' (Berger, 2001: 1083). In this sense the emerging global patchwork of resistance to neoliberalism, led by the WSF in Porto Alegre, may offer a way forward. The 2001 summit, for example, issued a call for mobilisation which urged a 'strengthening of alliances', a set of common actions seeking to reject the IFIs' discourses and building international mobilisation against NAFTA in particular (WSF, 2001: 122). The WSF has argued that globalisation is reinforcing a sexist and patriarchal system in which privatisation reigns and indigenous peoples and knowledge are not valued. Neoliberal globalisation, they argued, increases racism and does little about the debt crisis. The challenge for development geographies is to understand the spatial processes whereby exclusions and emancipations result from globalisation and global social, economic and political change. This may well involve challenging and confronting the normative and discursive nature of development itself.

BOX 7.5

Chapter-related websites

http://www.gatt.org/
GATT site.

http://www.wto.org/
WTO site.

http://www.theyesmen.org/
Includes spoof WTO site.

http://www.50years.org/ejn/
50 Years is Enough! Campaign.

http://www.econjustice.net/
World Banks Bonds Boycott.

http://www.weforum.org/ http://www.justact.org/
Youth Action for Global Justice.

http://www.portoalegre2002.org/
2002 World Social Forum.

http://www.id21.org/society/
Development research gateway.

http://www.focusweb.org/
Focus on the Global South.

http://www.warwick.ac.uk/csgr/
Globalisation and Regionalisation Study Centre.

http://www.weforum.org/
World Economic Forum.

http://www.anotherworldispossible.com/
Another world is possible.

http://www-heva.wto.ministerial.org
WTO ministerial documentation

http://www.worldbank.org/extdr/pb/globalization/
World Bank.

http://www.twnside.org.sg/title/siena-cn.htm/
Third World Network.

http://www.studentsforglobaljustice.org/
Students for Global Justice.

http://www.accnyc.org/
Anti-capitalist convergence against the WEF.

http://www.ipanet.net/http://www.miga.org/
MIGA.

http://www.tradejusticemovement.org.uk/
Trade for Justice Movement.

http://www.foeeurope.org/
Seattle to Brussels network.

http://www.privatizationlink.com/
Privatisation link.

http://www.iaigc.org
IAIGC.

http://www.ontheline.org/
On-the-line Productions.

http://www.vshiva.net/
Vandana Shiva.

http://www.zmag.org/
Zmag.

http://www.mozbusiness.gov.mz/
Mozambique Investment.

8
The Dissemination of Development

INTRODUCTION: DEVELOPMENT AS A 'GLOBAL MORAL IMPERATIVE'?

> I believe development as discourse has prevented us from seeing other very creative ways of addressing the poverty problem.
>
> (Yapa, 2002: 41)

> You are poor because you look at what you do not have. See what you possess, see what you are, and you will discover you are astonishingly rich.
>
> (African proverb, quoted in Ndione, 1994: 37)

If neoliberalism has become the centre-piece of development theology, then how do its preachers get their messages across and how can these theologies be deconstructed and resisted? This book has argued from the outset that development is fundamentally about ideology and the production and transmission of policy and discourses. As we saw in Chapter 7, there is also an important process at work whereby alternatives are often filtered out and in a way delegitimised by the major global development players. Standard approaches to development tend to focus only on the financial and material cargoes (e.g. in aid) that flow to peoples and places (Sogge, 2002), but what of the important flows of ideas that shape and structure our attentions towards the South or define our sense of its 'problems' and assign roles to 'us' or 'them' in these dramas? This chapter continues the concern with the diffusion of development ideologies in a 'globalising world' and suggests that one important way of viewing 'development' is to focus on the way in which its theories, strategies and ideologies are *disseminated* (through language as well as in terms of the material manifestations of development). The principal question here concerns the formulation of the rather pious 'motherhood and apple-pie image of development' (Munck, 1999: 200). It is therefore important to examine the ways in which discourses represent and place development as an 'uncontested human good' (Munck, 1999: 200). More importantly, as Yapa (2002) contends, some of these discourses of development have prevented us from seeing other, more creative ways of addressing the 'problem' of poverty and the question of how to eradicate it.

In particular the relationship between knowledge and power in development is also crucial here in that it is necessary to explore the constitution of 'socially important' knowledge of development. Further, it is useful to examine how power relations shape the definition of scenarios and situations, set the terms in which issues are discussed and understood and lead to the formulation of development goals, values and objectives in particular spaces and places. Particular forms of power then produce certain realities or, in Escobar's terms, certain 'domains of objects' and 'rituals of truth'. In important ways a discourse of development can determine how a country, a people and their past are represented. Discourses around the disciplining of weak, peripheral states in terms of the need for 'good governance' are a good example of this. Through debates about political democratisation and development, the good governance agenda comes to dominate the realm of what is said about 'developing countries', foreclosing other ways of thinking about politics and development. What we need to do here is to ask: Are there other, more important ways of addressing poverty and income inequality? If we have reached a situation of 'enter economism, exit politics' (Teivainen, 2002), how is it possible to go beyond it? It has become increasingly clear that neoliberalism dominates development thinking and practice, but the way in which this particular religion spreads its messages of progression and forwardness is far from straightforward.

The central objective of this chapter is to focus attention on the relationships between different groups in development and how they communicate knowledge and ideas between different cultural and economic spaces and across different spatial scales. In focusing on the power and knowledge of development it is necessary to also pay attention to the materiality of power relations, uneven development and income inequality in concrete and specific ways. As we have seen, stereotypical representations of 'Third World others' obscure the structural forces which produce these very geographies of inequality and unevenness. Actual political and economic processes that are producing different material geographies of inequality are masked and hidden behind the representations of agencies such as the World Bank which ignore the material production of poverty in countries where it has lending operations. Thus we also need to understand the *spatiality of power and discourse in development*: who is allowed to speak in certain kinds of discourses, where does this take place and with what implications for power relations? There are important social relationships between professionals, institutions and communities that we need to understand more fully. In this sense the dissemination of development is a two-way process which is never fixed in time or space but is rather fluid and dynamic, constituting peoples and places in a variety of contexts and settings. The spaces in between discourses, institutions and subjects in development are also important here, allowing us to pose further questions about the possibility of resistance to neoliberal discourses. There are also important variations here between the urban and rural spaces of development and between different gender and class groupings.

One theme of this chapter concerns the way in which the image of 'developing countries' is disseminated in western contexts through media coverage of humanitarian crises, which is particularly important in the context of debates about globalisation (see Box 8.1). Media representations therefore enframe the countries of the 'Third World' in multiple ways which often impact, through such processes as the 'CNN effect', on the nature and timing of humanitarian intervention. Why did none of the countries listed in Table 8.1 ever receive anything like the kind of western media attention or the level of humanitarian assistance that had been on offer to Kosovo in the same year? (Oxfam, 1999). It has been estimated that for every African life the world was willing to pay approximately US$10 in humanitarian aid, whereas for a Kosovan the figure stood at some US$600 (excluding peace-keeping and reconstruction costs). This theme of incomplete coverage was also explored at a gathering of journalists from both specialist African and mainstream British publications in 2001, brought together to discuss issues of representation and responsibility in covering stories about Africa and African peoples. The editor of *New African* magazine (Baffour Ankomah) bemoaned the 'heart of darkness' rhetoric which recurred constantly in western reporting, perpetuating notions 'from the ages of slavery and colonialism about why things happen in Africa' (Reporting the World, 2002: 1). Sensational pictures led the way with western audiences according to another journalist, with a consequent lack of understanding of the complexity involved in these issues. Racism often predominated in representations of the Rwandan genocide in 1994, for example, characterising this as a spontaneous uprising of tribal hatred rather than an organised political campaign. During the Indonesian intervention in East Timor in the mid-1990s, many reports had simply bought into a discourse of 'disintegration' and 'instability' used by President Suharto to justify his authoritarian policies (Reporting the World, 2002: 1). Accounts of violence in Indonesia were often presented through a range of 'fire' metaphors, as riots 'erupted' and tense situations 'ignited', masking the real roots of violence and their relationship to the regime in power. Journalists then talked of stretching the agenda 'sideways', making extensions into wider issues of political change and development. Ron McCullagh, Director of *Insight News*, called for journalists to devise new ways to make connections between the fate of African people and our own lives in the West (Reporting the World, 2002). One key question posed in the conclusions of the meeting is particularly relevant:

> What is 'our' role in this story?
> Is the message that these people will not be OK until our (benign) intervention, now in prospect? Or does the [media] report suggest that they would be OK but for our (malign) intervention? Reports about the continuing IMF dealings with

BOX 8.1

Reporting the World: from poverty to wildlife?

The organisation 3WE (2000) monitored 'international factual programming' in the UK and surveyed UK television between 1990 and 1999 and found that there had been a substantial decline in the volume of non-news/current affairs filmed outside of the British Isles. At the start of the decade, the largest category of 'developing country' factual program-ming concerned human rights, development and environmental issues at around 30 per cent of overall output. Programmes about religions, cultures and the arts in these countries were also significant at the start of the 1990s, representing about 20 per cent of overall output. In 1998 to 1999 these categories had been 'replaced by Travel and Wildlife programming which do not offer complete portraits of the developing world'. Some 60 per cent of programming concerned with 'developing countries' focused on travel (20 per cent) and wildlife (38 per cent) issues. Why, then, given the complexity of development issues we have been exploring, do programmers neglect key issues of poverty and inequality and their causes in favour of popular shows about wildlife which offer an 'incomplete portrait'? UK former Development Secretary Clare Short blamed the British media for 'dumbing down' international development issues in the mind of the British public, a challenge backed by experienced broadcaster Jonathan Dimbleby:

> Thus the so-called 'developing' world, where development is in many respects standing still or drifting backwards, has become a backdrop for cookery programmes and adventure holidays, or occasional 'exposés' in that voguish form of tele-reportage that titillates the viewer with tales of violent crime, drug busts or child prostitution.
>
> (*Guardian*, 1 June 1988: 9)

All the major UK channels reduced their programming on these issues considerably, yet the 3WE (2000) study found that the gap between the public service channels and new commercial channels had widened, with even fewer programmes devoted to development or environmental issues. According to Voluntary Services Overseas (VSO), some 80 per cent of the British public associates the developing world with 'doom-laden images of famine, disaster and Western aid' (VSO, 2001: 3). The VSO survey found that sixteen years on from Live Aid these images are uppermost in the popular British imagination of Africa and 'maintain a powerful grip on the British psyche' (VSO, 2001: 3). Similarly, research by the broadcasting trust 3WE (funded by Oxfam, Christian Aid, Comic Relief and other development charities) examined media representations of 'Third World' countries in Britain in 2001 and found that any serious examination of development had been abandoned in favour of 'reality' TV, holiday challenges and docusoaps (Vidal, 2002). The space for covering the lives, cultures and politics of the rest of the world had been closed off, with few programmes looking at society, the environment or development in the South. Earlier research by 3WE also revealed that between 1989 and 1999 peak-time factual programming about 'developing countries' fell by half on British channels 1–4. Consumer-oriented travel programmes increased to one-third of all foreign factual programming in 2000–2001 and reality programmes such as *Shipwrecked*, *Survivor* and *Temptation Island* comprised more than 10 per cent of all factual international programming on commercial channels in Britain. These shows were also popular in the United States, where the question of how the popular media shapes understandings of other cultures and economies is an important and very relevant one.

> Indonesia often include the phrase 'further assistance' [which should be] further interrogated – assistance for whom? What will the effect be of the particular IMF remedies being advanced as conditions for the 'assistance'?
>
> (Reporting the World, 2002: 3–4)

What is 'our' role in the story of global development and why are these 'remedies' not more widely explored and criticised by the media? What kinds of intervention are 'OK' and when? Words like poverty are disseminated partly through the mythology of development and its pharmacopoeia of remedies for poor countries. At the same time there are important material causes and consequences of poverty, rooted in particular places and spatial relations in a variety of societies that we also need to bring out here. The concrete and specific ways in which poverty becomes a material reality for billions of people around the world is precisely what is overlooked in so many neoliberal scriptings of international development. Through the variety of

TABLE 8.1 AFRICA'S FORGOTTEN CRISES

	Angola	*Ethiopia*	*Eritrea*	*Sierra Leone*	*Democratic Republic of Congo*	*Total*
Number at risk	2,000,000	5,000,000	Up to 500,000	1,500,000	500,000 to 1,500,000	10,000,000 approx.
Number displaced	1,70,000	315,000			750,000	4,625,000
Death rates per 1000 for under 5s	292	175	116	316	207	
Amount requested by humanitarian aid appeals in 1999	US$106 million	US$28 million	US$7.2 million requested by Eritrean government	US$25 million	US$38.8 million	US$205 million
Amount pledged by donors	US$57 million	US$26.6 million	n/a	US$9.5 million	US$4.3 million	US$97.4 million
Notes	3 million people were living in inaccessible, rebel-held areas. No data			70% of country accessible to humanitarian aid	Aid agencies restricted to major towns. No data	

***Source*: Adapted from Oxfam (1999).**

global and national development institutions that we have looked at so far it can be seen that each adopts its own myths of 'positive change' and what is required to bring this about, all of which has important implications for the way in which the poor are represented and development becomes a material reality.

This book has focused on the need to contest the *misrepresentation of poverty* and poor people, to challenge some of the myths that surround development issues and to understand how international development works both in theory and practice. In order to do this, we need to understand more fully how development is diffused and disseminated, how it is invented and popularised in different contexts and how it is imagined socially and spatially. Nederveen Pieterse argues that efforts to eradicate poverty (as with the MDGs) are often not considered by post-development writers and somehow 'slip off the map' of their critique, but, as we shall see, this has not always been the case. One way of responding to this critique (or at least anticipating it) is to focus on keywords such as livelihoods, social capital and social exclusion, and to examine the extent to which the conceptualisation of these ideas involves a 'North' to 'South' flow or diffusion. In the language of modernisation geographies, how much of a 'return-cargo' of ideas is there here?

It is interesting that the practice of poverty eradication strategies disappears from view in some accounts, in that the practice of development provides some of the clearest examples of this spreading of ideas and knowledge with material consequences. If we look at UNICEF's report the *Progress of Nations* (UNICEF, 2000) it is possible to examine some of the ways in which development agencies build a kind of branded image for themselves and justify interventions around this. UNICEF's concern with children and poverty is not in dispute here; we are simply seeking to interrogate further the ways in which such organisations construct development as a kind of 'global moral imperative'. In this way UNICEF set out to reclaim the 'lost children' of the world:

> We know where to find the lost children. They are in the tents and barracks of Africa . . . in the brothels of Asia, the slums of Europe and North

Figure 8.1 Spyglasses for security
Source: UN Department of Public Information

> America, the sweatshops of Latin America. Seeing their faces, even if only for a fleeting moment, how can we allow ourselves to forget them? Let us extend the gains now enjoyed by so many other children to this last, most isolated group. Let us be the ones who stand firm until all children lost in such dangerous obscurity emerge into a brighter future.
>
> (UNICEF, 2000: 11)

Children around the world, we are told, must emerge or be liberated from this 'dangerous obscurity' (since 'how can we forget their faces'?). The lost must be found and this should happen through donations to a UN development agency, UNICEF. Through these and other reports UNICEF's particular approach to development is disseminated, creating a vision and a discourse around itself and the grounds for intervening wherever this is deemed to be necessary, whether it be a brothel, a slum or a sweatshop. The image of this world of 'lost children' becomes clearer as they are seen to live in situations of danger, poverty and exploitation. These are the most isolated children in the world but they also live in Europe and North America, hence these are not simple geographies. Let us therefore 'extend the gains', the report continues; if only we all stand firm, these children can be rescued. The strategies of intervention adopted by UNICEF in these different contexts are quite varied however, but it none the less tries to create a kind of 'universal awareness' for its mission, through a certain kind a discourse about 'developing regions'.

When UNICEF talks about a 'global moral imperative' it is referring specifically to the emergency, relief and humanitarian nature of international development agendas, but the implications extend much further than these discourses. What is interesting here, however, is that where once it was seen as universalisable, the meaning of 'development' has often been narrowed in recent years to focus mainly on humanitarian relief for groups suffering the stigma of exclusion. Perhaps this is expecting too much of an organisation such as UNICEF, namely that it should open up its vision of development and consider the *causes* of the inequalities they importantly flag up. This narrowing of the meaning of development around humanitarian and emergency discourses (though far from being unique to one agency) raises all kinds of questions about the legacies of the past. It is also surely one of the 'gravest signs of the crisis of development' (Rist, 1997: 241).

If such importance is attached to global poverty reduction programmes by organisations such as USAID, CIDA, DFID, OECD or the panoply of UN agencies, surely we must be clear as to what we mean by poverty to begin with (White, 1999)? The development studies literature stresses the multi-dimensionality of poverty (considering this not just as material poverty but also as a lack of access/entitlements, political freedoms, health status or environmental quality, for example). Differences between absolute and relative poverty and temporary and permanent conditions of poverty are also important here (White, 1999). Are some kinds of poverty (e.g. income poverty and lack of material well-being) more immediate than others?

The term 'poverty' is more often than not employed in a wholly relative sense to refer to the poorest. Poverty is now increasingly seen, however,

Figure 8.2 Cartoon of UN poverty meeting
Source: CWS and Phore

as a phenomenon of many layers, determined by culture and power relations that range from the household to the nation. Some recognition has also been made that seasonal changes and short-term shocks affect people's movements into and out of poverty. Political violence can be a key part of this process for many people in the South, further emphasising the need to 'look more closely at the social and political milieu that engender poverty' (Grinspun, 2001: 7). Again this means paying attention to the specific local, national and global processes at work in the places and spaces of development. It also means bringing out the very real differences within and between multilateral and bilateral development agencies and the cultural and political gulfs between donors and recipients. Poverty is seen quite commonly as the 'antithesis of development' (O'Connor, 2001: 37), yet these are not always exact opposites since complex processes are involved. The term is seen as 'generally meaning a lack or deficiency' and focuses in most international development institutions on 'material poverty in respect of money, goods and services' (O'Connor, 2001: 37). The Bank claims to have had a new-found recognition of political power in recent years, but what role (if any) does it really have in eradicating poverty and in 'empowering' people, and to what extent is it prepared to take on in-country vested interests?

In human geography, many debates about poverty and its definition are much more concerned with poverty in the UK or the USA than in other countries of the world (O'Connor, 2001). Poverty is also often not considered in global terms by many geographers. There has, however, been a proliferation of poverty-related educational courses in geography and other disciplines:

> Global poverty is certainly not a fashionable subject for academic study, either in the countries most concerned or in the richer parts of the world. There is nothing to match the proliferation of institutes, degree programmes and journals of 'development studies' which have sprung up in the past half century (exemplified by the publication of the Development Studies Reader in 2001).
> (O'Connor, 2001: 38)

Poverty exists partly because, as Grinspun (2001: 3) has put it, its persistence is 'morally unconscionable', since the world has 'the resources and know-how to vanquish it'. Development therefore arises partly as a consequence of this 'moral unconscionable', of this desire to 'vanquish' poverty (through technology, resources and know-how). Organisations such as the UNDP now construct a variety of qualitative assessments of poverty, using household surveys and constructing 'poverty maps' to highlight disparities within national borders (Grinspun, 2001). Fighting poverty, as we have seen, is a highly political act therefore, involving 'underpinning assumptions' and a 'conditioning of mindsets' (Grinspun, 2001). How does this conditioning take place and how are these 'mindsets' and assumptions diffused and received? In addition to this focus on discourses, knowledge and power we can usefully draw upon deconstruction as a kind of methodological stance which allows us to understand the 'truth claims' that development texts make. Deconstruction is a method which interrogates the underlying and unarticulated presuppositions of a text and seeks to challenge a variety of western binaries such as First World/Third World, by finding traces of one element or one world within the other, thereby undermining dominant imagined geographies.

Many geographers are sceptical of such an approach, yet discourses on 'sustainable development' or 'people-centred' development or even 'green development' are clearly ripe for this kind of method of critique (Watts, 2000). There is a wealth of qualitative research documenting the variety of subject positions that are lumped together in the category 'poor'. The languages and buzz-words of development are also important here since they can mask continuity from colonial times in the representation of groups in ways that justify their access to resources and the exclusion of others from sharing in these resources. A key contention of this chapter then is that neoliberal development pathways are still rooted in colonial notions of the 'deserving' and 'non-deserving' poor and in a notion of poor countries as subservient to the more enlightened, all-knowing 'developed world' and its humanitarian concerns for its 'trustees'. It is argued here that the latest catchwords and languages of international development are producing new spaces of exclusion, particularly in the context of debates about globalisation, neoliberalism and development.

OBSCURING THE *CAUSES* OF POVERTY: THE POVERTY PROCESS ACRONYMS

> [F]ighting poverty is highly political. It entails a number of assumptions and perceptions that unconsciously condition mind-sets at every level of governance, from the household to the state. By doing so, poverty develops a web of relationships that become self-perpetuating well beyond the ambit of the poor themselves. For this reason; poverty cannot be dissociated from governance; they are two sides of the same coin.
>
> (Grinspun, 2001: 16)

One way of examining the flow of ideas between North and South concerns the bewildering array of poverty process acronyms used by the major multilateral and bilateral donors (see Table 8.2). The most famous and perhaps most ill-advised of these came with the reference to adjustment programmes or SAPs, but this was taken to a new dimension by the IFIs in the 1990s with the introduction of the Poverty Reduction and Growth Facility (PGRF) which replaced the much maligned previous system used by the Bank entitled 'Enhanced Structural Adjustment Facility'. In 1985 the external debts of African countries stood at US$95 billion, and at the end of 2001 that figure had reached a massive US$208 billion (World Bank, 2001a). In 1999 the World Bank and IMF decided to require countries to prepare a Poverty Reduction Strategy Paper (PRSP) as a condition for qualifying for concessional assistance and debt relief under the Heavily Indebted Poor Countries (HIPC) initiative. This initiative is unlikely to bring real benefit unless these reduction strategies can actually achieve strong and sustained economic growth (Jubilee Debt campaign, 2002). In effect, the Bank rebranded its ideology as a way of deflecting criticism and resisting change to the fundamentals of its ideology. This was reminiscent of the way in which the Bank 'officially' signalled the end of the long-standing notion of development as 'nationally managed economic growth' with the publication of the *World Development Report* of the early 1980s which focused primarily on the world market and began to measure 'participation in development' in these terms (Munck, 1999). At this point the Bank wrote out of the script any question of development fundamentally being about income *re*distribution. In the words of the IMF, the 'new' PRSP approach is 'results-oriented', focusing on the outcomes that will 'benefit the poor' (Jubilee Debt Campaign, 2002). It emerged towards the end of the 1990s, a decade in which the Bank shifted towards 'neoliberalism with a human face', and was also the result of the Bank's supposed adoption of a 'Comprehensive Development Framework' (CDF).

The 'comprehensive' approach was intended to involve consultations with 'stakeholders' in communities affected by Bank policies. The PRSP is, it is argued, 'partnership-oriented' and 'country-driven and owned'. This institution requires governments to recognise and involve civil society in the development process, but each context is very specific and particular, and there has not always been an open dialogue between the two in some cases. It is worth remembering, however, that the Bank looks to engender participation mainly as an end in itself, to produce the PRSP and then to legitimate this text before its audience and the international development community. However, what if 'civil society' groups do not wish to terminate their engagement in politics at this point? Some argue that the process of participation in PRSPs represents a kind of 'engineering of consent' for structural adjustment policies (see Box 8.2). In Bolivia the PRSP experience is often held up by international development agencies as a success story, with the country depicted as a mode structural adjustment performer; yet there were some harsh lessons and mistakes made in these processes which are much less frequently mentioned. The language of 'partnership' which emerged from colonial discourses of development is important here.

Once completed, the 'country-owned' PRSPs are assessed by the IFIs before relief is agreed and

BOX 8.2

Civil society, participation and the PRSP process in Uganda

In December 1999 the government of Uganda decided to revise its Poverty Eradication Action Plan (PEAP) and to formulate a PRSP as a precondition of its debt relief under the HIPC system. Civil society organisations (CSOs) were welcomed into the process by the Ministry of Finance, Planning and Economic Development (MFPED), where consultative meetings were held with government and World Bank officials. A civil society task-force was assembled and given the task of opening up the consultative process, compromising several NGOs such as Oxfam UK, Action Aid, VECO Uganda (Belgium), SNV (Holland) and MS Uganda (Denmark). The Ugandan NGOs included organisations such as Action for Development (ACFODE), the Ugandan Women's Network (UWONET) and several research institutions along with World Vision and the Catholic Medical Bureau. The Uganda Debt Network (UDN) became the lead agency for civil society participation in the poverty eradication and reduction strategy formulation. This is a coalition of NGOs (local/international), academics, research and religious organisations and individuals, and was formed in 1996, primarily to campaign for debt relief under the HIPC scheme. The UDN was also the lead organisation in Uganda for the Jubilee 2000 campaign for the total cancellation of the unpayable debt of all poor countries. Many organisations report that this was by and large a positive experience of participation where government officials were in 'close and continuous contact' with some civil society groups, but, as the reports neared completion, the space of civil society to really engage with policy was narrowed somewhat:

> Ugandan CSOs felt left out of the later stages of the process, when they were excluded from the discussions that turned the Uganda PEAP into the PRSP that was presented to the IMF and World Bank Executive Boards . . . there were fewer contacts with donors and more specifically the IMF and World Bank missions in the preparation of the IMF version of the PRSP document.
>
> (Jubilee Debt Campaign, 2002)

Participatory PRSPs are not really taken seriously by the IFIs. For some they appear to hold real promise as a process because they 'open a space for dialogue', according to Zie Gariyo, Co-ordinator of the Uganda Debt Network, between civil society groups and the institutions of the state. This then must be reflected in actual outcomes for popular participation to become more meaningful in countries such as Uganda (Jubilee Debt Campaign, 2002).

creditors dictate the terms of participation in the process while demanding a final seal of approval for the strategy. Their concern for good governance arises directly from colonial discourses. 'Ownership' today is also still a prerequisite of nations rather than individuals (Cooke, 2001). In Mozambique and Mauritania the IMF even said that rapid approval for debt relief would be forthcoming only if the government '*did not* put the PRSP out for public consultation' (Elliott, 2000: 25, emphasis added). A group of twenty-four southern countries published an important research paper on 'Governance-related conditionalities and the IFIs' in 2001 (Santiso, 2001). It questions whether the IFIs actually have the mandate or the competence to justify imposing governance-related conditionalities in the contexts where they operate. It raises concerns that IMF and World Bank country strategy papers are often full of 'faddish ideas', not realistic priorities. The failure to acknowledge a whole variety of external factors, including lack of transparency in the IMF and World Bank, are key barriers to improvement here.

At the UN Millennium summit in September 2000, 'the largest high-level development constituency ever' (Browne, 2001: xii), a variety of buzzwords and phrases were also used by world leaders to talk up their commitment to poverty eradication. Here it was declared that world leaders would 'spare no effort to free our fellow men, women and children from the abject and dehumanising conditions of extreme poverty' (cited in Foreword: iii). Once again there is this notion that children, men and women must be 'freed' (by this particular constituency) from the enslavement of abject and dehumanising conditions. The unjust nature of international trade and economics is thus quickly side-stepped, downplaying the centrality of crucial, structural processes of politics and economics. Many institutions talk about long-term targets and strategies for the 'alleviation' or sometimes 'eradication' of poverty, but how seriously are these commitments taken?

Evaluations of PRSPs are sometimes carried out, but their widening adoption has been criticised in some quarters, in that reduction strategies must always be linked to other macroeconomic trade liberalisation measures and a singular ideology of development. In Angola, for example, the UNDP-sponsored PRSP undercut local initiatives and was institutionally quite fragile, failing to build capacity as is so often promised (González de la Rocha, 2001). It is far from clear that civil societies and 'the poor' have been involved in and empowered by this process. Many of the former socialist countries of Eastern Europe have also had to grapple with this question of defining a threshold for social assistance (in a context of growing income and social disparities during the transition to capitalist, market economies). Because each of these measures is arbitrary in one way or another, they have also been quite controversial. None the less the existence of impoverishment in these contexts illustrates further that development is not simply a 'Third World' phenomenon but is something that organisations such as the OECD are slowly beginning to see as crucial for parts of Europe and the 'developed' world as well.

Focus on the Global South issued a report in 2002 assessing PRSP processes in Lao PDR, Cambodia and Vietnam based on interviews with World Bank officials and NGOs. The report concluded that the strategy papers came into conflict with long-established development plans, directions and debates in these areas. Even basic errors such as the failure to translate the strategies into either Khmer or Lao excluded many civil society groups (BWP, 2001b). There was also no transparent 'route map' to the process in Bangladesh, where the PRSP remained a condition of the receipt of future funding. Even the Bank's own Social Development Department acknowledges that information disclosure and consultation is primarily an urban process, confined to capital cities and particular social groups. Despite all the talk of widespread consultation, these processes are still driven by finance and planning ministries and exclude non-conventional NGOs such as community groups and some women's organisations that do not quite match the Bank's definition of appropriate forms of civil society. Bereft of any kind of gender analysis and often based on questionable data sources, how well informed are these accounts of development by the voices emanating from the places and peoples directly affected by these strategies? Perhaps these are static, 'document-centric' visions which bear little relation to views from below (BWP, 2002a). What about continuing participation *after* the strategy is completed? The macroeconomic situation and the entrenched political and economic processes which produce inequality in the national and international macroeconomy thus disappear from view and the core (neoliberal) assumptions remain unchallenged. The sense formulated here of joint learning and shared assessment across countries is a very shallow and superficial one.

When the UNDP published a study of poverty strategies entitled *Choices for the Poor: Lessons from National Poverty Strategies* (Grinspun, 2001), UNDP administrator Mark Malloch Brown pointed to the 'technical and political complexity of developing comprehensive anti-poverty policies' (Grinspun, 2001: iii). Words like coherence, consensus, choice, capacity and co-ordination seem to abound alongside phrases like sustainable, extreme, success and failure. *Choices for the Poor* follows a further UNDP initiative launched a few years earlier in 1996 to help poor countries develop national and local strategies for poverty reduction. Through this process (and the varied and multiple consequences it has in different contexts) we can see a particular ideology, strategy and theory of development being disseminated. The Poverty Strategies Initiative (PSI) became the UNDP's major means

Figure 8.3 Blowing a raspberry at UN commissions
Source: Cartoon by D. Cagle

of implementing its commitment to poverty eradication (made at the 1995 World Summit for Social Development).

It is useful to briefly consider here the experience of the PSI and PRSP processes in the Middle East, in the Lebanon and Palestine in particular. For much of the Cold War, the USA pursued a strategy of 'containment' in the Middle East, which was allied to wider discourses of danger about the Soviet Union. All manner of 'containment activities' were legitimated as a result of these discourses, involving support for moderate and conservative leaders in states such as Jordan and Lebanon, the overthrow of the Iranian government, support for Israel in the wars of 1967 and 1973, and the widespread distribution of arms through the region to aid struggles against the USSR (Pinto, 1999). When Lebanon's fifteen-year civil war came to an end in 1990 many lives had been lost and severe disruption caused to society by the conflict, damaging the infrastructure and resource base. As Jerve (2001: 304) puts it: '[p]overty and social inequality in Lebanon need to be understood in the context of these post-war traumas.' In some ways it is precisely here that the PSI/PRSP interventions seem to fall so far short, in failing to recognise and accommodate these specific postwar traumas. Violent conflict has continued not just in Lebanon but also in Palestine, where relations with Israel and the international community are a crucial factor in national politics. Unresolved territorial disputes and border closures, along with security problems, have made 'long-term national development planning virtually impossible' in this context (Jerve, 2001: 306). In both Lebanon and Palestine these reports can be engulfed in controversy. Critics of the regime in Palestine used the poverty reports there to press for changes in public spending by what was then the Palestine Authority (the state), but on the other hand the high poverty rates documented in the reports allowed the regime to lobby development donors for continued international assistance.

In Lebanon, a document called *Mapping Living Conditions in Lebanon* (UNDP, 1998) was also prepared in 1997 to 1998, identifying the spatial unevenness of impoverishment. A fractured state, dealing with many sectarian interests in Lebanon, can make policy responses and development planning rather complex. National development in Middle Eastern countries such as Lebanon and Palestine is also complicated by the fact that international agencies see borders and boundaries in ways which serve to naturalise complex ethnic and cultural differences between peoples, despite the intense contestation and struggle that surrounds and defines them. Thus the complex political and cultural struggles that forge particular nations and competing senses of nationhood are not seen as relevant in the all-inclusive, all-encompassing 'catch-all' models of the IFIs. The particular historical and political context of Palestine, for example, does not seem to be well understood by many development organisations. At the Copenhagen Summit for Social Development in 1995, the Minister of Social Affairs of the Palestine Authority felt it necessary to remind delegates that:

> Poverty in Palestine interlinks with factors different from those in other countries, where it is often attributed to structural social and economic imbalances. In Palestine, poverty basically interlinks with Israeli occupation.
>
> (Al Wazir, 1995, quoted in Jerve, 2001: 310)

The UNDP Programme of Assistance to the Palestinian People (PAPP) supported the preparation of a comprehensive situation analysis of poverty in Palestine, which resulted in the publication of the *Palestine Poverty Report* in 1998. An additional *Participatory Poverty Assessment* (planned with UK DFID finance) has also since been prepared and a National Poverty Commission (whose role remains unclear) has been set up. These kinds of reports often waver, however, between two competing and contradictory perspectives, from one based on a notion of poverty as a lack of something (of assets, basic services or income to sustain a minimum standard of living) to another exploring the relationship of a person 'to his or her social and physical environment' (Jerve, 2001: 312). Thus in the Palestinian context it is not clear if the material production of poverty is about a lack of assets or access or if this is also connected to the nature of the socio-cultural and geopolitical environment.

At the time of writing, all Palestinian cities are occupied and remain under Israeli curfew, preventing people's access to basic needs such as food and water, and to schools and hospitals. In addition, major human rights abuses have been committed by the Israeli forces in Palestine with impunity and without a great deal of resistance from the international community. Over the past eight years the Palestinian Authority has been enthusiastically preparing industrial zones to be built on the border with Israel (with the assistance of USAID) in the hope that joint investment between the two countries would cement a lasting peace, yet the Israeli occupation has today destroyed the Palestinian economy (Palestine Solidarity Campaign, 2002). In the late 1990s an emerging state apparatus in Palestine created some jobs, but unemployment remained high and many Palestinian businesses remained dependent on interactions with the Israeli economy. The Israeli economy has also suffered, as defence spending has risen by some US$2 billion a year since the Intifada began. Just to keep the tanks rolling in Palestinian towns costs some US$70 million a month (BBC, 21 June 2002). The US administration has offered to support a new Palestinian state but only under certain 'democratic' conditions established by Washington and without any clear timetable. The point here is that neoliberalism, and the strategies and acronyms used to convey it, seems unable to adapt and accommodate the specifics of certain kinds of context (e.g. associated with conflict and trauma) in particular places and communities.

These examples from the Middle East also highlight some of the ways in which, as Pieterse and Parekh (1995) argue, the region is commonly understood through a series of development-related metaphors in general and by a 'crisis discourse' in particular. Divisions within the Middle East are often papered over by accounts of the region's 'development', but these divisions are significant and have been historically produced (e.g. through the Gulf War or the Cold War) (Aarts, 1999). The end of global conflict, as elsewhere in the world, has thus led to the emergence of differentiated and conflicting national strategies for development which need to be carefully understood. Slowly but surely, from the mid–1980s onwards, governments across the region have begun diluting state-centred development strategies in favour of implementing 'liberalising' measures which coincided with an upsurge in the external debt of Arab countries (which had reached some US$160 billion in 1990) (Aarts, 1999). Arab states have been among the most reluctant to subscribe to the new neoliberal orthodoxies but 'one by one they have begun cautiously to endorse its principles and adopt its tenets piecemeal' (Aarts, 1999: 915). As a result the processes of economic and political liberalisation have been 'conspicuously partial' in the Middle East. Results so far have been very mixed and how relevant is the outward-oriented growth strategy which promises macroeconomic 'stability' and the creation of wealth and investment? High economic growth rates in the early postwar period of 1992 to 1998 have not been sustained, while declines in inflation rates and the return of foreign capital have not brought the predicted widespread benefits. What is important here though is the presumed universality of this method of assessing and responding to poverty, through the use of certain kinds of languages, strategy papers and acronyms.

POVERTY, LIVELIHOODS DISCOURSE AND SOCIAL CAPITAL

> What binds development NGOs together above all is talk.
>
> (Townsend, 2000: 15)

Another important and associated way in which we can explore the dissemination of development is to look at the increasing popularity of the concept of 'livelihoods' among many international donors and northern NGOs. The way in which the discourse on development is 'institutionalised' (Escobar, 1992) by multilateral and bilateral donors can be shown to partly condition what is ultimately written about development and what are constructed as its 'truths' (Escobar, 1995). The emergence of a Department for International Development (DFID) in Britain is a good example of how 'foggy aid-speak' (Sogge, 2002) can come to dominate international development practice today. An analysis of DFID approaches can tell us much about the ways in which aid and development act as a kind of 'feel good' or image industry that produces knowledge and ideologies which shape policy, norms and aspirations.

Former UK Development Secretary Clare Short spoke repeatedly of the need to 'mobilise the political will' which was seen as widely lacking, particularly among an apparently increasingly less charitable and more indifferent British public. Interestingly, DFID has been quick to associate itself with the millennium international development targets, committing itself to closely following the UN policy on aid provisions and the millennium development goal of a 50 per cent reduction in world poverty by the year 2015 (DFID, 1997). Even though DFID does not actually commit itself to cutting world poverty in half by 2015, it does want to play a part in the 'noble' and 'moral' task of reducing the number of people living in absolute poverty. Hewitt (2001) argues that DFID and New

Labour have successfully popularised the development agenda in the UK, generating more discussion among the public and adopting 'progressive' policies. For some commentators at least, there is more discussion and talk of development in the UK under the New Labour administration than under its predecessors (Hewitt, 2001).

The problem, however, is that this is 'fuzzy talk' which, in pointing to the future and 2015, camouflages actual political and economic injustices in the present that DFID seems unwilling to deal with and take on. In DFID parlance, aid and development are overburdened with far too much hope and hype in ways which mask other important forces such as unfair trade, undemocratic international development agencies, capital flight and 'brain drain', technological change and conflict. All of these are forces that aid is of course itself impacting upon and influencing. Thus none of this 'progressive' focus will successfully be brought to bear more widely as long as the 'market fundamentalism' of the IFIs continues and is endorsed actively by bilateral agencies such as these (Sogge, 2002). DFID has quickly become part of what Huntington once called the 'Davos Culture'. Thus it has rushed lovingly into the arms of the many institutions and professionals which live from 'development administration' (Rist, 1997). This is the kind of culture that has grown up around organisations such as the World Economic Forum which has usually met in Davos, Switerland:

> They know how to deal with computers, cellular phones, airline schedules, currency exchange and the like. But they also dress alike, exhibit the same amicable informality . . . and of course most of them interact in English. Since most of these cultural traits are of Western (and mostly American) provenance, individuals coming from different backgrounds must go through a process of socialization that will allow them to engage.
>
> (Huntington, quoted in Peet, 2002: 59)

DFID had no such problems in engaging with the neoliberal discourses on international development. DFID's legislative statements have been strong on rhetoric and presentation but weak on the substantive issues raised by policy objectives and on the means to implement them, deliberately setting out to align itself with the very latest and contemporary knowledge of development. Replete with the development buzz-words and phraseology of today (environmental conservation and sustainability, gender equalities, participation/empowerment, good state–market relations and 'good governance'), DFID has sought to position the British state as global broker of the most harmonious and 'sustainable' development drawing upon the very latest ideals for development, circulating within the 'development industry' of today (Power, 2000). The importance of the Bank's annual WDRs and other canons of neoliberal orthodoxy to the ideas drawn upon by DFID are somewhat alarming. Its position on these debates is strongly underpinned by an international neoliberal consensus associated with the major multilateral agencies and richest economies, suggesting that the recommended route to poverty reduction is by direct income growth among the poor rather than redistribution of wealth, either internally or internationally. In this way, there seems to be little about DFID that is 'conceptually new or operationally radical' (Gould, 1998: iv).

This lack of anything 'conceptually new or operationally radical to offer' is deeply characteristic of many debates about development today (Power, 2000). DFID is praised in many quarters, however, for its 'progressive approach', for example with respect to its concern for livelihoods analysis. DFID now claims to recognise the 'unremitting struggle to secure a livelihood in the face of adverse social, economic and often political circumstances' (Murray, 2001: 151). Rural livelihoods are multiple and complex (households may derive a part livelihood from farming; a part livelihood from migrant labour undertaken by household members in urban areas or other rural areas, and/or a part livelihood from petty trade or informal activities). Gendered livelihoods are also important here. Closely linked to the diversity of modes of livelihood at any one moment is the idea of diversification of livelihoods over time (Murray, 2001: 152). Through much of the 1990s in Africa, many people turned to non-farm sources of income in comparison to the previous decade (which was in turn linked to the introduction of SAPs, declining terms of trade and devalued currencies). Ellis (1998) points out that both rich and poor households can diversify their income in order to make ends meet and that migration has an impact on this process of diversification (see also de Haan, 1999).

One strength of this approach is that it does focus attention on income diversification in (neoliberalising) rural spaces and the societies that inhabit them, raising important questions about the impacts on farmers, rural markets and communities. Adjustment involved the dismantling of many state agricultural marketing boards and parastatals that had supported producers and their input requirements or had provided marketing facilities and controlled prices. The approach does focus therefore on responses to change in ways which are potentially useful and instructive, examining the erosion of communities or economies during the process of market and trade liberalisation. The approach also tries to be 'dynamic', committed to several dimensions of sustainability simultaneously and aiming to bridge the gap between macro and micro. It also moves beyond some of the expensive, large-scale surveys which generated masses of income and expenditure data but could ultimately say very little about the diversification of responses to poverty and of strategies adopted by people in particular places to counteract it (Bryceson, 2002). There is also a creative tension between several different dimensions of analysis and the approach does try to transcend binaries such as urban/rural, formal/informal, industrial/agricultural, raising some questions about intra-household and extra-household relations to a limited extent. Some scope for institutional change is also envisaged, while the approach is seen to recognise multiple influences, actors, strategies and outcomes. Some geographers have argued that the concept of livelihoods can provide more nuanced understandings of rural development and can focus attention on the importance of place (Bebbington, 2000). The sustainable (rural) livelihoods framework has been particularly influential in much recent research since its official adoption in Britain by DFID which notes that:

> A livelihood comprises the capabilities, assets (including both material and social resources) and activities required for a means of living. A livelihood is sustainable when it can cope with and recover from stresses and shocks and maintain or enhance its capabilities and assets both now and in the future, while not undermining the natural resource base.
>
> (DFID, 1999: Section 1.1)

A 'vulnerability context' is constructed which defines shifting seasonal constraints such as short-term economic shocks and rather vague longer-term trends of change, deploying various 'livelihood assets' or capital to construct the 'asset pentagon' (Francis, 2000). Capital is a social relation, however,

Figure 8.4 Fighting poverty through sustainable livelihoods: UNDP and the 'wheels of abundance'
Source: UN Department of Public Information

and not simply a 'thing', while landlessness cannot be understood as a uniquely distinctive attribute of the rural poor. Rather, capital and landlessness have to do with the 'working out of unequal social relations over time' (Murray, 2001: 154). The approach is seen as flexible, allowing for multiple realities, but it none the less represents a certain kind of discourse about poverty. Like numerous predecessors, the approach claims to be uniquely 'people-centred', based upon popular participation and embodying the realities of 'poor people' themselves. The whole notion of a 'vulnerability context' deserves to be torpedoed however: arguably rampant inflation, conflict, political instability and unemployment are also relevant and more important than allowed for in this framework. The language of 'multiplier effects' which seems to predominate here is quite bewildering, as is the somewhat bizarre notion of how it is possible to expand an individual's 'asset pentagons'. Inequalities of power and conflict are papered over either within or between communities in terms of ethnicity or class, for example. As Murray (2001) has argued, questions of international political economy often slip from view in the 'seductive rhetoric' of livelihood discourses:

> It is an essential pre-condition . . . to undertake a strong analysis at the regional or national level, and often at the international level also, of the political economy of change: key socio-economic trends, shifting political and economic and institutional pressures [and] the social relations of conflict and inequality that determine so many of

> the opportunities and constraints for different social classes within and beyond the population of immediate concern. The seductive rhetoric of 'participation' offers no substitute for independent rigorous analysis of this kind.
>
> (Murray, 2001: 15)

The principal tension here is that participation is central to the discourse of intervention that is based around the livelihoods framework and that there are therefore some important contradictions between these two objectives. If someone's livelihood is enhanced is it possible that another person's livelihood will consequently be undermined? Sustainable livelihoods debates are subject to the same criticisms and weaknesses that face discourses on sustainability more generally: *sustainable under what criteria, for whom and for how long*? (Elliott, 1994). Despite the occasional lip-service paid to public participation, preference is still given to 'control by scientists/technicians, national decision-makers and donors' (Sogge, 2002: 151). This means that while appearing to draw on local perceptions and know-how, there remains a persistent inability to really grasp how local people understand and define such things as livelihoods which are instead recast in a somewhat endogenous discourse encrypted by 'outsiders'.

The intellectual understanding of capital advanced by this approach is itself underdeveloped and impoverished, while the causes of poverty become issues of individual household attributes, rather than being seen as arising from unequal social relations between and within households and communities (Murray, 2001). Associated with this concern with livelihoods is the wider notion of 'social capital' adopted by some organisations such as DFID and more generally by the World Bank in recent years which can be critiqued for the extent to which attention is also shifted away from inequalities of power (Fine, 2001). This idea refers to the ability of people to work together for common purposes in groups and organisations and has been held up by the Bank as a testament to its close proximity to contemporary research. As we saw in Chapter 7, this concept is particularly interesting in that it raises questions about spatial patterns (Mohan and Mohan, 2002) and about the colonisation of the social sciences by neoliberalism. Although for the Bank this concept may adopt a wider notion of what capital is, it still ultimately opens up very little room for manoeuvre (Hart, 2001). It is through such concepts that other alternative ways of thinking about capital are actively delegitimated. The term is now used to discuss a wide range of issues and is often linked to discussions of civil society. Its use by the Bank is a good example of the way the organisation puts an economistic twist on so many aspects of social life by applying economic rationale and claiming to be 'holistic' in thinking about the 'human' implications of its actions. It also signifies the Bank's unchecked power to define truth and falsehood in development circles (Sogge, 2002).

The phrase 'social capital' figured especially prominently in the 2001 WDR *Attacking Poverty* in the sections on 'empowerment'. Ben Fine argues that social capital is part of the attempt by World Bank social theorists to be taken seriously by other economists. The concept of social capital is also problematic because, like the livelihoods approaches, it 'compromises with established doctrines, whilst absorbing and neutralizing more radical and coherent alternatives' (Fine, 2001, cited in BWP, 2001b). As Amin (2001b) has argued, it is possible to be 'duped' by the apparent changes of policy the Bank occasionally engages in, which claim to be part of the 'struggle against poverty' but do not really prioritise the welfare of ordinary

Figure 8.5 Drowning by numbers
Source: Bretton Woods Project

people at all. None the less, the combined effect of these changes has been a 'move towards multiple stakeholder approaches' and the partnership forged by states, capital and different groups of society (Mohan and Mohan, 2002). Thus the nature of relationships between plural actors is being drawn out to an extent, but the actual mechanisms which link the 'networks' and 'organisations' of social capital are much less well understood by these agencies. It may also be seen as quite a prescriptive approach, suggesting that if development fails locally it is because of a lack of the right sort of social capital, which in turn legitimates interventions in 'local' political economies as development agencies seek to 'build this' (Mohan and Mohan, 2002) in particular places.

One important way in which the notions of 'livelihoods' and 'social capital' have been popularised and institutionalised in development circles concerns the role of non-governmental organisations (NGOs). These organisations have been defined as a 'residual' category that includes a wide variety of formal associations which are not government led even though definitions vary enormously and many NGOs *do* receive finance and support from their home government. Between 1977 and 1978 the Canadian government agency CIDA's contribution to NGOs was equivalent to US$44 million, but by 1998 to 1999 it had risen to US$240.8 million despite those NGOs not having fared particularly well in alleviating poverty and protecting the environment (De Silva, 2002). These organisations can operate at local, national or international level but a key issue that has been raised in critiques of NGO activity concerns the closeness of their contact with 'local' and national cultures and knowledge. Although many have argued that the work of NGOs is often poorly monitored and sometimes imposes foreign solutions in delicate situations, it has been suggested that such organisations can provide an effective link between state and society, often working in 'very practical ways' (Edwards and Hulme, 1992: 14). NGOs have also often formed important alliances with social movements of various kinds.

None the less, since the 1980s and 1990s there has been an amazing 'explosion' in the numbers of NGOs and GROs (grassroots organisations) or CBOs (community-based organisations). One study which took a sample of only twenty-two countries found that NGO's generate US$1.1 trillion in revenue (Ryan, 1999). There are now in excess of two million such organisations in the USA alone, three-quarters of which have been established since 1968 (Watts, 2001). The 176 'international NGOs' of 1909 had risen to some 28,900 by 1993, virtually 90 per cent of which have been established since the 1960s (Commission on Global Governance, 1995). In and through this explosion of NGOs comes a further dissemination of development as certain ideas, practices and discourses begin to emerge around each one of them, representing a further 'institutionalisation' and 'professionalisation' of development (Escobar, 1992). For some commentators NGOs have mobilised little except 'compassion for children and victims of disaster' alongside large flows of income from donations (Sogge, 2002: 158) despite all the talk of 'capacity-building'. Many such organisations often seem more concerned with their own survival than with challenging poverty or tacking 'real' development issues:

> There is a great deal of *talk* about 'participation', 'listening to the poor' and 'partnership', and the donor organisations are often committed to these goals in principle. But the practice usually falls short of this talk.
>
> (Townsend, 2000: 1, emphasis in original)

Southern NGOs by contrast have much less power to influence the kinds of project that are undertaken and the way they are conceived. Townsend (2000) looked at the example of NGOs in India and found that even ideas which themselves originated in the South (such as the focus on microfinance) are taken up and promoted by northern NGOs which modify these ideas to fit neoliberal ideological preferences in a way which 'crowds out' locally derived alternatives and local conceptions of what constitutes development. What is interesting here is that all manner of NGOs claim to be engaged in the local, to be 'listening' to the poor and to be facilitating their participation, but in reality 'local voices are often the last heard' (Townsend, 2000: 4). In this way the 'knowledge economy' of the global development community of NGOs is dominated by a relatively small number of predominantly northern NGOs. This in turn is very closely shaped by global waves of 'fashions' in development thinking and new buzz-words such as 'livelihoods' and 'social capital' which arguably signals a lack of diversity in practices and ideas. The participation of southern 'partners' is thus to a degree conditional on the extent to which they can articulate such fads, fashions and buzz-words in their applications for funding and support. There is also something of a 'report culture' that has developed around many of these northern NGOs which focus on measuring and counting activities completed, 'progress' made and 'outputs' and 'indicators' rather than on deeper conversations and engagements with local communities and cultures. In India, many South Indian NGOs feel that money, power and information is concentrated in the capital cities to such an extent that they are geographically as well as socially excluded from participation (Townsend, 2000). It is difficult to generalise however across this mushrooming 'community' of NGOs but what we can say is that it is important to focus on the ways in which knowledge is traded, discussed and disseminated within this community and to examine how this relates to the knowledge of local peoples in particular places that are the subjects of these interventions.

Figure 8.6 Come to the UN: 'It's your world'. The UN Secretariat building in New York

Source: Designed by Gibby's Posters/UN Department of Public Information

Figure 8.7 Twenty years of South–South technical co-operation

Source: UN Department of Public Information, poster by Tracy Walker

DECONSTRUCTING THE 'DEVELOPMENT GATEWAY'

> [The Bank has become] a clearing house for knowledge about development – not just a corporate memory of best practices, but also a collector and disseminator of the *best development knowledge* from outside organisations.
>
> (World Bank, 1999: 140, emphasis added)

The UN *Human Development Report* for 2001 explores the potential of information and communication technologies (ICTs), touching upon the 'digital divide' and its implications for the South – recognising for the first time that this raises key issues about partnership and interconnections in an interdependent world. Much is made by international agencies of their capacity to close this imagined 'digital divide' (Abbott, 2001; Main, 2001; Mander, 2001; Winner, 2001). Allied to this is the World Bank's claim to have reinvented itself as a 'knowledge' bank, even though its notion of development continues to be apolitical and economistic.

What we need to examine here is the ways in which the Bank uses knowledge to legitimise and rationalise certain positions.

In particular the Bank's plans to build a Global Development Network (GDN) or gateway over the Internet are particularly interesting. The Bank hopes the gateway will become the 'premier web entry point for information about poverty and sustainable development' but for whom? Less than 30 per cent of the visitors to the World Bank's existing site come from outside the US (Patel, 2001). The gateway will have a mini-Internet bookstore (for those with access to credit cards), while news is provided by that well-known champion of grassroots knowledge in the South, *The Financial Times*. According to global agencies such as the Bank, technology now permits rapid access to information scattered all over the globe, and enables communications and interactions that transcend geographical boundaries. The 'material poor' need no longer be the 'information poor' according to some commentators (Schafer-Gross, 1995; Samoff and Stromquist, 2001). It is argued that if the people are considered 'information rich', they themselves can be 'empowered' to reduce inequalities of class or gender. In this way, technology is said to permit a faster take-off towards the good life such that people can not only close development 'gaps' but also begin to 'leap-frog ahead'.

This appears to be the philosophy behind the World Bank's reinvention of itself and the creation of the Development gateway, a very ambitious, far-reaching project costing US$69.5 million over the first three years and aiming to create a 'mega-website' to co-exist with the Bank's already extensive external website. Part of a rash of development-information portals on the web, it is rumoured that the idea first began with a meeting between Bill Gates and Bank President James Wolfensohn (Patel, 2001). The Bank is seeking to position itself as the world's 'mightiest think-tank' on development issues, the leading producer of doctrine and knowledge about how countries ought to develop, even aspiring to define the alternatives to its own visions (Sogge, 2002). Since the project began in 2000 to 2001, critiques of this 'Tower of Babel on the Internet' (Wilks, 2001) have been numerous. These have focused on the confused cacophony of voices in this project and the unrealistic 'visionary' plans put forward by people such as James Wolfensohn.

Interestingly, as a result of the riots and demonstrations surrounding its recent 'public' meetings, the Bank has been forced to reconsider its own information disclosure policy and issues surrounding the kinds of information made available in its own consultations. None the less, by the Bank's own admission there has been a continuing failure to provide local citizens and stakeholders with the information they need 'to meaningfully participate in decision-making about the projects affecting their lives' (Results Educational Fund, 2001: 2). There are thus major contradictions within and between the Bank's stated claims to be empowering people and allowing them to participate (through certain kinds of communication) in national or local development. By the mid-1990s the Bank was producing some 350 to 500 publications annually and by 2001 it had even gone into television as well as the Internet (Sogge, 2002). Much of its research arguably contributes more to legitimating what it does rather than to actually advancing understanding of the development issues involved.

Strangely, the Bank overlooks all of this in its claim that there are those who can and are taking advantage of the 'information era' and those that are *falling behind*:

> Knowledge is critical for development, because everything we do depends on knowledge. For countries in the vanguard of the world economy, the balance between knowledge and resources has shifted so far toward the former that knowledge has become perhaps the most important factor determining the standard of living – more so than land, than tools, than labor.
>
> (World Bank, 1999: 16)

Thus the Bank depicts a 'revolution' led by those in the 'vanguard of the global economy' and, by implication, 'forward-thinking' agencies like the Bank itself. A variety of institutions and agencies in the development business have come to a similar realisation that, as the Bank puts it in the 1999 *World Development Report,* 'knowledge *is* development' (World Bank, 1999: 130). Some of these agencies, however, hide behind the claim that the documents are 'country-owned' or 'country-driven' as a way of justifying their non-disclosure of certain kinds of information. The policy papers that are produced by technocrats within recipient countries can thus be understood as little more than a 'cardboard façade' to reflect ownership of dominant ideas (Sogge, 2002: 155). The World Bank in particular often talks about the need to protect the 'integrity of the deliberative process' in the formation of development policies, suggesting that some aspects of deliberations around adoption of its strategies need to be kept private, secret and protected. This is a massive contradiction, however, to the idea of the Bank as a 'gateway' to development knowledge when it deliberately holds back certain kinds of knowledge and information because they are seen to interfere with delicate deliberations around policy. Perhaps it is rather the other way around here in that the Bank 'does not want the public's deliberations to interfere with its own private deliberative process' (Results Educational Fund, 2001: 5). The way in which the Bank still penalises those who leak documents shows that the organisation is not keen on certain kinds of disclosure when it does not suit.

As usual, the 1999 WDR claimed to be inclusive of a range of perspectives but actually contained 'a sermon about what is to be done' (Samoff and Stromquist, 2001: 635). The Bank (thankfully) knows the way forward and preaches to its audience about what is to be done in the future. In 1995, in a statement about education policy, the Bank argued that its main contribution 'must be advice, designed to help governments develop education policies suitable for the circumstances of their countries' (World Bank, 1995: 14). Thus by invoking a concern for local circumstances in recipient countries the Bank constructs itself as responsive, tailoring solutions to the particularities of peoples and places. What is also interesting here is the Bank's sense that knowledge is a 'thing' that can be acquired, borrowed, appropriated or sold. The Bank's role is therefore one of 'disseminator' of 'best' knowledge and 'best' practices, but who classifies them as such? In this way knowledge becomes rather like a microwave, an automobile or a radio: a thing that can be produced or traded, exported or imported. According to the Bank, developing countries will remain 'importers rather than principal producers of technical knowledge' for some time (World Bank, 1999: 24).

Thus the Bank defines the roles for senders and receivers of this knowledge and views this scenario as something which is likely to remain the case 'for some time' into the future. There is often a slippage in the Bank's use of terms (as with other global institutions) between the words 'knowledge' and 'information', for example, which seem almost to fuse together. 'Knowledge gaps' and 'information problems' blur into each other quite often in the Bank's reading of these debates however (Samoff and Stromquist, 2001). Information (of a very particular kind) has thus come to be seen as *the principal determinant* of national economic growth and development. As the World Bank explains in various sections of the WDR, poor countries 'lag' or 'fall' behind in a number of ways in this regard:

> 'Poor countries – and poor people – differ from rich ones not only because they have less capital but because they have less knowledge' (1999: 1).
>
> 'Indeed even greater than the knowledge gap is the gap in the capacity to create knowledge' (1999: 2).
>
> 'The need for developing countries to increase their capacity to use knowledge cannot be overstated' (1999: 16).
>
> 'Countries that fail to encourage investment in the effective use of global and local knowledge are likely to *fall behind* those that succeed in encouraging it' (1999: 25, emphasis added).

The Bank therefore is depicted as a global and local manager of knowledge for development, looking to communicate information to planners and decision-makers and subsequently to the recipients of 'development' knowledge. What is missing from this account, however, is a sense of how context and agency can transform information into knowledge (Samoff and Stromquist, 2001). Thus every individual human being/agent or subject thinks their place in the world and transforms various forms of information into knowledge every day. The power, discourses and knowledge of the World Bank thus create *models of subjectivity*, which can, as Escobar (1995) has shown, be seen partly as an effect of its networks of power–knowledge. All subjectivities are gendered of course and are the product of a matrix of habits, practices and discourses (McDowell and Sharp, 1999). Although the concept of subjectivity is a complex one, it raises questions about agency, action and authorship in the making of development knowledge and about the situated and partial nature of *knowing* development.

Information has to be repackaged, translated, contextualised and communicated through complex pathways that are overlooked by the Bank and its 'corporate memory' of neoliberal best practices. How is this information passed on, generated and appropriated? In this sense, '[h]uman agents, multi-step communication flows, status-based transmission barriers, and more, all disappear from view'

(Samoff and Stromquist, 2001: 643). The Bank also adopts a rather static and linear model of the policy-making process. How is the same information viewed differently by, say, a government official as opposed to a community activist? In a way there is an ideological arrogance in this, assuming that it is unimportant for 'developing countries' to assess the value of others' experiences for their own setting. Of course the Bank will be quick to respond that 'indigenous knowledge' is also important to its work (major sections of the 1999 WDR are given over to this), but this is again seen as something which can simply be appended to the Bank's own knowledge of development. In practice, this 'knowledge' is simply not used by the majority of Bank staff in their everyday practices: 'What is old, or more often, what is deemed to be old, though in fact it may be quite recent, becomes romanticized and at the same time fossilized' (Samoff and Stromquist, 2001: 647).

The knowledge of the 'indigenous' thus is sometimes 'romanticised' and fossilised by the Bank which often has a limited sense of what is old and what is new. How 'empowering' is this knowledge from the Bank for the world's poor? Even when indigenous communities 'produce' the knowledge, they remain dependent on those who generate the technology employed to distil and distribute that knowledge. Where poor countries and their citizens are not technically excluded (and these areas are still vast), they therefore may sometimes be 'priced out of participation' because of the uneven distribution of global resources (Samoff and Stromquist, 2001: 650). As early as the Bank's 1996 annual meeting James Wolfensohn had talked of the 'light' of knowledge 'enlightening the lives of people everywhere', representing knowledge as something which 'easily travels the world' and the only hope for billions of people 'that still live in the darkness of poverty' (Wolfensohn, quoted in Patel, 2001: 2). The parallels here with colonial representations of enlightenment and non-western darkness are striking; perhaps the only difference is that the Bank has its own Gospel.

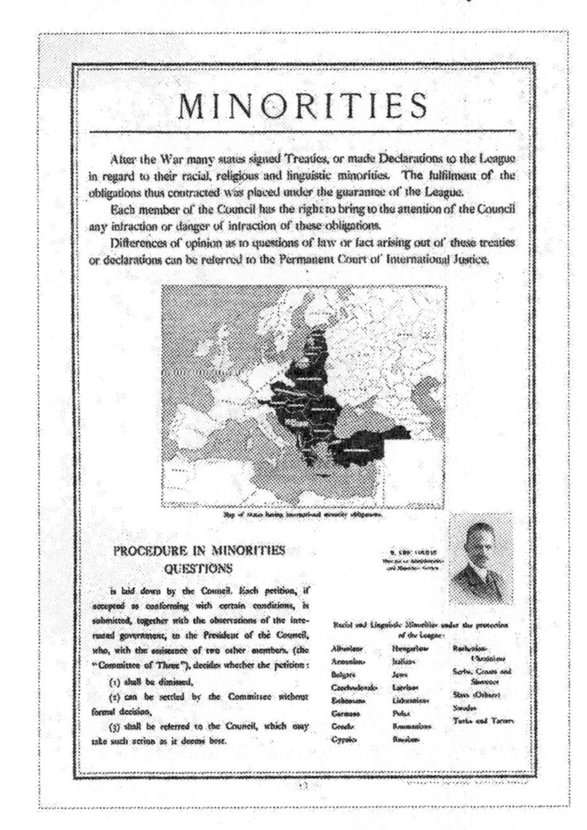

Figure 8.8 Racial, linguistic and religious minorities: League of Nations poster, *c.* 1938

Source: UN Department of Public Information

For Appadurai (2000: 3) there has been a widespread failure to recognise the significance of 'globalisation from below' (see Chapter 9), which we can argue is certainly true of the World Bank's discourses on global development. This relates to the 'growing disjuncture between the globalisation of knowledge and the knowledge of globalisation' (ibid.: 3) and thus we might support the call for a 'new architecture for producing and sharing knowledge about globalisation' (Appadurai, 2000: 18). This is a crucial argument here in that it suggests that what is not needed is yet another global development portal on the Internet (there are already some much more critical and useful sites) but rather an entirely and different new global architecture for producing and sharing knowledge about development. As we have seen, the Bank often crowds out the search for alternative theories, strategies and ideologies, as in the case of South Africa. The Bank disseminates knowledge about development based around its own ideology and notions of effective development strategies, but how desirable is this when it actively crowds out the scope for change, for thinking outside of the development 'box'? (Freire, 1970)? One interesting feature of the World Bank's website which came online in 2002 was a subsection entitled *Ten Things You Never Knew About the World Bank* (available in French, English, Spanish and Japanese). This site articulates an argument that 'the World Bank's priorities have changed dramatically' in recent years and seeks to remind us that the organisation is varied and dynamic:

1. The World Bank is the largest external funder of education.
2. The World Bank is among the world's largest funders in the fight against HIV/AIDS
3. The World Bank is among the world's largest external funders of health programs.
4. The World Bank strongly supports debt relief.
5. The World Bank is one of the largest international funders of biodiversity projects.
6. The World Bank works in partnership more than ever before.
7. The World Bank is a leader in the fight against corruption worldwide.
8. Civil society plays an ever larger role in the Bank's work.
9. The World Bank helps countries emerging from conflict.
10. The World Bank is listening to the voices of the poor.

(http://www.worldbank.org/tenthings/)

For all the rhetoric about 'listening' and 'partnership', the idea of funding and funders ('the world's largest') and the idea of the Bank as a 'leader' with a 'large role' that seeks to 'help' the poor, remain important in this vision. All but one of these ten 'facts' we never knew starts with the words 'The World Bank . . .'. Shortly after this appeared on the Bank's wedsite, the US campaign group Global Exchange published a pamphlet setting down ten reasons to abolish the World Bank (Danaher, 2002).

CONCLUSIONS: SOLUTIONS IN SEARCH OF PROBLEMS?

> The answer is needed urgently. While there has been more progress with poverty reduction in the past 50 years than in any comparable period in human history, poverty remains a *dire global problem*.
>
> (World Bank, 2000b: 2, emphasis added)

This chapter has focused on the way in which poverty is constructed as a 'dire global problem' through a series of languages, discourses and buzz-words which enframe the people and places of the South in particular ways. Words such as these have major consequences, framing the construction and implementation of policy and covering up the continuity with their institutional predecessors (Cooke, 2001). Thus when these institutions use such phrases as 'ownership' and 'partnership', these are not neutral, unproblematic and ahistorical, but in some ways serve to illustrate the legacies of earlier, imperial visions of development administration and trusteeship. These keywords and languages are thus caught up in a kind of 'semantic conjuring' (Rist, 1997: 238) where an image and concept of 'under-development' is constructed and disseminated, defining and legitimating the terms of the debate. A good example of this is the way in which structural adjustment became unspeakable in the late 1980s and early 1990s, while the Bank is apparently now considering renaming its adjustment lending as 'development policy support lending' (BWP, July–August 2002). This allows them to claim that the performance of these strategies is improving and should be retained as lessons are learned and reflected upon. In this process development becomes primarily a technical enterprise with an 'illusion of neutrality' (Cooke, 2001: 18). This is a complex process involving a whole range of actors from private aid agencies, NGOs and policy think-tanks to bilateral and multilateral agencies such as USAID or DFID. Agencies such as USAID have even created think-tanks especially for the purpose of promoting neoliberalism and market fundamentalism (Sogge, 2002).

In many ways the central argument here has been that dominant discourses of development have 'prevented us from seeing other very creative ways of addressing the poverty problem' (Yapa, 2002: 41). It could even be argued that development, as currently conceived by those at the commanding heights of the development industry, is about sending out solutions in search of problems (Sogge, 2002). Particular notions and maps of the 'routes' and 'directions' countries *should* take and follow are prescribed and disseminated in a number of different ways through a range of different discourses and institutions. Modernisation and 'globalisation' are themselves examples of buzz-words which are disseminated through the development industry and have come to be internalised in various ways. This is not to downplay the very real differences between bilateral donors (such as USAID, CIDA and DFID), but to try to understand more fully how the ideological space for thinking creatively about development is narrowed by these organisations and their visions, cultures and practices. Indeed in some ways it is the tensions and differences between these agencies which can trigger critique and counter-ideas.

The World Bank regularly uses phrases like 'country-owned' and 'country-driven' in its discussions of the development strategies it is involved with, but to what extent is this really true? As we have seen, the links to enlightenment ideals about truth and light and to colonial representations of the 'other' are important here:

> The parallels with the first chapter of St John's Gospel are striking. . . . The Bank's Gospel in developing countries has its precedent in Early Modern Europe and colonialism. The dispatch of a battalion of consultants from head office isn't called a 'mission' for nothing.
>
> (Patel, 2001: 2)

Similarly phrases such as 'participation' are often bandied about by development agencies with reference to a variety of spatial scales but the extent to which the 'voices of the poor' truly have an impact on the deliberative process of these agencies remains quite limited. As we shall see in Chapter 9, when these strategies have been called into the question by social movements and international coalitions of various kinds, we are told that there are no alternatives. Thus (neoliberal) development is constructed as unquestionably a force for good, as sweet as apple pie and as inherently progressive as motherhood. Although there is much that geographers can engage with concerning debates about social capital and livelihoods, it is worth recalling that these are the latest in a long line of simplistic labels that have been placed on people and places that are far from homogeneous. The websites of some of the major global institutions sometimes assemble a range of individual and community 'stories' to illustrate successes and proximity to 'cutting-edge' ideas. The point here is not to call into question the authenticity of these stories but rather to focus on the way they are conveyed in order to disseminate a particular image of development. In a way this process can define and rank the categories by which development is measured and understood, but it can also assign value and even label particular identities (Sogge, 2002). In this respect debates about postcolonialism are particularly relevant in that they help to destabilise the myths that surround poverty and its eradication and go some way towards illustrating the variety of subject positions that are lumped together whenever these agencies talk about 'poor people' as if they represented some sort of uniform, homogeneous collective confined to a singular space.

It is also necessary to call into question the repeated exiting of politics and political economy from the debate and the tendency to downplay and misinterpret the complex realm of political and economic struggles around 'poverty' and its variegated cultural and historical meanings. The bewildering number of languages and acronyms of the PRSP and associated processes cover up the real political and economic roots of struggles in societies undergoing neoliberalisation and simplify very contested and dynamic political spaces such as those of Israel or Palestine. Similarly, the media can offer simplifications of the political and economic struggles waged by people across the South by ignoring complex issues in programming schedules and focusing instead on stereotypical images of hunger, famine and deprivation. In this way the media shape our understanding of 'distant others' and the social and political worlds they inhabit.

The United Nations Development Programme today talks about 'human poverty' and about 'a denial of choices and opportunities for living a tolerable life', while the World Bank frequently refers to 'income poverty' or to 'living on less than a dollar a day'. Then there is 'absolute' poverty – those below a defined poverty line or threshold (usually defined

TABLE 8.2 POVERTY PROCESS ACRONYMS

PRP	Poverty Reduction Strategy
PRSP	Poverty Reduction Strategy Paper
IPRSP	Interim Poverty Reduction Strategy Paper
PRGF	Poverty Reduction and Growth Facility
PRSC	Poverty Reduction Support Credit
SAL	Structural Adjustment Loan
SAPRI	Structural Adjustment Participatory Review Initiative
CDF	Comprehensive Development Framework
CAS	Country Assistance Strategy
NSSD	National Strategy for Sustainable Development
ESAF	Enhanced Structural Adjustment Facility
NEPAD	New Economic Partnership for African Development
SPA	Strategic Partnership for Africa
PEAP	Poverty Eradication Action Plan

in US currency) – as well as 'relative' poverty – where someone is poor in relation to those around them. Not only do various types of poverty exist but there are also a number of particular ways of speaking about and representing poverty, not all of which find expression in DFID, World Bank or UNCTAD reports. As Jerve (2001) has argued, most poverty research treats the urban and rural spheres as distinct sectors, despite the spatial continuum that often exists between them, which precludes any easy or sharp demarcations. Static poverty concepts and measures fail to capture the cyclical movements of people and the transfer of resources that bind urban and rural spaces together. How can poverty be understood when these crucial issues are so often neglected and how can broad-based economic growth ever be created given this common unevenness?

It is crucial not to misunderstand this argument however – we are not simply calling for a different way of speaking about poverty, content to cast off the notion of 'poverty' as something that becomes only a vehicle for the dissemination of development. As Patel (2001: 1) reminds us, the World Bank's talk matters 'more than your average newspaper or magazine'. The crucial point here is that very real political and economic processes of exploitation, unevenness and inequality are obscured in many of these visions and representations of 'Third World development'. In addition, we have seen that a great deal of time and money is spent by the Bank in particular on 'investing its products with the look and feel of impartiality', using a number of established western academics to 'manufacture knowledge under the Bank's brand' (Patel, 2001: 3).

Thus the Bank and other agencies seek to brand their visions and approaches and to manufacture knowledge which supports their visions. It is therefore important to recognise that such representations are also bound up with the process of perpetuating these relations of impoverishment and inequality and are thus related directly to the material production of poverty. Recently, governments have begun to use the term 'social exclusion' as a useful tool for describing what poor people experience. This is useful, as long as it is not an

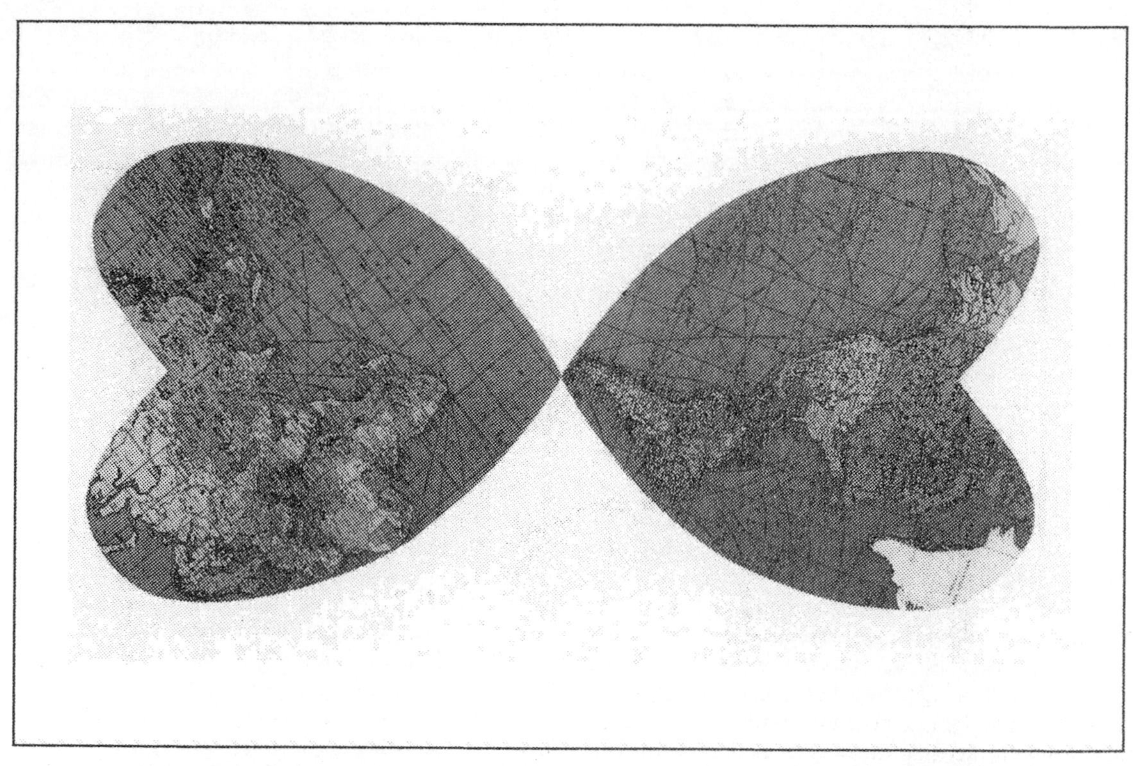

Figure 8.9 Whole World (also known as 'Tolerance flag') by Robert Rauschenberg for the fiftieth anniversary of the WHO, 1948–1998
Source: UN Department of Public Information

excuse for failing to invest in development assistance but could be read as yet another example of a 'northern' concept which has been transferred from North to South. Will there come a time when there will be a return cargo of ideas, a South–North transfer of poverty concepts and definitions?

If some people, areas or communities are 'socially excluded', then we need to know precisely what they are excluded from and the extent to which this forms the basis of development action and intervention. Arguably, the dissemination of development is itself a highly exclusionary process, but this should not allow us to lose sight of a concern for the 'included' and the reasons for their inclusion. The interesting point about all these definitions is that they only really define the poor: there are far more words for poverty than there are for wealth. Yet it is the poor who are often then seen as the 'problem' – a problem that 'will always be with us' and one that requires institutions at the vanguard of the world economy to vanquish its 'darkness'. In the foyer of the Bank's Washington HQ stands an (expensive) luminous sign which reads 'Our dream is a world free of poverty' but why not dream of a world free of inequality? Is it possible to dream of a world free of the need for the World Bank?

The concept of social exclusion does not always foreground the economic roles and relationships at the household and community levels that are performed mainly by women. Yet women all over the world bear the brunt of poverty, partly because of the extra burden of responsibilities they have for the household and partly because they lack access to land, credit and employment. Labels and measurements are useful tools for dealing with poverty, but sometimes they can detract from what being poor really means in particular places. In the records and statistics of World Development Reports there is a silence about the ways in which poverty is *materially created and produced* by specific political and economic processes at work in each place and locality. There are also important silences here about the doubt and despair that come with being poor, despite the Bank's attempts to 'listen to the voices' of the impoverished. What peoples in places of 'social exclusion' least need, it would appear, is yet another label (Van der Gaag, 2002); nor do they need institutions that claim to hear their voices while simultaneously silencing them and systematically disguising the material causes of their impoverishment, which are local, national and global.

The Human Development Report of 1991 was quite a significant turning point in that it introduced the idea of the Human Development Index (HDI). Although repeating the standard normative definitions of development and the usual claims to focus on people, it did break (to some extent) with the sacred cow of economic growth and it did distinguish that wealth was not created or accumulated simply as an end in itself. Quoting Aristotle, the Report reminds us that 'Wealth is evidently not the good we are seeking, for it is merely useful and for the sake of something else' (UNDP, 1991). For many global development institutions wealth creation is of paramount importance, but what is this 'something else' for which wealth is useful and why is this questioned so infrequently by these institutions? Amartya Sen, one of the consultants working with UNDP at the time this Report was prepared, has questioned the dangers of the ascendancy of economics in development thinking and the myopic views that this has led to in some quarters, arguing instead for increasing 'human choice' and placing 'freedom' at the centre of the debate (Sen, 2001). Thus age-old enlightenment ideals resurface once again in contemporary development thinking, just as they have at repeated junctures during the post-war 'invention' of development as an arena of theory and state practice.

In the past fifty years poverty has fallen more than in the previous five hundred years, according to the UN and the World Bank. Over the course of the twentieth century some three to four billion of the world's people are said to have experienced substantial improvements in their quality of life according to these agencies and their narrations of progress. Since 1960 in particular, we are told that child death rates in developing countries have been halved and malnutrition has declined by more than one-third. However, such figures hide the fact that the absolute number of poor people is *increasing* as a direct consequence of neoliberal strategies and the resulting liberalisation of development. Between 1987 and 1993, in the heyday of structural adjustment, the number of people with an income of less than a dollar a day actually rose by almost 100 million according to the Bank's own statistics (World Bank, 1995). World development institutions are increasingly coming to realise then that these concerns with poverty are not exclusive only to the 'Majority World' but are also very relevant in the 'developed world', in Eastern Europe and in the countries of the former Soviet Union where the average incidence of income poverty increased sevenfold between 1988 and 1994 (Van der Gaag, 2002).

The 'champagne glass' of global income will not be shaken up or turned on its head very easily, nor is it likely that world poverty will be 'eradicated' by half by 2015. The most industrialised countries today have some 147 of the world's 225 richest people, while globally the gap between rich and poor is increasing all the time – the three richest people in the world, for example (including Microsoft CEO Bill Gates), have assets that exceed the combined gross domestic product of some forty-eight countries of the South. We may well ask why the concentration of wealth in the hands of a few is never really an issue in poverty eradication campaigns and discourses. The urgency of addressing what the UN calls the 'scandal' of poverty is not hard to see (and is not being disputed here) but important doubts remain about whether this is truly being recognised at the highest levels, particularly since links have been made between poverty and international terrorism in the light of the terrorist attacks of 11th September 2001.

Poverty is supposedly very much on the international agenda today; the number of papers, statements and targets has mushroomed, and ideas of social capital, participation and livelihoods have become the new buzz-words. Every development agency now claims they want to involve poor people in the debate about what should be done, to enhance poor people's livelihoods or quality of life. It remains to be seen if theory becomes practice however, and if this will ever be more than just 'talk'. If poverty is to be eradicated, it must be more than talk and empty rhetoric as has so often been the case in the past. Many people experiencing poverty have another, even more radical and important point to make. One of the most extraordinary things about listening to people in poverty in different parts of the world is that so many are concerned not just about inequality, but about the way money has become the measure of all things (Van der Gaag, 2002). Is money and wealth creation what we should value and aim to achieve through development, and what happens if we use other ways of valuing people; surely it is the poor who are rich? Governments which aim for economic growth alone often fail to address fully the *causes and consequences of poverty*. The emphasis also has to be on issues of equity and distribution, not just economic growth *per se* but on issues of its social and spatial unevenness. What is so disconcerting about the World Bank is that it has maintained a remarkably consistent view of development as economic growth (rather than (re)distribution) throughout almost its entire history.

It is important therefore to consider the ways in which development institutions (including NGOs, global institutions such as the World Bank and bilateral donors such as DFID) disseminate a particular notion of development, power and knowledge. Certain visions of indigenous knowledge have emerged and particular regions of the world are invented as 'exporters' of knowledge and expertise, with others seen as long-term recipients. People experiencing poverty are often aware that any 'help' – whether from governments or charities – works best when those being 'helped' are organised into workers' movements, trade unions or women's networks, for example. It is also therefore necessary to find a range of new approaches to development co-operation which are based on genuine partnership, rather than empty, rhetorical commitments to co-operation. Poverty eradication requires not just an escalation of aid transfers but rather a systematic and simultaneous reform on a whole variety of fronts as well as the establishment of international coalitions for change which begin with the needs and interests of citizens rather than corporations. For too long development has been posing as a solution when in fact it may well be a large chunk of the problem. Through quiet encroachments and collective insubordinations of various kinds, people of the South are beginning to show their belief in and passionate desire for alternatives to 'neoliberalism with a human face'. Many observers have pledged

Figure 8.10 Earth Summit: recycling platitudes
Source: CWS and Gable

to avoid using the development gateway, for example (Patel, 2001), and, to use a phrase borrowed from the Southern African People's Solidarity Network (SAPSN), are now seeking to support alternative sources of knowledge, pursuing instead a strategy of 'constructive disengagement'.

BOX 8.3

Chapter-related websites

http://www.neweconomics.org/
New Economics.

http://www.unctad.org/
UNCTAD.

http://www.devint.org/glossary.htm#Glossary
Contains material on acronyms and poverty process languages and shorthand.

http://www.brettonwoodsproject.org/faq/gloss/
Bretton Woods glossary.

http://www.imf.org/np/term/index.sap?index=eng&index_langi
IMF.

http://www.aidc.org.za/
Alternative Information Centre on Development.

http://www.gdnet.org/
GD Net.

http://www.realworldbank.org/ http://www.worldbank.org/gateway/
World Bank.

http://www.g24org/publicat.htm
G-24 organisation.

http://www.unicef.org/pon00/gmi.html/
UNICEF.

http://www.globalissues.org/Geopolitics/
Global Issues site on geopolitics.

http://voiceoftheturtle.org/gateway/
World Bank Bonds Boycott/SAPSN link.

http://www.ibt.co.uk/
International Broadcasting Trust (IBT).

http://www.reportingtheworld.org/
Reporting the World.

http://www.unido.org/
UNIDO also sees itself as a 'global forum' for 'generating and disseminating knowledge relating to industrial matters' (UNIDO, 2001: 3).

http://un.org/works/
The UN Works' website tries to portray the image that the UN works (effectively) for development, education, culture, human rights, health, labour and so on. The home-page explains: 'a little girl is back home. A soldier starts a new life. A village finds prosperity. These are the stories of how the UN works.'

http://www.worldbank.org/tenthings/
Ten things you never knew about the World Bank.

http://www.globalexchange.com/
Ten reasons to abolish the World Bank.

9

'Theorising Back'

views from the South and the globalisation of resistance

> The whole planet is witnessing a series of very varied and apparently unconnected social tremors . . . Indirect relations affect hundreds of mlIions of people who may not be aware of the link that binds them to the world economic system but nonetheless suffer its disastrous consequences in their daily lives . . . the chain of cause and effect is not obvious and requires exposition.
>
> (Houtart, 2001, vi)

> It takes a singularly disengaged person not to realise that there is widespread dissatisfaction with the process and products of globalisation. . . . Rarely though, can resistance be easily pigeon-holed.
>
> (Parnwell and Rigg, 2001, 206)

OPENING A SPACE IN THE HISTORY AND GEOGRAPHY OF INTERNATIONAL DEVELOPMENT

This chapter continues the focus on where and how knowledge and theories of development have been produced and considers the recent shift in perspective towards a concern with redistributing power through popular 'participation' and 'empowerment' approaches. In what ways do development theories and practices express centralising tendencies and why have 'grassroots' and 'bottom-up' approaches been important? Questions of social and cultural identity (particularly to do with gender and ethnicity) are discussed here, since it is necessary to understand how the places and spaces of development are not socially homogeneous but fractured by a range of important identities and through *resistances* of various kinds. Apparently unconnected 'social tremors' around the world in recent years have highlighted the links that bind peoples and places to the world economic system. The word *resistance* refers to 'any action, imbued with intent, that attempts to challenge, change, or retain particular societal relations, processes and/or institutions' (Routledge, 1997: 69). This chapter seeks to examine a wide range of resistance to neoliberal ideologies and free trade mythologies, exploring the connections that can and are being made between peoples and places around the world for the promotion of support and solidarity networks. In an earlier chapter we explored how dependency theorists had sought to 'theorise back', to countenance liberal assumptions about growth and progress. For David Slater, the dependency writers deployed a geopolitical imagination that sought to prioritise autonomy and difference in order to 'break the subordinating effect of metropolis-satelite relations' (Slater, 1993: 430). This chapter seeks to extend this concern by focusing on insurgent resistance groups and the ways in which they are also attempting to strike or answer back and are similarly deploying a geopolitical imagination in order to challenge western assumptions about development from afar and from below.

As we have seen, many development practices have resulted in the impoverishment and marginalisation of indigenous peoples, women, peasant farmers and industrial workers and in a deterioration of certain economic, social and ecological conditions. In addition, 'western' values have sometimes been emphasised at the expense of traditional and indigenous knowledge systems, while certain groups have been excluded and marginalised in the decision-making process (such as people with disabilities). All this has led to the formation of

Is this what Democracy looks like??

Seattle, Washington, November '99: Police attack peaceful protesters with tear gas and plastic bullets at the protests against the WTO.

Prague, Czechoslovakia, Sept. 2000: Police beat and silence people demonstrating against the IMF and World Bank.

Quebec City, April 2001: Activists protesting against the FTAA are met by 6000 riot police who attacked them with over 1000 plastic bullets and 6000 canisters of tear gas.

Genoa, Italy, June 2001: Protester murdered in cold blood by police at protests against the G8 Summit. Other peaceful demonstrators are brutally attacked by the police.

Figure 9.1 This poster concerning a meeting in Ottawa (2002) highlights the recent protests in the cities of Prague, Genoa, Seattle and Quebec

Source: Adbusters

'myriad social movements which articulate struggles for political autonomy, and cultural, ecological and economic survival' (Routledge, 1999: 77).

Many of these struggles are important for geographers to study since they are often place-specific in character and we need to understand why and where they arise. The various forms resistance takes are not always public and overtly political but can be subtle, small-scale struggles and resistance grounded in day-to-day practices, located within the space of the household (Mohanty, 1991; Radcliffe, 1999). Resistance may also occur around struggles over access to resources or may be the product of various tensions and conflicts within society, reflecting cultural, ecological, political and economic concerns (Routledge, 1999). These varied *terrains of resistance* are not only based around particular places and localities, however, but may also be global in character, transcending the boundaries of nation through globalised resistance to ideologies such as neoliberalism. Orthodox accounts of resistance have focused on a dominating power source (such as the state) and public forms of political resistance, seeing power as being held by particular people or institutions which is then used to repress identities and knowledge. This is understandable since power does accumulate in certain sites, producing an uneven topography of power. Resistance is a form of power in its own right however, and even the most powerless and marginalised people have the capacity to resist through a number of weapons.

Power is thus diffused through society in a web of relations. In thinking about the ways in which to conceptualise and visualise the changing geographies of governance and resistance it is important to keep in mind this sense of the relational and diffuse nature of power (McDowell and Sharp, 1999). Important alternative geographies of development emerge from resistance of various kinds that can inform our concern to rethink and reconceptualise approaches. In Chapter 7 we explored the shifting geographical scales of governance in relation to globalisation and neoliberalism and what we need to do now is examine how these scales and forms of governance are being transformed (if at all) by networks of resistance and protest. If there is a move away from the dominance of nation-states in thought and practice towards the emergence of networks (Agnew, 1998), how does this relate to the contested nature of development and its geopolitical spaces? In the context of a globalising world an effective theorisation of power can 'no longer ignore the significance of space' (Slater, 2002: 255). Geographers then are well placed to map these changing *cartographies of struggle* around identity and society as well as themes of economic justice and can attempt to understand the intersections of politics, culture and economy as they unfold in particular spaces and places on the ground. In Chapter 6 on postcolonialism we saw how struggles over cultural difference were related to the meaning and imagination of contemporary development, but it is also necessary to extend this concern to struggles around economic and political justice.

Social movements can be very diverse and are far from homogeneous, including squatter movements, women's associations, human rights organisations, neighbourhood groups, indigenous rights groups, youth groups and self-help organisations of various kinds. In a way, these movements arise, most importantly, as a result of tensions and contradictions

Figure 9.2 This 'think different' poster from Aqitart highlights street protests
Source: Adbusters

in the development process both locally and globally and have occasionally moved beyond issues raised by urban struggles to embrace wider communities and priorities. Here we return to Escobar's concern with popular struggles to *defend* places and ecologies. In many ways social movements are the visible signs of deep transformations and of the growing emphasis on culture and identity politics, but also of wider shifts away from the state-led developmentalisms of the past. They have 'helped to lay to rest the myth of totality and [have] reinforced the notion that society has no centre' (Munck, 1999: 206). Thus there is no developmental society ready and able to automatically correspond to the mission of particular plans and administrations. Many of these groups therefore emerge in response to the problems and limitations of development and can involve the formation of identities and solidarities based around class, gender, kinship or neighbourhood. Indeed, social and cultural identities are always a product of struggle and contestation over meaning. Many are multidimensional, addressing issues of poverty, gender, culture and ecology in a simultaneous fashion and many take place within the realm of 'civil society'. As geographers, therefore, a contribution can be made in that:

> we can contribute to the understanding of struggles for survival in different cultural contexts. Given that we live within an increasingly interdependent world, we might also consider ways of attempting to contribute towards these struggles, to make our contribution towards an environmentally sustainable and socially just world.
>
> (Routledge, 1999: 83)

Geographers can focus on the 'everyday' practices that construct identities and forms of governance (Thrift, 2000). What we have seen in recent years across the South is that resistance is assembled from the materials and practices of everyday life. In Argentina this was powerfully illustrated when millions of people took to the streets with pots and pans from their kitchens to communicate their protest. In many parts of the South, women have been particularly adept at making precisely this point about the everyday implications of adjustment on their lives by domesticating their struggles, by bringing the private into the public sphere. With respect to development the emergence of resistance is based partly around the demise of embedded forms of statism and 'developmentalist' notions of nationhood. Important challenges from ethnic and gender identities have sought to counter state-centred views of development, creating new identities and political spaces in the process. We have seen that in the context of globalisation, deterritorialisation raises important questions about the power and role of the state and its relations to place-specific identities (Flint, 2002). These identities and places have challenged local, state and global practices of development. Places have both 'real' and 'imagined' significance in this regard in that all struggles over politics, the economy and culture are grounded in particular places and in interpretations of those places and their relations with other areas (Holloway and Hubbard, 2001). Since the early 1990s new transnational forms of resistance have been playing an increasingly important role in making visible and combating the inequities of the post-Cold War international order (Boehmer and Moore-Gilbert, 2002).

Stories about the power of development have often been told rather problematically, giving the impression that powerful people simply impose their ideas and values on the groups below them (Holloway and Hubbard, 2001), but we need to understand the more subtle and diffuse relations that exist between people and places at all scales and levels of society, from the level of the individual household all the way up to global relations. What we need to examine is how these complex power relations are changed and reordered by resistance of various kinds and in various localities. What effect, if any, does rejecting the disciplinary power of states and global institutions have on the theorisation and practice of development in different places at different times?

BOX 9.1

The collapse of the Argentinian economy and the *cacerolazos*

In the middle of a fourth year of economic recession and with the official unemployment rate running at 24 per cent many Argentinians took to the streets in December 2001 to protest against the government's latest economic policies which restricted people's access to their own savings. In the context of further cutbacks in social services (MacEwan, 2002), presidential and ministerial resignations followed along with the largest debt default in history – when the new government defaulted on $155 million of the total national debt. Falling incomes and rising unemployment contrast directly with the picture of Argentina painted by the IMF (for which the country was the 'poster-child' of conservative economic policies) during the 1990s (MacEwan, 2002: 2). Much attention was focused on Argentina in 2001 as the national economy collapsed and many began to wonder how the country would manage its financial and political crises. The country was struggling to meet its debt repayments and did not find immediate support from the IMF, which delayed debt restructuring until the government agreed to a further reduction in budgeted social spending (already down to a minimum). A variety of different political and economic errors were committed not only by Argentinian institutions but also by the IFIs, which reacted slowly and unsympathetically. Incredible pressure was placed by the Fund on a quite fragile democracy. Yet the situation in Argentina illustrates the more general failure of IMF prescriptions for long-term macroeconomic growth and stability. The 'growth' the Bank had trumpeted in the 1990s was built on rising levels of international debt and short-term revenue injections from periodic sales of government-owned industries. The good times had thus been built on remarkably weak and short-term foundations. US-based firms often benefit from these privatisations but what use have they been for the one in five Argentinians who are now unemployed? As MacEwan (2002: 4) has argued, privatisation is especially problematic 'when it only replaces an inefficient government monopoly with a private monopoly yielding huge profits for its owners'.

In response to the collapse of their economy millions of Argentine people took to the streets to demonstrate their rejection of political parties, and politically aligned trade unions. A *cacerolazos* was organised, a pots-and-pans protest, involving many women. In addition, self-mobilised assemblies appeared all over the country. *Cacerolazos* were held in London as well in January 2002. This could be seen as part of a tradition of resistance led by women in Argentina of popular struggle against slavery during the time of conquest or against the disappearance of family members during the dictatorship in the 1970s and 1980s. These demonstrations highlight further the need to reclaim freedoms and political spaces from the new enclosures of neoliberal development discourses. In addition, a militant movement known as the *Piqueteros*, a large portion of them unemployed women, have pushed the authorities for negotiations about the need for public works employment and subsistence programmes during the economic crisis (Brecher and Costello, 2002). Discontent came to a head as the government accepted even more austerity programmes from the IMF and set about imposing a state of siege to curb the protests.

Thus it can be argued that:

> Taken together, such a perspective on the relations of place and power might appear to imply that geography is 'carved out' by the rich and the powerful, who occupy places of distinction and prestige while frequently relegating marginalized groups (and behaviours) to areas typified by disinvestment and decay.
>
> (Holloway and Hubbard, 2001: 208)

The objective here is not to carve out a mythical map of universalised exploitation for all 'Third World' countries that are thus confined to a space of 'decay', but rather we must seek to grasp the real and imagined connections and relations between people and places. For Foucault, power was something that was exercised from innumerable points and also something that comes very much 'from below'. Prior nationalist modes of resistance (which had ended colonialism and western hegemony in parts of the South between 1945 and 1975) had proved generally ineffective and incomplete in some way. None the less, the formation of anti-colonial identities in the struggles for national independence during this time came about partly as a result of the interrelationships between different peoples, involving places that were often 'far from "home" ' (Boehmer and

Figure 9.3 World march of women in Mozambique (2000)
Source: Frelimo Forum Mulher

Moore-Gilbert, 2002: 17). In many cases these movements for national independence were the outcome of a significant popular struggle (as in India and Mozambique, for example). What is meant here is that challenges to colonialism were never simply about one singular space of identity but rather were the product of interactions and interrelationships between different social and cultural formations.

New approaches to the theory and praxis of citizenship have begun to emerge, through demands for greater collective rights or more inclusive democracies around the world. Addressing the post-Cold War predicament of development will involve, in important ways, the efforts of global social movements to 'construct a new systemic alternative that builds on the failures and successes of progressive initiatives in the twentieth century' (Berger, 2001: 1085). A key question that this chapter seeks to address is how best to understand the transnational alliances that have been forged between different constituencies around the world of development discourses. The symbolism of the space of the 'Third World' in mobilising and legitimating struggles around development is important here. An extension of this view is to seek to understand the symbolism of places and the identities and struggles that emerge around them.

If there has been a failure to acknowledge the power and significance of 'globalisation from below' (Appadurai, 2000), how well placed are social scientists to embrace such a form of globalisation, in the context of apartheid in the academy (which restricts opportunities for researchers from the South) and of a continuing 'cognitive colonialism' (Parnwell and Rigg, 2001: 206) in the theory and practice of development? Do the required political and economic structures exist for us to engage with these changes from 'below'? There is also a risk here of misinterpreting and romanticising the resistant and of inflating dominance. None the less, social identities and changing social movements represent a crucial component in any study of power (Slater, 2002). Geographers do have a role, however, in mapping the changing spatiality of power and the geographical expression of networks of governance and resistance. In particular we can highlight the ways in which discourses of 'good governance' seek to contain the political and to 'manage' these new networks of resistance. There is also the danger that resistance may be misinterpreted:

> The emphasis on resistance [in post-structural interpretations of development] is, in some sense, welcome and appropriate, but to phrase it categorically as resistance to state interventions, or oppositions to modernisation, seems unhelpful: for while explaining some phenomena, others become harder to explain when resistance is essentialised in this way.
>
> (Bebbington, 2000: 513)

Thus it is important not to romanticise resistance or define restrictively only certain kinds of acts as resistance when in fact this can be multiple and varied. Resistance can be violent and authoritarian in nature and is not always inherently progressive. Identities are complex and plural and thus there are also varied forms of identification. The term 'archipelagos of resistance' (Slater, 2002: 261) seems to capture this need to focus on how collective and cross-cultural identities are formed around challenges to the power of development's discourses.

These arise at particular points and in particular places, and are very important in how sense and meaning is made of development as an idea or organisational framework of practice. Individuals and groups create their own geographies, using places in a whole range of ways that were not always envisaged or intended by planners and administrators (Holloway and Hubbard, 2001). Dam construction in Vietnam, China or the Lao People's Democratic Republic, for example, may be viewed through the lenses of local people, national governments, multilateral agencies or NGO activists (Parnwell and Rigg, 2001). We thus cannot essentialise the 'local' as uniformly progressive or oppose this to an oppressive, reactionary 'global' (Kiely, 2000). The term 'anti-globalisation' is a case in point, papering over the varied and multiple complaints that have emerged from different locales at different times and across a diversity of political ideologies. Rather, it is necessary to tease out the complex interweaving of sovereignty, governance and territory (Slater, 2002). This comes back to the need to move away from the simplistic image that there is some sort of 'evil genie who organizes the system, loading the dice and making sure the same people win all the time' (Rist, 1997: 122).

The first section of this chapter deals with the question of empowerment, exploring the notion of people's power and the use of other keywords such as 'emancipation'. The second section focuses on the notion of 'geopolitics from below' and discusses further the crisis of developmentalism in some states of the 'South'. The third section raises questions about the extent to which networks and coalitions of resistances can be 'grassroots' as well as 'global' in terms of their reach and implications. A final section looks at the specific case of resistance to neoliberalism in post-apartheid South Africa.

EMPOWERMENT FROM ABOVE OR BELOW?

People are unquestionably central to the development process and an integral element in all development strategies, yet we have seen that this has not always been the case. In looking at how development pathways 'from below' have emerged from different acts of resistance, it is possible to see how people challenge the meanings and practices of development, through the formation of their identities, solidarities and resistance. The survival needs of people have often been overlooked and there has frequently been a failure to consider the implications of development policies at the level of individuals, households and communities. Development policy is thus often likely to be undermined by this failure to view the household and families in a holistic manner (Potter *et al.*, 1999).

Inequalities also manifest themselves at this level of the household, meaning that not all households have the same level of access to income, assets (e.g. land) or local and national power structures. It is important therefore for development organisations to understand the different roles and responsibilities of men and women at the household level. Development strategies assume 'people' and 'communities' to be homogeneous and passive rather than differentiated and dynamic. Development organisations have also made erroneous assumptions about the distribution of power and decision-making at the household level, for example taking it for granted that there is an equal division of labour between men and women within households and communities (Power, 2000). As we have already seen, national statistics often conceal marked variations which exist within and between specific regions, between rural and urban areas and between men and women. The contributions of women in particular to household income and welfare needs to be understood much more fully (Moser, 1993; Kabeer, 1994). Women have also been subject, in different regions and settings, to the multiple failings of development programmes and discourses, but also:

> to [the] misuse of international development funds . . . to the impact of structural adjustment firings on marital relations; [women] have experienced the demolition of their homes and the destruction of the fragile balance of coping strategies that formerly made it possible for them to pay school fees for their children. And they *actively, efficaciously and indignantly resist such development* . . . we are only beginning to record [this story] across the globe.
>
> (Perry and Schenck, 2001: 257–258, emphasis added)

Women have been absolutely central to the very formation of effective resistance around the world and the record of these important stories does need to be much fuller and more able to inform development policy than is currently the case. The recent conflicts between 'Third World' and 'First World' feminists are also interesting here. The global and international nature of feminism has been characterised partly by struggles and differences between 'First World' and 'Third World' feminisms over what constitutes a human right (Radcliffe, 1994) and around the problematic notion of a 'Third World woman' (see Chapters 5 and 6). Western feminism thus has problems coming to terms with the question of difference and interpreting struggles of women in the 'Third World'. Mohanty (1991) critiques the essentialism of feminist theory, seeking to highlight the problems involved in downplaying cultural and economic differences separating women in supposedly common struggles and concerns. This work also cautioned about the use of generalising assumptions in referring to resistance. The variability of economic and cultural backgrounds from which feminist writings emerge complexifies the search for common understandings, definitions and declarations. Again this would seem to highlight the importance of being sensitive to place and an understanding of difference in the formation of international coalitions and transnational networks. The literature on gender and development suggests, among many other things, the different forms male and female resistance can take and how women in particular have been able to challenge and overcome male bias in the development process in their everyday rejection of and resistance to development in a number of different contexts and settings. This also involves a concern for masculinity and the ways in which this is expressed through, or contested in, resistance of various kinds.

As we have seen, mainstream theories of development, although taking many different forms during the twentieth century, have often been centred on the primacy of the nation-state (Hettne, 1995; Kiely and Marfleet 1998). The state has been seen typically as the primary agent of development, particularly since the 1960s when many countries attempted ambitious large-scale modernisation projects with the backing of international development institutions and commercial lenders. The formation of new states, particularly in Asia and in sub-Saharan Africa, followed at a time when development ideals were becoming increasingly popular worldwide. The 'age of great optimism' imbued these emerging states with a sense of the importance of development in the postcolonial era. This was originally underwritten by a degree of popular support and legitimacy for newly formed states and their visions and strategies, but this was severely curtailed in the 1980s and 1990s by indebtedness, reduced commodity prices, a growing dependence on international donors, and escalating recession and protectionism in the global economy.

This chapter suggests that a different focus, on peoples and places, on solidarity and struggle, is useful and important here. State-centred accounts of development ignore the multiple ways in which development is (re)made every day, on the ground or at the grassroots level by men, women and children around the world. There is therefore an important connection we need to establish here between the subjective identities of development and the changing nature of sovereignty. The important point is to focus on the ways in which the authority of the state is being deligitimated and 'hollowed out' as a result of resistance and struggles in different (yet often similar) places. There are no universal definitions of the state to offer here and we must remember that states do not behave in any predetermined way. That would be to subscribe to some of the 'Thirdworldist' mythologies which surrounded neocolonialism and the notion of states acting like puppets on a stage with prearranged roles and constumes. Ideology also has an important bearing on the way the role of the state is viewed in development. Reflecting the World Bank's neoliberal framework, the World Development Report (WDR) views an effective state almost exclusively 'through the lens of economic efficiency', with states seen as providing either 'barriers' or 'lubricants' to free market economic reforms. Not surprisingly, the Bank often 'locks society out' (BWP, 2000: 43) in its definition of the state.

The term 'empowerment' is now used regularly by the Bank and other sections of the 'development industry' (Ferguson, 1990) to describe its commitment to popular participation in decision-making for development. Although there is much that is laudable about this agenda, it seems particularly shallow and rhetorical, particularly when articulated by the (rather undemocratic) World Bank. Facilitating 'people's participation' is now on the agenda of most international development institutions but, for Majid Rahnema, the increasingly widespread acceptance of the idea of participation suggests that it has been severely diluted and has lost some radical political potential:

> governments and development institutions are no longer scared by the outcome of people's participation . . . there is little evidence to indicate that

> the participatory approach, as it evolved, did, as a rule, succeed in bringing about new forms of peoples' power.
>
> (Rahnema, 1992: 118–123)

These 'new forms of peoples power' relate, in part, to local experiences of governance and development and the politics of everyday challenges to established institutions. Escobar (1995) also seems somewhat disillusioned with the development concepts of 'participation' and 'empowerment', pointing out that development planners and politicians have often tried to manipulate experiences of participation to suit their own ends. Throughout the 1990s there was a growing recognition of the need to involve local people and communities as agents of their own development 'if only because of the manifest failure of the main theoretical perspectives on development' (Thomas, 2000: 48). The problem remains, however, that participatory processes are often undertaken 'ritualistically' by development agencies and NGOs that have turned out to be manipulative, or harmful to 'those who were supposed to be empowered' (Cooke and Kothari, 2001: 1).

While much is made of realising human potential in development (even increasingly by those with economistic interpretations), 'people-centred' visions of development have not always been as 'empowering' as they have claimed to be. Empowerment speaks of making people the agents of their own development, while doing little about the *causes of inequality*, or about the need for transforming the nature of ties with international development institutions and their ideals. Empowerment implies redistributing power and transforming institutions, but in many cases the principal structures of political and economic power have remained relatively untransformed. The idea of empowerment and its increasing acceptance in development circles is partly a consequence of the rising influence of non-governmental organisations (NGOs), which have often been highly localised and focused on particular issues. There remains little evidence, however, that NGOs are managing to engage in the formal political process successfully 'without becoming embroiled in partisan politics and the distortions that accompany the struggle for state power' (Edwards and Hulme, 1995: 7). Moreover, it remains 'wildly optimistic' to see this vast universe of NGOs as anti-capitalist, radical and emancipatory (Watts, 2001). The extent to which such organisations are liberating in their interventions is questionable, since:

Figure 9.4 Did someone say something?
Source: Cartoon by Maddocks/ID21

> Some of the emancipatory zeal *has* been harnessed by the NGO community but the very existence of dubious hybrid entities such as BONGOs (Business-oriented NGOs) or GRINGOs (government-run NGOs), for instance, exemplifies the extent to which the porous boundary between state and civil society can substitute market-oriented individualism for the radical autonomy of community empowerment.
>
> (Watts, 2001: 177)

The relationship between the state and NGOs is complex but these organisations *can* have an impact on the possibility of democratic development in many societies, particularly in situations of political instability or conflict where they can sometimes provide critical linkages in the absence of effective or disputed state administrations. This, however, assumes that these organisations fully understand the roots and nature of conflict and conditions of instability in which they operate, when this is often far from being the case. In addition, why should NGOs prop up a state that is ailing, thus allowing state administrations to abdicate responsibility for providing basic social services at all times? In some ways therefore it is possible that many development agencies fail to recognise and understand the causes and conditions of political conflict and their wider impact on development (P. White, 1999). To be critical about the sense of 'emancipatory zeal' that NGO communities can harness for development is not to reject the ultimate objective of redistributing power but rather to raise questions about the missionary zeal with which this term is embraced by some northern organisations.

GEOPOLITICS FROM BELOW AND THE CRISES OF STATE-CENTRED DEVELOPMENTALISM

> The very concept of dialogue can turn out to be an ideological weapon in the hands of those who hold power.
>
> (Houtart, 2001: viii)

A crisis of state-centred development strategies and statist 'developmentalism' has been at the heart of a number of recent challenges to development theories and practices that are emerging from below. There is a danger here, however, of drawing a veil over the varied nature of these struggles, 'downgrading' their specificities by exploring their commonalities (Slater, 2002). The places and spaces of development are not socially homogeneous however, but are fractured by a range of important social and cultural identities and through resistances, of various kinds. As we have seen, there have been various forms of resistance to neoliberal ideologies of development, sometimes grouped together as 'globalisation from below'. This also includes the emergence of 'geopolitics from below':

> In contrast to official political discourse about the global economy, these challenges articulate a 'globalization from below' that comprises a 'geopolitics from below' – an evolving international network of groups, organizations and social movements
>
> (Routledge, 1998: 253).

As we have seen, development knowledge is constructed from positions of power and privilege and we need also to understand how the actions of states, elites and international institutions are contested by those who face domination and subordination as a consequence of development knowledge and practices. These varied histories of resistance may be termed 'geopolitics from below', in reference to challenges directed against the hegemony of certain kinds of visions of development. The region of knowledge that later came to be known as 'geopolitics' first emerged in the metropoles of rival imperial powers and then proliferated in Latin America and among the 'Superpowers' (Atkinson, 2000; Dodds and Ó Tuathail, 1996). As we have seen, geopolitics stems partly from the recognition that geography is fundamentally about power and the authority to organise, occupy and administer space:

> Geography is about power. Although often assumed to be innocent, the geography of the world is not a product of nature but a product of the struggles between competing authorities over the power to organize, occupy, and administer space.
>
> (Ó Tuathail, 1996: 1)

It is precisely these struggles over space and the authority to organise, occupy and administer space that the notion of 'geopolitics from below' seeks to flag up. Following Ó Tuathail and Agnew (1992: 191) we can interpret geopolitics both as a self-conscious tradition (a corpus) of writings and institutions and as the discursive practice by which 'intellectuals of statecraft "spatialise" international politics in such a way as to represent it as a "world" characterised by particular types of peoples and dramas'. Similarly it is useful to examine how various groups, concerned with development, seek to 'spatialise' international economic and political change and to represent this as a world characterised by particular peoples and dramas. In order to understand these resistances we need to adopt a broader and more 'holistic' view of politics, going beyond the 'state-in-society' approach of comparative politics which focuses on the social and political interactions between states and societies.

As Agnew (1998) shows, geopolitical imaginations or ideas do not simply exist 'out there' in texts or policy documents but emanate directly from and feed into practice and social actions. What is also important here is the 'anti-geopolitics' that emerges from below through movements of resistance that seek to challenge the claims of political classes, states and other institutions of development (Routledge, 1998: 245). This can take myriad forms and can operate at a variety of geographical scales, including colonial anti-geopolitics, which points to the armed liberation struggles fought against colonial powers and the movement of decolonisation, and 'Cold War anti-geopolitics', which involves dependency theory, the rejection of US containment discourses, the emergence of an anti-nuclear peace movement and the resistance to socialist ideologies in Eastern Europe. Then there is 'New World Order anti-geopolitics', which refers to the contested discourses surrounding the Gulf War and US involvement in the Middle East, the emergence of the Zapatistas and transnational resistance to the myths of free trade (Routledge, 1998). The relationships between these geographies is important here since it raises questions about the networks of power which operate around development and the various different scales at which this occurs.

In terms of the economy, social movements often articulate conflicts around the productive resources in a society such as forest or water resources, calling for new services, new forms of access and more equitable distributions. Many of these movements seem on one level to express 'political struggles' for power and resources, but also articulate cultural struggles over identity and attachment to place. In some ways these movements have stopped expecting everything to come from the good will of those in positions of power (Rist, 1997) and instead organise themselves collectively, inventing new forms of social and spatial linkages and new ways of securing an existence. The social and geographical diversity of these struggles makes it difficult to generalise, but exclusion does form a common ground and this has even been claimed as the basis of the autonomy of these movements.

The *Movimento dos Trabalhadores Rurais Sem Terra* (MST) or Landless Rural Workers' Movement in Brazil is a useful example of the organised struggle over access to socio-economic and political resources. Forged around a mass social movement that was first formed in 1985, the MST comprises many of those dispossessed in the course of Brazilian state-centred 'development'. The MST is thus made up of croppers, casual pickers, farm labourers and those dispossessed of their land by commercial agriculture or mechanisation (Langevin and Rosset, 1997; Routledge, 1999; MST, 2002). The movement has targeted Brazil's vast estates that lie unused by private landowners, where illegal land invasions were organised to retake possession of these crucial resources, and the land is then resettled, and communities established. Symbolically, long marches to the capital city, Brasilia, have been organised, often in opposition to the government of Fernando Henrique Cardoso. The principal objective of the 'sem terra' organisation is to carry out land reform from below and to challenge the elite's grip on so-called democratic rule. In some ways the MST has created many new co-operatives and is formulating an 'alternative rural development strategy', one which challenges the authority and neoliberal strategies of the state. In the MST's manifesto to the Brazilian people of August 2000, the movement argued that the country's problems had become more acute under Cardoso and were made worse by the President's attempts to 'modernise' rural areas (MST, 2000).

Figure 9.5 'We have a chance to lift millions out of poverty'
Source: Oxfam

The MST manifesto also talked of 'building another project for Brazil', a popular project, one where people 'take the reigns of economic policy' (MST, 2000: 2). There is widespread support for these objectives from many Brazilians, who share a concern for democratising popular access to resources such as land. The country's largest commercial agricultural enterprises comprise only 1.6 per cent of all farms, yet they hold some 53.4 per cent of all agricultural land (MST, 2002). Challenging the global economic and political order imposed by the WTO, IMF and World Bank in Brazil, this movement does have a quite extraordinary capacity to 'mobilise the excluded' (Rocha, 2002: 12) and has reinvented itself around a more ecological focus in recent years, starting environmental education programmes and awareness campaigns. In this sense, one reading of this organisation is that it might be understood as among the 'torch-bearing front runners in the global movement towards greater sustainability' (Rocha, 2002: 3). Around the notion of 'landless' peoples, a variety of cultural identities have been asserted, as these debates are also very much about nationhood, belonging and values and meanings of national citizenship. The movement now has community radio stations and its own newspaper. The MST has also set up a National Collective for Gender. Today more than 250,000 families have won land titles to over fifteen million acres as a result of the MST, transforming Brazil's development and politics. It organises in states across Brazil, although it is estimated that some 4.8 million farmers still have no access to land. The MST has also set up schools and food co-ops.

The state-centred character of politics and development has also been challenged in Mexico through critiques of neoliberal strategies and the role of the Mexican state in forging development. According to Esteva, hundreds of associations across the country are struggling to 'regenerate' their local space (Esteva, 2001). One of the most frequently cited examples has been that of the *Ejercito Zapatista Liberación Nacional* – the EZLN or Zapatistas that emerged in the state of Chiapas. The Zapatistas seek to resist NAFTA and is made up of predominantly indigenous (Mayan) peoples. Chiapas is among the poorest states in Mexico and its resources have been ruthlessly exploited. In addition, land ownership in this state has become more uneven as a result of the free trade agreement, while the neoliberal principles behind it have had many negative impacts on agricultural prices and markets. In their struggle, the Zapatistas have occupied the capital of Chiapas State and many other provincial towns, and have made good use of the media in disseminating knowledge about the uneven process of development there and in building resistance to this. In March 2001 the Zapatistas travelled from Chiapas through thirteen different states, arriving at the Zócalo – the central square of the capital city in Mexico – to demand a place in the constitution.

What was interesting about the march was the extent to which displays of sympathy and solidarity came from various parts of the Americas. It also took place around a particular locality within the capital city with all the symbolism that this act involved. In addition, these struggles have simultaneously articulated ecological, cultural and economic objectives, waging their conflict on several fronts and seeking to form alliances with similar organisations around the world. It is precisely this kind of connectivity, linking peoples and places around resistance, struggle and solidarity, that geographers interested in development must focus on. In 1998 Subcomandante Marcos, spokesperson of the Zapatistas (EZLN), said that as a result, his movement was little more than a 'symptom' of something much bigger:

Figure 9.6 Subcommandante Marcos discussing constitutional reform concerning rights and indigenous culture, April 2001
Source: EZLN (2002)

> Don't give too much weight to the EZLN; it's nothing more than a symptom of something more. Years from now, whether or not the EZLN is still around, there is going to be protest and social ferment in many places. I know this because when we rose up against the Government we began to receive displays of solidarity and sympathy not only from Mexicans but from people in Chile, Argentina, Canada, the United States and Central America. They told us that the uprising represents something that they wanted to say, and now they have found the words to say it, each in his or her respective country. I believe the fallacious notion of the end of history has finally been destroyed.
>
> (Marcos, quoted in Ainger, 2001: 2).

At the risk of giving 'too much weight' to the Zapatistas here, what is particularly interesting about Marcos and the Chiapas struggles is this key concern with people and how they are able to make and rewrite history, to change its course by popular pressure 'from below'. The Zapatistas first appeared in the southern-most Mexican state of Chiapas as the Cold War came to an end, challenging the liberal economic order that prevailed there and the developmentalism of the Mexican state. A key aspect of their politics has been their efforts to create a space for new understandings of politics and citizenship to those offered by the dominant liberal vision which emerged among Mexican political elites in the post-Cold War era (Berger, 2001). They seek to articulate a new form of politics that operates simultaneously at local, national and global levels. Moreover, the Zapatistas have articulated an *alternative geography of Chiapas*, highlighting the economic, ecological and cultural exploitation of the indigenous Mayan peoples and peasants for the enrichment of national and international markets (Routledge, 1998). Such resistance is a response to local conditions (those existing in Chiapas, Mexico and North America more generally), but they also acknowledge that these outcomes are in part the product of global forces at work in other spaces and places and therefore in need of transnational responses. What we need to examine here is how the Zapatistas' struggle articulated something that other movements of resistance 'wanted to say' and how common languages of resistance have weaved webs of relations between places that link and connect these common struggles.

The retreat from state-guided developmentalism in a number of other 'paradigmatic' southern countries (such as India, Mexico and Brazil) has led to an increasing number of resistance movements to national, regional and local strategies (Berger, 2001). This is not to downgrade the specificities of these countries' histories, cultures and resistance groups but rather to further interrogate the commonality of developmentalism as the core of these related, but different, crises of state legitimacy and authority. In postcolonial India, for example, resistance has emerged around a whole range of themes relating to social and cultural identities and the question of national economic justice. This has been evident at a variety of spatial scales, and each one produces a kind of 'reinvention of India' (Corbridge and Harriss, 2000). Struggles around the introduction of GM crops or the construction of a dam, among others, have attracted the attention of both the national and international media. The Narmada River Valley project, which spans the

Figure 9.7 Zapatista guerrillas in the Lacandon Forest
Source: EZLN (2002)

states of Madhya Pradesh, Maharashtra and Gujarat, is a case in point. This project envisaged the construction of some thirty major dams along the sacred river and its tributaries involving the flooding of fertile lands and the submergence of long-established towns and villages, leading to many evictions. Two of the major dams are already built and in recent years protest has centred on the third Sardar Sarovar reservoir scheme and has been organised through the *Narmada Bachao Andolan* (Save Narmada Movement (NBA)). This dam is supported by a US$450 million loan from the World Bank, and threatens massive ecological damage and the displacement of hundreds of thousands of people, many of them poor and indigenous.

The NBA has brought together individuals, groups and organisations from across and beyond India who have demanded an end to the project by disrupting construction, blockading roads and through mass demonstrations (Routledge, 1999). Protest has spread through the valley and has drawn international attention to the cultural heritage of this region of India, and to the important spiritual connections to the place that these evictions seek to sever. The protests have thus simultaneously been about ecological and cultural survival and have led to important debates about regional political autonomy. In this way the geopolitical imagination of India is also important here, as are the ways in which different actors in Indian society seek maps of meaning, relevance and order and to project them on to the contested political universe (Chaturvedi, 2000). Jawaharlal Nehru's imaginative geography of postcolonial Indian development, which had a singular and monolithic vision of national unity, failed to 'displace the widespread sense of attachment to "place" in India' (Chaturvedi, 2000: 220), trying to substitute this with a vision of the national collective.

In many parts of India the territories in which people live have important historical and cultural symbolisms of attachment to places and localities. There has, however, been a growing feeling among some groups of alienation from state-centred discourses on development and visions of national citizenship and identity, particularly during the 1990s. In India the idea of decentralised democracy is evident in the way rural people have demanded control over their rivers and forests. Here many people have sought to resist the 'centrifugal tendencies' of the state and the particular renditions that state-centred discourses on development involve (Chaturvedi,

2000: 230). The new networks of international resistance mean that different movements are learning from each other, sharing tactics and resources. Thus NBA leader Medha Patkar has also sought to explore commonalities between struggles in the Narmada valley and those of more 'global' protests against the IMF and World Bank in cities such as Prague and Seattle (BWP, 2000). According to Medha Patkar, this involves going beyond western models:

> If the vast majority of our population is to be fed and clothed, then a balanced vision with our own priorities in place of the Western models is a must. There is no other way but to redefine 'modernity' and the goals of development.
>
> (Patkar, quoted in NBA, 1991: 1)

This then is an important part of the work of organisations such as the NBA which are thus seeking to formulate new goals and meanings for the

Figure 9.8 Women protesting against the construction of the Maheshwar Dam, July 2000
Source: NBA

Figure 9.9 Adivasis from the village Nimrani at a rally, April 2000
Source: NBA

Figure 9.10 Medha Patkar with children at a peace meeting, Satyagraha, August 2000
Source: NBA

development process and to redefine 'modernity' as a consequence.

'WE DO NOT NEED YOU TO SAVE OUR FORESTS'

Protest, tremors and 'social ferment' around free trade and development issues has also arisen in Thailand, where many thousands of rural people have gathered at the gates of Government House in the capital city (Bangkok) to protest their exclusion from the Asian miracles and success stories that the IFIs are so fond of reminiscing about. These have included villagers affected by big dam projects, small farmers and fisher folk who came together to create a rural coalition – the Assembly of the Poor – comprising the excluded. Protestors went right to the heart of the capital city to erect a makeshift 'village of the poor', where they camped out in the smog and the traffic for ninety-nine days, surviving by growing vegetables illegally along the banks of the city's river (Ainger, 2001). They declared at the time that:

> Rivers and forests, on which the survival of rural families depends, have been plundered from the people. . . . The collapse of agricultural society forces people out of their communities to cheaply sell their labour in the city. . . . The people must set up the country's development direction. The people must be the real beneficiaries of development.
>
> (quoted in Ainger, 2001: 2)

The message from the assembly was clear: that

the 'development direction' must be shaped by and for the benefit of the people. In Thailand the incidence of poverty has fallen in some areas of the country but not in others (Dixon, 1999), and growth has often not extended far beyond the geographical limits of Bangkok. Inequality has thus increased in Thailand as a whole (Jansen, 2001). The highly uneven pattern of growth is distorted both regionally (in favour of the metropolitan region) and sectorally (in favour of certain industries). The rapid growth of the Thai economy has also had considerable environmental costs, with a loss of forest cover and biodiversity, traffic congestion and air and water pollution (Rigg, 1995). It was even suggested, by Walden Bello of the Bangkok-based Focus on the Global South, that loose environmental controls actually attract foreign investors who face stricter controls at home (Bello *et al.*, 1998), leading to the 'Siamese tragedy' of Thai development. New protest groups are starting to emerge across Thailand however. In addition, there has been the emergence of something of a 'new localism' (Pasuk and Baker, 2000) which is now challenging many of the assumptions that had underlain the orthodox development doctrine of unfettered capitalism, particularly since the melt-down of 1997. In some ways this new localism:

> has drawn heavily on populist, nationalist, romantic and Utopian visions of traditional, harmonious and self-reliant (typically rural) communities, and prevailing anti-development or sustainable development discourses.
>
> (Parnwell and Rigg, 2001: 207)

At the local level, what is emerging in Thailand are genuine and serious attempts by people 'to resist, mould, interpret and, indeed, vivify the forces of globalisation' (Parnwell and Rigg, 2001: 207). A small number of other Asian countries also enjoyed high growth rates until the well-documented 'Asian crises' began in 1997. The 'secret' of the Asian success stories depends on individual ideological persuasion and has been much debated, given its dependence on interactions between business and the state (Haggard, 1998). There is no 'East Asian miracle' however; rather there are 'several, different stories of East Asian countries' (Jansen, 2001: 352). The Thai 'Assembly of the Poor' came together again in 1998 to join the coalition of protest movements against the IMF bail-out programme in the wake of the Asian financial crisis, and again when thousands converged on the Asian Development Bank meetings in Chiang Mai in May 2000. Some of the protesters reportedly carried a tombstone on their back on which were inscribed the words: 'There is a price on the water, a meter in the rice paddies, dollars in the soil, resorts in the forests.' This notion of 'dollars in the soil' again seems to resonate with the resistance of other excluded peoples around the world which have often been centred on issues of access to ecological resources. One study in the late 1990s concluded that some 30 per cent of Thai children apparently believed that the IMF was in fact a UFO! (Ainger, 2001).

Few if any of the major institutions – the WTOs, the World Bank, the IMF – have been able to meet in recent years without being accompanied by visible signs of protest. The 2002 G-8 meeting in Western Canada, for example, spent some US$300 million on security measures and threatened the use of lethal force against protestors (BWP, July/August 2002). These amounts nearly outweighed new 'commitments' secured at the meeting on African development. This also involved policing the peaceful parallel conference called the 'G-6B' (Group of 6 billion) which met at the same time. 'IMF riots' – over the price of staples such as food and fuel – have been occurring across the South since the 1970s.

Some commentators have referred to an emerging grassroots network or to 'the multitude' (Hardt and Negri, 2000). This is seen to represent the inversion or the mirror image of a stratum of concentrated power 'from above', where the market is king. This 'multitude' embodies the real world below, all that cannot be reduced to the status of a commodity to be bought and sold in a global marketplace, focusing on human beings, nature, culture and diversity. In fact, it is problematic to talk of an 'anti-global' movement at all, in that these different movements and multitudes, as we have seen, combine a geopolitics from below which is very much about the global and the power of global institutions. This involves, as Hardt and Negri (2000) suggest, a challenging of the idea that 'the global surfaces of the world market are interchangeable'. Just as the geography of modernisation and modernisation surfaces was once seen as interchangeable, so neoliberalism is based around a similar assumption that free trade is good for all, irrespective of particular cultures and economies. Against the religious orthodoxy and the model of a 'single economic blueprint', where the market rules, this 'multitude' represents diverse, people-centred alternatives. In

Figure 9.11 World Economic System: 'Error' poster
Source: Adbusters

the Zapatistas' words: 'One no, many yeses.' Against what Shiva (2001) calls the 'monoculture' of economic globalisation there is thus a growing demand for a world where many worlds fit (Ainger, 2001: 2). This is an objective which must also be central to any rethinking of development geographies, which seeks to contest singular blueprints and orthodoxies and carves out a space for representations of other worlds. The call for 'people-centred development' is far from new but its widespread (mis)appropriation by neoliberal agencies has devalued the idea of popular participation and empowerment through development. New approaches to the democratisation of development are unfolding across the world, and geographers need to acknowledge these struggles and their significance for theorising development.

Vía Campesina is one particularly important example of an international peasant union uniting farmers, rural women, indigenous groups and landless peoples and is one of the most innovative new forms of international networking that has emerged so far. Created in 1992 (when peasant leaders from Central and North America met in Nicaragua), this international movement campaigns on a number of issues such as food sovereignty, agrarian reform, credit and external debt, technology, women's participation and rural development. It is especially interesting in that it has a very wide geographical coverage and holds meetings every three years, rotating the location of meetings

BOX 9.2

The World Forum for Alternatives

The manifesto of the World Forum for Alternatives, chaired by Samir Amin in Dakar, Senegal, was published in March 1997 and its core assumptions were reiterated at the 'Other Davos' in 2001 (Amin, 2001b). The idea of the forum is to bring together social movements and intellectuals in collective analyses of global issues and in the search for alternatives. It was born as an idea in 1996, before taking shape in Cairo in March 1997. Initially it sought to target the World Economic Forum meetings in Davos, Switzerland, but has since produced several new organisations and networks, setting up working groups on social movements and alternative modes of socio-economic organisation. Its 1997 manifesto included the following statements:

- It is time to reclaim the March of History.
- It is time to make the Economy serve the peoples of the World.
- It is time to break down the wall between North and South.
- It is time to confront the crisis of our civilisation.
- It is time to refuse the dictatorship of Money.
- It is time to replace cynicism with hope.
- It is time to rebuild and democratise the State.
- It is time to recreate the citizenry.
- It is time to Salvage Collective Values.
- It is time to Globalise Social Struggles.
- It is time to build on People's Resistance.
- Now is the time for joining forces.
- A time for Creative Universal Thought has arrived.
- The time to rebuild and extend Democracy is here.

***Source*: WSF, 2001: 122–123**

between member regions. Again this organisation also targets the World Bank, the IMF and the WTO and seeks to build international coalitions.

The movement declared 17 April 2002, for example, an international day of protest in order to 'delegitimize neo-liberal globalization' (*Via Campesina*, 2002: 1). It also took an active role in the Porto Alegre WSF summits and has introduced some important gender position papers used to inform discussion and policy within social movements and social forums. These position papers often refer to the 'enslaving dependence' of TNCs and the impoverishment of women through adjustment reforms. Members of the movement include the Landless of Brazil (MST) and the radical Karnataka State Farmers' Association of India. With a combined membership of millions, it represents one of the largest single organisations of people opposed to the WTO. What is interesting about this movement is that some of the first points of resistance to global capitalism seem to be appearing from those people whose livelihoods still depend directly on everyday access to natural resources (Ainger, 2001). In response, movements of natural resource-based communities have created coalitions of the dispossessed, excluded and marginalised. For example, members of a network of Indian *adivasi* activists invaded World Bank offices in New Delhi and plastered its walls with cow dung. At the time they declared:

> For the World Bank and the WTO, our forests are a marketable commodity. But for us, the forests are a home, our source of livelihood, the dwelling of our gods, the burial ground of our ancestors, the inspiration of our culture. We do not need you to save our forests. We will not let you sell our forests. So go back from our forests and our country.
>
> (quoted in Ainger, 2001: 5–6)

At the risk of over-simplification, the message that seems to issue from such movements is that we don't need 'you' (the World Bank, the IMF or the WTO) to save 'us' or our forests and communities. Neoliberal institutions often efface or deny the history of struggles in countries such as Mexico or India, refusing to acknowledge the diversity and inspiration of their cultures and heritage. These institutions also often overlook people's historic and cultural attachments to place. The National Alliance of Peoples' Movements in India, galvanised by the incredible energy of the NBA, unites over a hundred mass organisations, from fisher folk to farmers. Common to all the diverse struggles that the NBA represents is the fight for people's control over their own lives and resources. Sanjay Sangavi of the NBA describes this as 'the emergence of a new politics of environmental socialism in India' (Sangavi, quoted in Ainger, 2001: 6). The potential political force as grassroots groups like these begin to link up internationally is immense. 'Our resistances arise separately, but we are beginning to recognize one another' (Ainger, 2001: 1). Medha Patkar of the NBA movement in India was on the streets of Prague in September 2000, protesting against the World Bank and the IMF. At the time she argued:

> It's not about the First and Third World, North and South. There is a section of the population that is just as present in the US and in Britain – the homeless, unemployed people, on the streets of London – which is also there in the indigenous communities, villages and farms of India, Indonesia, the Philippines, Mexico, Brazil. And all those who face the backlash of this kind of economics are coming together – to create a new, people-centred world order.
>
> (Patkar, quoted in Ainger, 2001: 4)

This is an interesting contention then that the labels North/South, developed/developing, rich/poor or First World/Third World are in a way irrelevant. Within London and Los Angeles are all of these dualities and more. Poverty and unemployment link people, places and communities in diverse locations such that the fate and future of Mexican farmers is not entirely disconnected from that of American farmers, for example, in that free trade agreements and ideologies dominate political and economic interactions between these and many other countries around the world. Resistance at the 'grassroots' level is not new however. The question is: can diverse, dispersed movements everywhere manage to construct bonds of solidarity and support? In establishing ties these movements can manage to support each other and 'together they will be able to change the course of contemporary history' (Chomsky, quoted in Ainger, 2001: 6). Allied to an increasingly popular opposition to global institutions and neoliberalism in the North, 'bonds of solidarity' can be internationalised and come to transcend the boundaries of place and nation. This comes back to legacies of 'Thirdworldism' and the idea of a project of linked resistance. These sorts of bonds and connections between countries of the South are an issue in postcolonial theory where a fundamental question concerns the postcolonial thematic of accepting and accommodating difference and recognising shared or common histories of exclusion and marginalisation. Can contemporary solidarities challenge the shape and contours of established geographies of development? Again the need to rethink the local and global is paramount in that it is necessary to examine how the context of 'local' struggles can be theorised in a way which allows for the fact that the local is not totally bounded but is interconnected with global processes and ideologies. In order to begin to answer this question we can explore the international roots of South African resistance.

NEOLIBERALISM, DEMOCRACY AND RESISTANCE IN POST-APARTHEID SOUTH AFRICA

> Neoliberal development discourse is utterly unsuited to the conditions prevailing in post-apartheid South Africa.
>
> (Peet, 2002: 66)

The liberation of South Africa from centuries of apartheid in the early 1990s is a useful example of the importance of popular power, social movements and international coalitions in bringing about radical change. Apartheid was centred upon a kind of 'governmentalisation of geography', where racial segregation created supposedly separate spaces of development according to an individual's racial identity. The deracialisation of these spaces was a long and complex process. Soon after the end of centuries of apartheid segregation in South Africa, the 'sharks of global capitalism' quickly came to encircle the country (Haffajee, 2001). Many South Africans have thus been frustrated with the severe limitations placed on policy under a neoliberal development framework and within the wider neoliberal development agencies that dominate contemporary development thinking (Peet, 2002). After the country's first two multi-party elections, the African National Congress (ANC) has made some quite drastic reorientations of policy as a result of external financial and political pressures in ways which are increasingly being resisted by many people in many places across this 'rainbow nation'.

The ANC itself was built around grassroots networks of organisations and individuals that were searching for alternatives and came together around cultural identities which were commonly excluded from society and the majority of political and economic spaces. The Reconstruction and Development Programme (RDP) was the first attempt by the ANC to map out a strategy of development in post-apartheid times and aimed to be 'people-driven', 'integrated' and 'sustainable' in its conception. Its goals included one million houses, universal and affordable electricity, a national health scheme and social security. By 1999, following the adoption of the RDP, some three million people were provided with safe drinking-water from taps within 200m of their home (Peet, 2002), and there were other advances in housing and health-care, for example. None the less, since the mid-1990s there has been a growth in the pressure exerted on the ANC to curtail the more radical dimensions of the RDP's objectives. The ANC has since then changed quite substantially the course and character of its ideology from that of the liberation movement which took power on a wave of euphoria in 1994. The RDP began as a radical social-democratic policy document based on the Freedom Charter and centred on human, infrastructural and economic development. In 1996, however, the ANC was forced by powerful investors and the IMF to adapt itself to the 'realities' of the global economy with its new Growth Employment and Redistribution strategy (GEAR).

This newer programme shifted the emphasis from growth through redistribution towards redistribution through growth and has not been popular with some of the unions, which have pointed out that GEAR is more a concession to the corporate world than it is to working-class South Africans (Peet, 2002). International economic elites helped to 'discipline' and shape this programme and its heart is neoliberal, placing macroeconomic targets such as low inflation and a low budget deficit (3 per cent) at the apex of policy formation, and relegating development goals to second place. Since then, health, welfare, education, electrification and housing budgets have been slashed (Bond, 2001). Income disparity has actually increased since the end of apartheid and unemployment stood at 37.6 per cent in 2000 (Peet, 2002). Land redistribution has also been slow, despite the highly organised network of social movements that were in place at the end of apartheid.

For South African activists such as Trevor Ngwane the reason is that '[t]here's been a shift in policy from a redistributive policy to a trickle-down policy', where the benefits of 'growth' must trickle down rather than be actively redistributed. A veteran anti-apartheid activist born in Soweto, Ngwane was expelled as a local councillor of the ANC for Pimville in 1999, disciplined after objecting to the government's World Bank-influenced development model for Johannesburg involving privatisation (known locally as 'corporatisation') of public services such as electricity and water. The struggle against apartheid in South Africa is so recent that a proud culture of resistance is still latent in the townships, and it is this that is feeding the groundswell of resistance at the grassroots level (Haffajee, 2001). Many meetings have been organised which have been allied to the Anti-Privatization Forum, of which Ngwane is secretary, a national forum that links a range of organisations which oppose various forms of privatisation and which assist with community struggles against them. Ngwane has protested in Washington against the World Bank, joining the World Social Forum in Porto Alegre, Brazil, and the World Economic Forum meeting on South African soil in June 2001. In conjunction with local and international academics, radical groups, trade unionists and others, this new movement is nascent but has potential, and has been particularly effective in the protests against the global pharmaceutical giants and in favour of affordable AIDS drugs, for example. Of the ANC, Ngwane argues that the organisation is:

> a shell of its former self. It has no mass politics; it only prepares for power struggles . . . but most people are demobilized, cynical; they are leaving the stage. . . . Our problem now is to provide a political home for these people, but there isn't a consensus of how we relate to the state. We are a young democracy remember. . . . The ANC in power is very unresponsive. This is their big mistake. When people elect you, you've got to be there for them. . . . When the next election comes in 2004, there will be pressure from the left for a more coherent approach.
>
> (quoted in Haffajee, 2001)

This is a view shared by Patrick Bond (2000a,b) who argues that half of the World Bank's US$200 million in loans to South Africa went to expand white consumers' access to electricity, which was denied to virtually all black South Africans until the 1980s. The apartheid debt inherited by the ANC in 1994 was around US$25 billion, but because of power relations prevailing at the time, and a fear of offending foreign lenders, Nelson Mandela and his advisers agreed to service the loans. In response a group of activists formed Jubilee South Africa, demanding total cancellation by creditors in the USA, Switzerland, Britain and Germany. Led by the Archbishop of Cape Town Desmond Tutu, by Njongonkulu Ndungane and Mandela's official biographer, Professor Fatima Meer, the Jubilee movement also demands reparations from financiers who supported apartheid and colonialism throughout the region. The World Bank and Citibank are other key targets, and Jubilee South Africa (working with other coalitions from around the world) has helped catalyse the World Bank Bonds Boycott, reviving the international solidarity tactics once used to encourage disinvestment from companies doing business in apartheid-era South Africa. This campaign highlights the fact that because the Bank raises its funds by issuing bonds on the private financial markets, it is possible to bankrupt the Bank by boycotting the bonds that support it (Hari, 2002). Many US cities have already joined the boycott, which has been attacked by Bank President James Wolfensohn and UK former Development Secretary Clare Short as 'ill advised'. This is very reminiscent of the ways in which many South Africans boycotted the payments for rent and services provided in the townships during the apartheid era. In a similar way international coalitions remind the development and commercial banks that ordinary people can very quickly bring powerful organisations to their knees and force the pace and direction of change in important ways.

Johannesburg hosted the World Conference on Sustainable Development (Rio-plus 10) in August and September 2002 which fast became a focus for resistance to national and international development discourses and the disciplinary power of the IFIs. In many ways hundreds of citizen mobilisations around the world, from Seattle to Cochabamba, from Prague to Harare, have provided the inspiration for this (Bond, 2000a). Local activists in the healthcare, water, environment, economic justice, community, women's, youth, church and labour movements were all involved in the meeting and did not hesitate to remind visitors that racial liberation in South Africa has come at a huge socio-economic cost:

> For Ngwane, the metaphor of the anti-apartheid struggle – such as has inspired the World Bank Bonds Boycott campaign – also applies to 'decommodification' struggles over land, air, water and everything in between, uniting grassroots progressives against common enemies and around 'rights-based' demands that put people before profits. And perhaps Rio-plus 10 will be where we break the chains of global apartheid, not least the neoliberal policies foisted on the new South Africa as part of its Faustian compromise with globalization.
>
> (Bond, 2000a: 88)

Patrick Bond also focuses on the economic arguments that were used to persuade ANC leaders of the benefits of what he calls a 'social contract capitalism' (2000b: 53) and a 'narrowing of economic discourse'. In this it should be remembered that the development aims of the GEAR and RDP programmes were initially progressive (although the ANC has subsequently failed to meet many of its targets). The IFI-imposed strictures on public expenditure have limited this however. This comes back to the question of the damage done by the World Bank in ideology and policy spheres, where redistributive policies have been overlooked in favour of 'essentially status quo arrangements' (Bond, 2000a: 155).

The shape of alternatives to a dying Washington consensus is perhaps less clear however, except to say that alliances of 'progressive' forces on a global (but not globalising) basis are assuming increasing significance. The nature of these types of alliance needs to be specified much more clearly but can be investigated through an analysis of resistance emerging around specific products and processes (see Box 9.3). The World Bank's strategies of cooption and persuasion of key think-tanks, foundations and government officials are also interesting here. International solidarity led to the demise of the apartheid regime, but the freedoms of South Africa are being curtailed today in a different sense because of neoliberal policies. Solidarity among people who have been 'pushed to struggle in defence of their standard of living' (Ngwane, quoted in Haffajee, 2001) involves connecting movements based in Soweto, the township that was a symbol of the struggle against apartheid, to a range of other local, national and international struggles and coalitions. A key question concerns the extent to which the community of 'experts' now writing South African development discourse are willing to respond to these struggles and the important questions they raise. Neoliberalism may well be 'utterly unsuitable' in this respect (Peet, 2002: 66).

Another issue concerns the importance of gender in some of these emerging networks of South African resistance. Morrell (2001) talks about how shortly after the ANC election victory of 1994 many commentators had negative views of men, promoting stereotypes and isolating one or two aspects of masculinity and assuming they are universal and commonplace. Much attention is still focused on the inferior position of women relative to men but is shifting slowly from a concern with women's access to resources to include issues of masculinity. Masculinities are fluid (geographically and historically) and are not fixed but are constantly being broken down, defended/protected and remade. Gender relations are fundamentally about power and also about the 'patriarchal dividend' or the advantage that accrues to men through the subordination of women. Ethnicity and class can influence how this 'patriarchal dividend' comes to men, how they understand their masculinity and how they seek to deploy it.

Predominantly, gendered power relations have left a legacy whereby women are more likely to be disadvantaged relative to men, have less access to resources, benefits, information and decision-making, and fewer rights both within the household and in the public sphere. Thus far, then, these concerns and the struggle for gender equality have been narrowly perceived to be a 'women's issue' and gender programmes designed with a sole focus on women. But if men generally benefit from gender power relations, can we continue to ignore their roles in the struggle for gender equality? The development of gender programmes that involve men and the role they can play in a movement towards more gender-equitable development has been relatively slow in South Africa. The involvement of men in the transformative process required to attain gender equality has a number of entry points, some of which are: ending gender-based violence, commencing human rights and peace initiatives, and poverty reduction strategies (Moser and Clark, 2001; Moser and Norton, 2001). Much of the literature continues to be 'gender-blind', portraying men as perpetrators (in defence of their nation) and women as 'victims' of violence (of

BOX 9.3

Straightening the bent world of the banana trade

One particular agricultural product around which resistance to global injustices has begun to emerge is bananas. Bananas are in many ways strangely symptomatic of the wide range of injustices present in the globalisation of international trade: unacceptable working and living conditions for producers, environmental disaster forced by toxic chemicals and intensive farming methods, suppression of independent organisation through trade unions and the disproportionate economic and political power of the handful of multinational corporations which supply the markets for bananas in northern countries (http://www.bananalink.org.uk/). Bananas are in turn linked to many of the international trade rules that we have examined in previous chapters which consequently shape the lives of both consumers and producers. Some 55 million tonnes of bananas are produced each year in Africa, Asia, the Caribbean and Latin America due to the suitability of climatic conditions. The vast majority are consumed locally but about 20 per cent enters world trade markets. Brazil and India are the two major producers in an ever more crowded global market but use growing levels of chemical inputs to help them compete. UK NGO Banana Link campaigns to make these connections more explicit through a variety of educational resources on line through their website and through various forms of activism. Among other things, the organisation issues calls for support and solidarity with the actions of unions fighting the exploitative practices of TNCs such as Del Monte. In Colombia, Dole's attacks on the rights of workers were highlighted by the organisation in conjunction with the Colombian union SINTRAINAGRO which was seeking to organise against Dole's flagrant disregard for ILO conventions. Dole is a company that claims (before consumers) to be fair and just in its production, but organisations such as Banana Link are exposing this, illustrating the on-the-ground realities of production and suggesting in the process what consumers can do to change these practices. This organisation also works towards a 'sustainable banana economy' and attempts to mobilise the British public via campaigns and actions. Links have been made with ethical trade movements connected to other Fair Trade goods such as coffee and cocoa but also with a wide variety of other national and international campaign groups.

'Fair trade' bananas from Costa Rica entered the UK (through 1000 Co-op stores) in January 2000 – but elsewhere Fair Trade bananas account for only 8 per cent of the total share (Schwarz, 2001). The Fair Trade network of European and North American activists now works in more than thirty producer countries, paying an agreed minimum price (even where the world market price has fallen) and also providing a 'social premium' to help poor producers. Ninety different Fair Trade products (such as coffee, tea, honey, sugar, chocolate, cocoa, fruit and fruit juice) are now available in UK supermarkets and are responsible for an annual turnover of some £23 million a year (Schwarz, 2001). The experiment remains vulnerable to world market fluctuations, however, and is not always the clear-cut alternative to the capitalist system it is constructed to be. Do social premiums always reach the poor and is this simply 'fairer' trade rather than truly fair trade *per se*? In the UK over 75 per cent of banana sales are now through the major supermarkets (Chambron, 2000). UK consumers can enjoy organic bananas, red bananas, apple bananas, home-ripening bananas and sweet baby bananas (among other variations) which pander to a variety of consumer needs and desires.

Liddell and Donovan (2000: 6) estimate that of the huge profits generated for these companies only tiny shares reach producers and only 12 per cent of the total revenues remain in the producing country. Local companies will sometimes be used by some of the TNCs to disguise this link with oppressive working conditions, offloading pressures to local firms (who in turn make decisions about unionisation). Fair Trade and local producer co-operatives hold some hope for the future but not without major reform in the way that international trade is organised. Greater connections are being made which make more explicit the long-standing relationships between producers and consumers – many of the latter are often blissfully unaware of the injustices in this trade which need to be peeled back and exposed. According to Anne-Claire Chambron (1999, 2000) bananas are the 'green gold' of TNCs. Connections are also being made among producers, from South Africa to Jamaica, where people are beginning to mobilise around these injustices.

abduction and abuse). The causes, costs and consequences of violence are gendered however. Women should not be seen as 'objects' in the process of resisting development but as active agents. We thus need to be alert to the *power relations of gender* as well as to local constructions of masculinity and femininity.

CONCLUSIONS: A GLOBAL FABRIC OF STRUGGLE?

> Over the past decade a transformation has been taking place as the threads of local movements are woven into a new global fabric of struggle. They [social movements] are beginning to understand that unless they can organize transnationally, they're dead.
>
> (Ainger, 2001: 5)

A fabric of struggle linking cities such as Seattle, Melbourne, Prague, Quebec, Genoa and Washington is beginning to be sown which raises important questions about the order of twenty-first-century century world development and the need for democratising the major global development institutions. In January 2000 the World Economic Forum was again under siege at its base in Davos, Switzerland. Throughout the year, women and men, farmers, workers, the unemployed, professionals, students and indigenous peoples expressed their commitment to struggle for rights, freedoms, education and employment. What we have seen is that each resistance movement, in its own way, operates to an extent outside the structures of state power in attempting to (re)create models of some kind of direct democracy, often based around the notion of some kind of 'alternative' development. For the Zapatistas, the outlawing of the common ownership of land (fought for by Emiliano Zapata, folk hero of the Mexican revolution that began in 1910) was the crucial turning point. In Ecuador, the IMF-imposed dollarisation of the economy was the match that ignited a variety of popular protests and struggles by indigenous peoples seeking to explore the capacity to develop collectively (Ainger, 2001).

In Thailand and India, it was the plundering of natural resources and the absence of popular participation in development planning. For South Africans, post-apartheid 'trade-offs' with the global economic elite in the context of escalating income inequality has also led to new waves of resistance. For much of Southeast Asia it was the IMF austerity measures imposed on their shattered economies after the financial crisis of 1997. Thousand of activists across Mexico mobilised millions of people, radicalising communities as they went and calling for changes from within and 'from below', while articulating counter-visions and strategies. In some ways, radicalised development geographies can seek to focus on these expressions of 'geopolitics from below' and the alternative notions of development that are gradually emerging from them. Although some of these struggles are avowedly about specific issues of land and ecology, they are caught up with wider issues of territory, identity and statehood and represent important practical exercises in 'theorising back'. In a way, people's identities are constituted through these struggles and in relation to the forces they seek to resist. Activists have sought to shatter the image of world institutions as uncontested forces for good, disputing the power of neoliberal agencies such as the IFIs, the WEF or WTO and rejecting their tendency to 'interfere' in national development policy debates. The World

Figure 9.12 'Get on board for Genoa'
Source: Globalise Resistance

Social Forum 'Call for Mobilization' in January 2001 captures the reasons why these struggles are necessary:

> we are fighting against the hegemony of finance, the destruction of our cultures, the monopolization of knowledge, mass media and communication, the degradation of nature, and the destruction of the quality of life by transnational corporations and anti-democratic policies.
>
> (WSF, 2001: 122)

The WSF in particular has begun to emerge as a global space and an alternative international assembly for the discussion and networking that is taking place on a people-to-people basis between grassroots organisations from around the world. What is especially interesting about the forum is its concern to contest the monopolisation of knowledge about development by the IFIs and its insistence that other ways of knowing are possible, with indigenous sources seen as crucial to development thinking and practice. In 2002, the second WSF brought together 51,300 participants, representing 4,909 organisations from some 131 different countries (Brecher and Costello, 2002). The Forum has its critics, however, who argue that it has not yet come up with a 'blueprint' for global social reform for all its global dialogues and attempts to avoid the imposition of new 'Third World' elites. Again there is this persistent notion that there is a single recipe

Figure 9.13 A shirt urging people to Jam the World Trade Organisation
Source: Adbusters

book, a simple blueprint that can be formed for all 'developing countries'. The Porto Alegre call for social movements (WSF, 2002: 1) had already noted that 'diversity is our strength, the basis of our unity'. In addition, it is not especially difficult to see what these organisations are 'for': they seek democracy, the abolition of indebtedness, freedom from poverty and violence, and gender equality and rights (among other things). What connects these various social tremors to earlier waves of 'Thirdworldism' is a common concern for self-determination, especially the rights of indigenous communities with respect to land and resources.

Another key commonality in many of these struggles is that they centre upon those people who are excluded from development plans and strategies or do not benefit directly from 'progress' in their societies. The term 'multitude' points up the fact that these people are seen as the expendable, the invisible people who global development ignores and those at the heart of what the Zapatistas call the 'fourth world war' (Ainger, 2001: 5). New coalitions of the dispossessed are uniting not just within countries, but internationally. Furthermore many older social movements, principally labour and trade unions that have existed for some time, are also being drawn into these processes of mobilisation (the success of which will also ultimately depend on their participation). Thus a variety of social movements, as we have seen, have demanded land, constitutional recognition and, perhaps most importantly, *meaningful participation* in development planning. They represent a 'troublesome forest that walks', a 'stream that joins other streams to become a river' in the words of Subcommandante Marcos (quoted in Ainger, 2001: 5). In a variety of spaces, from rural to urban, from the South to the North, unrest against the global political and economic order is quickly spreading. These disparate threads are the early stages of a movement that is beginning to reconstitute the global economic landscape, reshaping the way development is played out in the twenty-first century. In addition, from these disparate threads of global struggles against neoliberalism are emerging important new forms of resistance, building solidarity and support networks that unite diverse locations and struggles. The campaign to make drugs available for AIDS patients at reasonable prices is a good example of a struggle that made connections between local and global in terms of mobilising for wider popular access to healthcare. There is some evidence that these kinds of campaigns are yielding some (although often rather limited) results, forcing changes in trade rules and the operations of profit-seeking TNCs (Brecher and Costello, 2002).

The free trade agreements signed in the Americas in recent years such as the FTAA and NAFTA and the summit meetings where their progress is discussed have been particularly important sites of struggle in the internationalisation of resistance. Protesters, aggressive and passive alike, have been met with clouds of teargas, water cannons, pepper spray and rubber bullets. All kinds of groups and coalitions have come together around these meetings however, often involving large numbers of local trade unionists (as in Quebec) and public employees. At the alternative 'Peoples' Summits', organised around these meetings, the voices of dissent against the modern-day 'enclosures' of free trade have been clear and articulate. Movements across Latin America also mobilised for one of the largest, most ambitious self-organised referenda ever attempted, the *consulta* popular. At the end of 2001 fourteen countries in the Americas were thus asked to vote on the principles of the Free Trade Area of the Americas. Broad-based social movements organised at a variety of spatial levels have been keen to enable this kind of popular consultation and are at least trying to illustrate its usefulness and necessity. The red banner of the MST, now a familiar site at many demos in Brazil, seems to capture why this is important. Depicting MST workers with tools, their flag symbolises the rootedness of their work in the everyday struggles of workers and labourers. Most recently, the MST has begun to work through organisations such as the CLOC (Latin American Co-ordination of Peasant Organisations) and has also begun exploring connections with African organisations since early 2000 and the beginnings of the WSF meetings.

The final communiqués which have been issued from a number of the 'People's Summits' surrounding free trade meetings seem to suggest that 'another world is possible' and also desirable in the context of neoliberalism. In December 2000 a meeting of campaigners in Dakar (Senegal) assessed Africa's debt crisis and the human effects of SAPs. Condemning the neoliberal model, the participants (many from the Jubilee movement uniting North and South) highlighted the ecological aspects of debt and called for its total cancellation alongside a rejection of the PRSPs and SAPs developed by the IFIs. The Dakar manifesto issued at the conference favours globalisation based on *solidarity among people*, with priority to basic human needs, calling for reviews of external borrowing and stressing the importance of promoting 'home-grown' solutions to development rather than a 'universal model'.

Future resistance may become more effective if it relates more directly to particular ideologies of development, such as those codified in the World Bank's successor to structural adjustment, the Poverty Reduction Strategy Papers (PRSPs). This was a classic case of new wine in old bottles, as the 'responsive' Bank was seen to replace the infamous adjustment policies of old, but the logic remains unchanged according to southern civil society groups such as Focus on the Global South and Jubilee South. The views, the perspectives, the mode of analysis and even the menu of options brought by each country delegation remain much as before. Focusing on poverty reduction in the way that these agencies do arguably narrows the space for a discussion of other development issues and alternatives/ models. Although the blueprint approach of the past may have been reworked, 'civil societies' are not universally the same and depart from different political and economic contexts and starting points which defy easy generalisation about their intended role in policy 'dialogue'. Many texts on development uncritically praise the role of civil societies without questioning their legitimacy, effectiveness and capacity to represent their constituencies.

Various statements are made by the major development institutions about broadening and deepening participation in policy formation in favour of 'civil society', but real and active citizen engagement is often a myth. The World Bank often believes that any issue can be resolved if you throw enough statistics and references at it. Opening up the process of report-writing (e.g. on the WDR) would be the clearest statement yet from the Bank that it truly embraces ideas of widening participation. This chapter has drawn attention to the need to focus on varied and multiple forms of resistance to the imposition of this 'universal model' in order to understand the capacity for promoting these kinds of alternative, 'home-grown' interpretations of development. After an initial period of shock and confusion in the wake of 11 September 2001, campaigners have once again resumed protests against the IFIs and multilateral trade and economic organisations, restarting their campaign against this universalising model. In order to change these agencies and to undermine the power and authority

of their ideologies, new coalitions need to be explored (as with the banana trade), linking worlds through wider senses of connectivity, linking people and places, poverty and consumption, producers and consumers, North and South. One of the principal obstacles to such connectivity is that:

> transnational resistance today itself depends on goods (from education to means of communication like mobiles or computers) which are unequally distributed throughout the world, allowing certain groups (notably in the 'north') far greater access to the circuits of resistance, encouraging them – if only by default – to 'represent' less well-resourced constituencies.
>
> (Boehmer and Moore-Gilbert, 2002: 19)

Important questions need to be asked here about who is representing whom in the process of organising transnational resistance and about who is claiming to speak for the less well-resourced constituencies of the unequal world in which we live. Postcolonial studies have enormous potential in this regard. This can help to point up the problems involved in assuming that this unequal world speaks with one voice or is unable to represent itself. There is a need to move away from what Pieterse (1998) refers to as the tendency in post-development and postcolonial writings to focus on resistance but not on empowerment. This comes back to the notion that postcolonialism offers a useful critique but few alternatives, avoiding complex questions concerning development defined as liberation and emancipation. In addition, transnational resistance will need to form structures and organisations which are stable and lasting if alliances are to be formed that can be meaningful (e.g. with trade unions). Again this is dependent partly on recognising the strength that comes from diversity and the particularly important contributions of women to traditions of struggle and resistance in the South. At the same time, existing forms of social organisation (such as trade unionism) need to adapt to new forms of political and economic reality to revitalise their politics and to form alliances with other kinds of groups and organisations. This will obviously take some time and patience, particularly when it will also necessarily raise complex issues of partnership.

Figure 9.14 'Stop Sweatshops' demonstration in New York, March 2001
Source: The author

A key objective for many post-development writers is to seek to 'delegitimise' the institutions of global governance and the idea of development more generally. As we have seen however, many critics point to the absence of alternatives and criticise post-development writers for romanticising the local or calling for some kind of prejudicing of the local over the global. The point rather should be to consider how democracies which emerge from the (often romanticised) 'grassroots' organisations can be linked to wider struggles and changes at a variety of geographical scales. None the less, surely the initiatives of those excluded from development are preferable to 'an anyway impossible transformation of international structures' (Rist, 1997: 245)? There has clearly been a flourishing of grassroots organising and as a result the very ideas of democracy and development are being reinvented. As a result there is also a need for new ways of thinking about subjectivity and political agency. It is not therefore a question of local vs. global but rather one of rethinking the entire meaning of these terms, the identities that are constructed around them and the ways in which they come together in different times and places (Massey, 1994; Massey and Jess, 1995). In a globalising world the intersection of politics and economics in a number of places and spaces around the world is increasingly important and must be a key focal point for challenging prevailing inequalities of wealth and power (Wills, 2002).

In this chapter we have looked at a number of countries such as Argentina, South Africa, India, Brazil, Mexico and Thailand, each of which stands as a different yet similar example of the grounded and place-specific nature of resistance around common themes of exclusion and impoverishment. Lasting coalitions for promoting global social change and reform will require further connections to be made between struggles for identity and

economic justice, bringing together peoples and groups to find and cement common causes and alliances without masking the differences between them. Post-development writers and social movements share similar rejections of development and understandings of the crises of neoliberalism and in this sense there is surely scope for much greater dialogue between the two. In addition, a postcolonial geography, which involves close attention to identity, culture and subjectivity, will have an important role to play in the way we seek to understand place-specific identities and practices. There is now a much wider recognition among social scientists of the multiple and unstable nature of identities and their constant reconstruction in the development process (Munck, 1999). Geographers are well placed to map these changing cartographies of struggle and to grapple with the intersections of politics, economy and culture 'as they unfold on the ground' (Wills, 2002: 96). Rethinking geographies of development may also enable us to demystify development and to focus on *growth through equality* rather than equality via growth (Wallerstein, 1994). In this sense, Wallerstein (1994) argues that development can be a 'lodestar' for the hopes and dreams of social movements rather than just a distant illusion. Quiet encroachments and collective insubordinations of various kinds and in various places have shown that there *are* alternatives and new concepts of growth and progress are being fashioned through practices of various kinds every day.

BOX 9.4

Chapter-related websites

http://www.globalizethis.org/
Globalise This!

http://www.attac.org/
Anti-globalisation organisation ATTAC.

http://www.wtowatch.org/
WTO Watch, a trade observatory on WTO, globalisation and sustainability.

http://www.southcentre.org/
The South Centre.

http://www.oneworld.net/campaigns/trade/
One World.

http://www.twnside.org.sg/trade_1.htm/
Third World Network.

http://www.brettonwoodsproject.org/topic/knowledgebank/
Bretton Woods Project.

http://www.developmentgap.org/
Development Gap.

http://www.fairtrade.org.uk
Fair Trade resources.

http://www.bananalink.org/
Banana Link.

http://www.jubileesouth.net
Jubilee South.

http://www.g6bpeoplesummit.org/
G-6B Summit.

http://www.rowmanlittlefield.org/
Some useful information about social justice activism.

http://www.ifg.org/
International Forum on Globalisation, a network of activists and intellectuals.

http://www.forestpeoples.org/
Briefings on World Bank forest policy.

http://www.worldbank.org/poverty/empowerment/
World Bank empowerment issues.

http://www.mstbrazil.org/
MST Rural Landless Workers' Movement in Brazil.

http://www.viacampesina.org/
Vía Campesina.

http://www.lanic.utexas.edu/la/region/indigenous/
Latin American resources on indigenous peoples with very good links.

http://www.ezln.org/
Zapatistas.

http://www.narmada.org/
Friends of Narmada.

http://flag.blackened.net/revolt/mexico.html/
Contains good archive resources on the Zapatistas.

10
Conclusions
resisting the temptations of remedies, mirages and fairy-tales

> Once you grasp this, once you understand that neoliberalism is not a force like gravity but a totally artificial construct, you can also understand that what some people have created, other people can change.
>
> (George, 2001: 7)

TOWARDS 2015

The CIA report, *Global Trends 2015*, warns of the dangers associated with the polarisation between winners and losers in the first fifteen years of the twenty-first century. It talks about the forecast global economic boom (compared with that of the 1960s and 1970s) and argues that there is a growing global middle class which is now 'two billion strong' (CIA, 2000: 22). The basic trend towards the expansion of international trade is seen in the report as both inevitable and inexorable and we are told that there are real risks of emerging market countries and economies 'falling' behind, save for a few possible 'breakout candidates' (CIA, 2000: 22) such as China and India that might manage to escape this prescribed scenario. Great swathes of the globe are seen as characterised by 'endemic' corruption and political instability which is not regarded as the product of foreign interventions but of *internal* breakdowns or autocracy (particularly in the Middle East and North Africa). Growth will be uneven, it is recognised, and liberalisation and globalisation will inevitably 'create bumps in the road' which may be highly disruptive. The possibility of instability and turmoil spreading (from Russia to Brazil) is also recognised in the report, with the US economy identified as 'the most important driver of recent global growth' (CIA, 2000: 24).

Middle-Eastern states are rather problematically portrayed by the CIA as sharing a singular, common view of globalisation (which sees more challenges and opportunities), while sub-Saharan Africa is cloaked in the now familiar guise of endemic disease:

> The interplay of demographics and disease – as well as poor governance – will be the major determinants of Africa's increasing marginalization by 2015. . . . Conditions for economic development in sub-Saharan Africa are limited by the persistence of conflicts, poor political leadership and endemic corruption, and uncertain weather conditions.
>
> (CIA, 2000: 45)

This marginalisation, however, is not seen as related to the functioning of the world economy, to the neoliberal strategies of organisations such as the World Bank or to the action of powerful states such as the USA. These countries 'lag behind', so this argument follows, simply because of their failure to pursue certain kinds of reforms or even because of the weather rather than as a result of the failures of multilateral institutions and their preoccupations with 'free trade' and markets. We have seen that the IFIs disseminate an economistic view of development and adopt rather problematic and highly contradictory concepts of indigenous knowledge, participation and empowerment. Like many other actors in the global development industry today, the Bank claims to be working in partnership with poor countries, 'facilitating' strategies which are 'country-driven' and 'country-owned' wherever they go.

Figure 10.1 Smart aid or smart bombs?
Source: CWS and Zetterling

Surely the best way, however, to illustrate a commitment to decentralising power and formulating more localised 'ownership' of development would be to move the organisation's capital from Washington, DC?

It has been estimated that the World Bank generates US$2 billion in economic activity for the Washington, DC region every year (BWP, May/June 2002). These figures include some US$851 million paid to nearly 14,000 employees, consultants and contractors and some US$217 million paid for goods and services. Clearly if the Bank left to go to another world capital there would be a 'hue and cry' about the loss of jobs and income, but it would confirm the Bank's willingness to base itself and its ideologies in other kinds of places and spaces. If the US Congress again threatens to suspend its financial support for the Bank, campaigners for reform of the IFIs should lobby to move the organisation since 'after decades of interest payments many Southern capitals could use the cash' (BWP, May/June 2002: 6). Strangely, at the G-8 summit in Western Canada in June 2002 reform of the IFIs failed to even make the agenda.

Figure 10.2 Time for the UN to act to sustain global development
Source: Adam Lee and *The Straits Times* (Singapore)

What is particularly interesting about the largest global development agencies and their vision of progress and positive change is the way in which they are based on particular kinds of imagined geographies of global difference and notions of 'backward', 'underdeveloped', 'developing' or 'Third World' countries. The debates of the 1960s and 1970s about radical geography seem to offer important beginnings here in formulating a critique of these visions and imaginations in that they incorporated a call for new relations between what Buchanan (1977b: 366) called the 'director' and 'directed' societies:

> Social, economic and ecological crises in countries such as the USA or Japan, conventionally regarded in the past as demonstration models of the success of orthodox GNP-oriented development . . . these things are strengthening the argument of those who believe a complete re-thinking of the development issue is long overdue. And until this re-thinking is done the export of dubious models of development to the Third World is simply an act of irresponsible arrogance.
>
> (Buchanan, 1977b: 374–375)

Thus arguably this 'irresponsible arrogance' continues to shape development thinking today and as such we might consider rejecting the formulae of what Buchanan called the 'dirty word' of development and its prescriptions aimed at 'Third World patients'. Radical geographers such as Buchanan have refocused attention on individual human

beings as agents, envisioning 'development as liberation' (an idea popular in Latin America when he was writing). Others have asked important questions about development 'for whom' and 'for what', criticising the role of 'experts' and the technocratic solutions they put forward, while refusing the idea that development was about economic growth only and about 'becoming some sort of replica of the developed societies' (Buchanan, 1977b: 366). Buchanan and others have also noted the confusing and misleading nature of development and its persistent failure, in some conceptions, to get to the root causes. Today this deception is more extreme than ever.

Nearly forty years after Weisskopf (1964) described the myth of a growing GNP as a dogma, a shibboleth, a golden calf and a centre of worship, few measures of growth, change and development seem able to capture the diversity and multiplicity of meanings and objectives involved in development debates. Much is said about population growth in the 'Third World', but the consumption of four-fifths of the Earth's resources by one-fifth of its population 'remains largely unproblematised' (Yapa, 2002: 36). There are thus severe ecological limits to the universal attainment of the good life predicted by certain models of development (with its assumption that all consumers aspire to 'western' affluence). Even the 'champagne glass of inequality' referred to by the UN points to certain kinds of western consumption. The point here is that the poor are so often 'examined, judged and found wanting because the contents of their consumption basket are modest' (Yapa, 2002: 36). The issue then is not that GNP, for example, as a measure of 'progress' is fundamentally flawed (as has been more than amply demonstrated in the literature) but rather that it is so frequently and unconsciously presumed that access to per capita income will lead to particular patterns of consumption and to certain kinds of goods associated with the historical experiences of a few.

In an important book entitled *Toward 2000*, Raymond Williams (1983) looks at the linkages between the cardinal points East–West and North–South, acknowledging the extent to which, despite its limitations, the concept of the three worlds has dramatically focused attention on the 'appalling facts of contemporary poverty'. Although the 'three worlds' schema involves dramatic simplifications and a number of simple pictures and images of development, these are important beginnings:

> Whether we are thinking of 'the Third World' or of 'The South', it matters very much whether we are seeing a blocked and generalised poverty, or a more complex system . . . there are not only radical but operative differences between the 'newly industrialising countries' (from South Korea to Brazil) and the OPEC oil producing countries, the strategic-mineral and cash-crop economies and . . . the most desperately poor and disadvantaged peoples.
>
> (Williams, 1983: 203)

Thus the image of a 'blocked' and 'generalised' world of poverty 'matters very much' to the way we distinguish economic and political difference between countries. Major internal variations of

Figure 10.3 Poster advertising a 'Buy Nothing Day' (November 2001)
Source: Image courtesy of www.adbusters.org (2002)

Figure 10.4 Aids in Africa is swept under the cartographic continental carpet
Source: Cartoon by M. Keefe for Aids Africa

income and power further complicate these dramas and simplicities. Writing some twenty years ago, Williams noted the sustained and 'often reckless' development of an international credit economy (with commercial banks lining up to provide capital to 'boost' and 'stimulate' growth), and pointed out how the pressures for a continuing adaptation to 'externally conceived development' were immense (Williams, 1983: 207). In this way it may be seen that development is inherently ideological rather than the natural, unfolding, evolving process it is sometimes seen as. This in turn 'naturalises' the process by which hierarchies are created and countries come to be naturally seen as 'underdeveloped' or 'developing' according to predefined and prescribed models and histories. This also involves a naturalisation of borders and boundaries between 'developing countries', a downplaying of their fluidity and contested nature as markers of cultural and historical difference between states. In the global capitalist economy, but also in many actual and proposed alternatives to it, the 'idea of growth' is continually taken for granted 'as the sovereign remedy for all existing economic inequalities' (Williams, 1983: 213). Thus there are 'sovereign remedies' which reign at particular moments in time and for many countries of the world through the disciplinary power of the largest agencies. Thus what is needed is a *deliberalisation* of development and a denaturalisation of the nebulous phenomenon that is neoliberalism, often seen as natural a force as gravity itself (George, 2001).

The changes involved in rethinking development practice are so substantial and resistance from existing interests will be so certain and powerful that nobody can suppose that this will be anything but a very long and complex struggle (Williams, 1983). New forms of alliance and political and transnational labour struggles have begun to emerge but they remain disparate and in some cases incoherent. According to the WSF, people are organising resistance and engaging in struggles in order to create alternatives to such a scenario:

> Some are rebuilding knowledge on the basis of experiences of struggle, some are trying out new economic forms, some are creating the basis of a new kind of politics, and some are inventing new cultures. *It is time to build on people's resistance.*
> (WSF, 2001: 122, emphasis added)

This book partly takes its lead from the WSF, the Other Davos, in that it also aims to amplify the voices of those who are protesting against the structural injustices of the current economic system in order to raise awareness of the possibility of thinking and planning the future differently. The Other Davos refers to a meeting of social movements from different parts of the world that met in Davos in January 1999 at the time of the WEF meeting there, laying down some ideas and guidelines for the construction of networks of solidarity and collective action. These included the MST from Brazil, a trade union group (PICIS) from South Korea, the National Federation of Farmworkers Organizations (FENOP) from Burkina Faso, the Women's Movement from Quebec and the Movement of the Unemployed from France (Houtart, 2001).

GEOGRAPHIES OF NEOLIBERALISATION

Neoliberalism concentrates its efforts in three main areas: free trade in goods and services, the free circulation of capital and freedom of investment. Neoliberal agencies are very good at defining the economic 'winners' but have far less to say about the losers: 'to whom nothing in particular is owed' (George, 2001: 15). In some ways neoliberal discourses have changed the very nature of politics, where exclusion is normalised and naturalised as acceptable and as part and parcel of transformation or the inevitable 'bumps on the road' to progress. At the centre of the debate therefore is a liberalism which sees competition as fundamental to successful development, a competition between nations, regions, places or firms and between individuals. Competition works, neoliberal advocates argue, because it separates the 'sheep from the goats, the men from the boys, the fit from the unfit' (George, 2001). The public sector has a limited role to play here since it is seen as unable to obey the central law of competing for profits. Alternative concepts of growth, progress and even development itself may well be forthcoming and result from these conversations and dialogues. As we have seen, new ideas and perspectives are being fashioned every day in and through place-based identities and practices by peoples across the South in multiple ways where development concepts and practices are continually resisted and remade.

The IFIs continue to concentrate on building the same model of a 'steady state' in the imaginary

BOX 10.1

NEPAD: a 'new' partnership

The New Partnership for Africa's Development (NEPAD) represents a contemporary pledge by African leaders 'to eradicate poverty and to place their countries on a path of sustainable growth and development'. NEPAD attempts to integrate existing bilateral and multilateral financing programmes, aiming to build a 'strong and competitive economy for the continent'. The reaction of civil society groups to the initiative has been mixed. Many have lauded the attempt to forge a new relationship with development partners and its focus on African ownership and reliance. According to the Third World Network (TWN) this 'new relationship' represents little more than a rehashing of neoliberalism. More worry-ingly, according to Trevor Ngwane, a South African activist, 'no civic society, church, political party, parliament or demo-cratic body was consulted in Africa' (cited in BWP, 2002b: 2). At the African Social Forum in Mali in January 2002 there was a consensus against NEPAD, reflected in the attitude of the Bamako declaration produced at the meeting.

A meeting on 26 June 2002 in the Western Canadian mountain resort of Kananaskis further discussed NEPAD, in addition to the progress made so far with the MDGs and the global war against terrorism. Private sector representatives from Microsoft, Coca-Cola and Chevron, among others, recently met to sign a 'Dakar declaration' to 'set up structures under which they can cooperate with NEPAD' (BWP, 2002b: 2). We can question, however, whether the emphasis on this model of governance will allow African states 'sufficient space to articulate indigenous development strategies and political development models' (Santiso, 2001). That NEPAD stood accused of being 'top-down' in its conception among African statesmen such as Thabo Mbeki has not augured well for its likely future success. The US$64 billion plan is unlikely to work, according to many of its African critics, because it does not involve the consent of the people and calls for no real change to existing structures and institutions. During a three-day African forum for discussing NEPAD held in Nairobi in May 2002, keynote speaker Professor Adedeji Adebayo accused 'the West' and international institutions of frustrating efforts to resuscitate African economies: 'Helping Africa is always a refrain and gratuitous rhetoric at UN sessions and international conferences' (quoted in Sammy, 2002: 2).

Many critics are concerned that NEPAD is also still partly based around a number of presidential authoritarian regimes that now claim to be democratic, such as that of former President Daniel Arap Moi in Kenya. Adedeji has had personal experience of how African initiatives (such as those formulated by the UN Economic Commission for Africa, ECA,) are stifled by donors and financial institutions. Alternative formulations of adjustment programmes were drawn up by the ECA (albeit with their own limitations), but these were also swept aside as 'impractical', costly or damaging to trade and enterprise. Adedeji's final comment at the meeting in Nairobi was to say that 'Africa requires building anew; not rehabilitation or reconstruction' (quoted in Sammy, 2002: 3). The suggestion here is that other Africas are possible, that altogether different and more radical changes are not just a possibility but a necessity for any such 'partnership' to be effective. More specifically, NEPAD is weak and inconsistent on questions of gender relations and is silent on issues of inequality and discrimination. These ought to have been a more central focal point of NEPAD objectives (in addition to arising more generally 'from below'). Similarly, not even the Bank's attempts to reduce the incidence of poverty by 'statistical fiat' (defining as poor only those with incomes of less than two dollars a day – one dollar for those in extreme poverty) have succeeded in disguising either the extent and scope of the problem, or its root causes (Petras and Veltmeyer, 2002: 286).

At the 2002 G-8 meeting, NEPAD was again referred to as the 'Marshall plan' for Africa, with British Prime Minister Tony Blair reiterating his promise to 'heal the scar' of African poverty, to 'rescue' Africa through poverty eradication, leading to a 'new start' for the continent. US President George Bush had already offered a US$5 billion boost in US foreign aid from 2004 if 'developing countries' pledge to meet human rights and corruption conditionalities. Canada pushed the US administration to come up with its half of the US$8 billion promised jointly by the USA and the EU in Monterrey, Mexico, earlier in 2002. This money was promised originally as a response to Kofi Annan's call to the Monterrey delegates to pay US$50 billion to help meet the 2015 millennium development goals.

space of the Third World (as obtains in the core) which, as we have seen, is deeply problematic for all sorts of reasons. Governance-related conditionalities from the donor and aid community overwhelmingly dominate the agendas of the international development architecture. Consequently (and not for the first time) there has been a tendency to compare the capabilities of postcolonial states with those of 'western' countries. The popular diagnosis which follows is that development aid and assistance can perform a transplant of the conditions and processes at work in donor countries and in the 'civilised' West. Once again postcolonial states are rudely abstracted from their specific socio-political contexts, as their histories are denied and overlooked in search of the holy mantra of 'good governance'.

A postcolonial critique of these discourses would seek to illustrate how the pressure for 'good government' is partly a legacy of the late colonial period and of emergent discourses of developmentalism in the age of decolonisation. Writing in 1967, John Lee argued that the national development 'project' emerging in Britain between the 1930s and 1960s (and its emerging concern for 'good governments') was an outgrowth of the late colonial era in Asia and Africa (Lee, 1967). During this period, the idea of development was used increasingly as a framework for metropolitan policy interventions which claimed to be improving living standards while relegitimating Empire (Berger, 2001). Britain could advise India on what was constituted by 'good governance', so it was argued, since the metropole was truly enlightened and its democracy firmly entrenched. The idea of good government here was central to the creation of modern subjects. Hence, as Gupta (1998) argues, some strands of decolonisation discourses in India held that national development involved 'mimicking' the historical trajectory of the former imperial ruler. This assumption that development involves a mimicry or replication of something that already exists has been common and is related very much to the way in which the 'metaphor of development gave global hegemony to a purely Western genealogy of history' (Esteva, 1992: 9). Thus postcolonial geographies of development are partly about challenging the hegemony of these 'purely' western histories.

A whole variety of patronising references are made continually in contemporary development discourses to 'strong' and 'weak' states and even 'failed states'. At the beginning of the twenty-first century a language of friends and 'foes' has been (re)invented, along a continuum or order and 'stability' that is defined from without, by 'western' political leaders. By no means should we be tempted to see the 'western' state as a finished project however, especially given all the problems and contradictions in the idea of being 'western'. There are many problems then with representing the dominant idea and practice of development as western instead of something more specific such as 'liberal' (Berger, 2001). Deconstructive engagements with development discourse have also shown how difficult it is to move outside and extricate ourselves from the idea and image of development as 'western'. As Mohammed Ayoob (1995) argues, the assumption is that security-building in the 'Third World' is necessary because of the lack of 'adequate stateness' that is seen to be found in the non-western world. New conceptions of security are required that move beyond the simplistic labels of 'core'/'periphery', 'First World'/'Third World' but rather seek to blur and disrupt these artificial geographical imaginings and boundaries. This is not a simple or straightforward project since so many Cold War discourses of geopolitics balkanised knowledge about international politics, annexing them into a range of separate disciplinary specialisms, including development geography:

> Hence the importance of opening analysis up to the different processes of state formation and historical circumstances constitutive of various post-colonial states, thereby considering different forms rather than obscuring diverse trajectories of state formation.
>
> (Bilgin and Morton, 2002: 73)

All too often the diversity of paths to development or 'trajectories' towards the good life are obscured by the reconstituted 'Sinatra Doctrine' of neoliberalism, founded as it is on the classical tenets of (European) Enlightenment rationality. In this sense we might follow the suggestion of Jean-Francis Bayart (1991), who has argued that it is necessary to dispense with the idea of the 'Third World' by focusing instead on the *specific historic trajectories* of postcolonial states and explorations of their interactions with different societies and cultures. More direct connections need to be made between human security issues (as in the US relationship with Colombia) and wider development issues and concerns, relating to social, cultural and economic processes. In this sense it becomes possible to turn the IFIs' notion of failure around on itself. As Bilgin and Morton (2002: 63) have argued, rarely is the question asked 'Who has failed the "failed state?" ' Perhaps, therefore, rather than focus on 'failed states', increased attention should be given to the 'failed universalisation' of the 'imported state within the post-colonial world' (Bilgin and Morton, 2002: 75).

This then is a key issue here: rather than representing the absence of 'development' as the consequence of a failure to obtain adequate statehood in the eyes of a select band of 'democracies' (themselves forged over much longer time periods), the objective therefore must be to ask questions about why the universalisation of a particular model of statehood or development has failed. There is such enormous and apparent faith in the idea that political democratisation is the corollary and the foundation of economic liberalisation and social justice, yet these 'democracies' often have little to offer in terms of the active redistribution of wealth. Moreover, it is as if the World Bank and IMF themselves are seen to be in no need of a dose of their own medicine, as if they are wholly democratic, pluralistic and open institutions engaged closely with civil societies and communities everywhere. The reality is surely altogether different, with all the silences and absences in their annual writing of global development 'progress', such as in the WDR, which ignore a whole range of crucial debates about sovereignty, political participation, class, gender and disability (among others). No account is seen as necessary of how a postcolonial state is formed or constituted and limited consideration is given to each country's historical involvement with international political discourses. These issues are dismissed as too abstract and 'complex' to enter into the intellectual universe of the IFIs.

Figure 10.5 The IMF: Helping the poor from above
Source: Andrzej Krauze, *Guardian*, 28 March 2002

Concepts of 'rogue' states (also known as 'outlaw' and 'backlash' states') have dominated a great deal of the international development communities' thinking on the Middle East, a trend which has intensified considerably in the wake of the terrorist attacks on New York and Washington in 2001. Except for Cuba and North Korea, most of the 'rogue' states defined by successive US administrations are Muslim (Iraq, Libya, Sudan, Syria), with Iran being the only non-Arab country among the group. The rationale for the emergence of the rogue-state doctrine stemmed 'more from fears of budget cuts following the vanishing of the Soviet threat than from serious security concerns' (Zoubir, 2002: 33). The USA has accordingly expected the rest of the world to isolate these rogues by severing commercial ties, imposing multilateral sanctions and embargoes and hampering the military and technological potential of these states. Middle-Eastern and North African countries can thus be forced to cooperate with the US geopolitical agenda, 'including pro-Israeli interpretations of the Middle East peace process' (Zoubir, 2002: 37). UN sanctions on rogue states (such as Libya) have severely affected the development of these countries, as many Libyans,

Figure 10.6 So many evildoers, so little time!
Source: Beattie: Atlantic Syndication

Cubans and Iraqis would attest. UN sanctions have also taken a 'tragic toll' in Libya, costing the country an estimated US$19 billion and leading to some 21,000 deaths since their imposition in 1992, a US$5.9 billion shortfall in national agriculture and a major decline in the Libyan economy and in standards of living (Zoubir, 2002: 38).

Countries that dare to reject US hegemony in their region or are opposed to US and other foreign interests and presence are isolated and punished in this way, yet somehow are expected to liberalise and develop their economies. In announcing a new tied package after the UN Summit in Monterrey in 2002, President Bush argued that 'liberty and law and opportunity are the conditions of development' (cited in BBC, 22 March 2002). The relationship between declining volumes of US foreign assistance and the rising budget for the Department of Homeland Security is complex but important in rethinking 'development' and its geographies. It follows that if the emphasis is on smart bombs rather than smart aid, international development will continue to occupy a low priority for the US state unless it relates directly to the security of the US homeland (Zuza, 2002). US aid programmes to the Middle East will become the object of increasing scrutiny and attention in the years ahead however. It remains to be seen how this relates to the war on terrorism which may mean that the USA will focus its spending on foreign aid around unilateral concerns and priorities. It also remains unclear how these wars against terrorism will affect the World Bank's conditionality policies and Bretton Woods financing more generally.

Figure 10.7 Peace in the Middle East, 1989
Source: UN Department of Public Information

Figure 10.8 The Parikrama begins with Didi offering prayers to the Goddess Narmada. The Parikrama is an old tradition in which people living on the banks of the Narmada River get an opportunity to come out of their homes and mix with people from various communities and get to know the different facets of their culture
Source: NBA

How do these countries go from rogue and outlaw status to adequate statehood in the eyes of the world's most powerful countries? The answers are far from clear but will continue to pose recurrent dilemmas for the meaning of international development in the twenty-first century. The neglect of Middle-Eastern geographies of development therefore needs to be halted. As Chaudhry (1997: 29) says of Saudi Arabia and Yemen, in 'general discussions of political economy' these are 'two of the least studied cases on record'. By widening the compass a little it is possible to engage this 'region' in a wider variety of debates about the meaning of development and its principal agents. The case of Saudi Arabia, for example, offers a perspective on the origins or invention of these ideas about the [oil] firms as 'agents of development' (Vitalis, 2002: 186). By looking at such cases there is much we can learn about oil politics or neocolonialism today (relevant also in the Niger Delta, for example), about 'emerging markets' and the formation of states and state planning agencies. How indeed is it possible to understand the world economy or the role of the USA in world politics and in shaping international development discourses *without* reference to such countries? It is crucial, however, not to ascribe to the Middle East a false unity when there is such political and economic diversity. There is thus no single Islam but several, just as there are different Americas. 'Islam' and 'the West', as Said (2002) argues, are inadequate banners to follow blindly and they must not be allowed to obscure interdependent histories of injustice and oppression. The events of 2001 and the emergence of the global war on terrorism illustrate how development discourses are dynamic and influenced by changing geopolitical circumstances.

A range of perspectives on development have been explored here, including those of the Asian and African leaders who came to power in the 1950s and 1960s, taking over the newly formed colonial development projects established by imperial powers in the dying days of colonialism and inheriting the long-standing machinery of the colonial states. The role of the state is crucial to rethinking development today and this means considering the possibility of other ways of imagining community:

> the terms on which the newly sovereign nation-states in Asia and Africa were both consolidated, and then incorporated into the wider global order, ensured that the '*state*', often a direct inheritance of the colonial era, was the '*poisoned gift of national liberation*' at the same time as 'the nation' became the prescribed, if not '*the only way to imagine community*'.
>
> (Berger, 2001a: 225, emphasis in original)

Thus the imagination of development and the imagination of the nation have been closely intertwined in a variety of spaces and places since 1945. Even in the Soviet bloc of socialist regimes, the nation was seen as the natural unit of development and the state was seen as the key in the struggle for modernity, to the extent of trying to achieve a 'mimicry' of the economic achievements of the capitalist powers (Dirlik, 1994). What is particularly interesting about many debates concerning national and international development today is that many people are claiming exclusion from the imagined communities of development as the basis of their cultural and political autonomy. In this way marginalised groups can make a virtue out of necessity by profiting from the fact that they are not allowed to share in the booty of 'development' (Rist, 1997). In this way we can draw on the experiences of social movements, exploring the forms of social and spatial linkages that have been developed as new ways of securing an existence.

In a way these varied resistances have often highlighted the 'blindspots' of development discourses and the overwhelming dominance of economistic approaches. This means that organisations such as the World Bank proclaim, for example, to be cognisant of gender, disability and other 'social' relations but only ever in a rather limited, shallow and quite 'instrumental' sense:

> The only arguments that the Bank puts forward to rationalize its interest in gender issues are generally of an economic order: investing in women is profitable, hence justified. *Carried away by its focus on the economic*, the Bank has undertaken econometric analyses in several areas on the rate of return on investment in programmes promoting women.
>
> (Bessis, 2001: 19, emphasis added)

Thus the Bank (and many other sections of the international development business) is fond of econometric analyses, becoming carried away in the process. PRSP processes, for example, have been characterised by a lack of transparency and accountability (to use the Bank's favoured parlance) as these crucial documents have generally been scripted by an 'inner circle from the Finance Ministry and the Bank' (BWP, May/June 2002: 5). 'Civil societies', as the IFIs like to call them, have been (tokenistically) asked for their input on the targets necessary to ensure poverty reduction but they are far from being meaningfully involved in truly open discussions about *how* to achieve these aims and objectives or about the desirability of alternatives in each country. This is always off-limits for the Bank and the Fund. Linking bilateral and multilateral aid to each country's 'compliance' with these IFI-scripted poverty reduction strategies has exerted enormous pressure on recipient states. As we have seen, a key aspect of these debates is the need for 'ownership' and 'empowerment'. If this was really meaningful in the eyes of the international development community, then voting power at the IFIs would be based on democratic rather than economic considerations, such as population size (Dowden, 2002). The Bank or Fund would then be dominated by China or India and might operate quite differently as a result.

In 2002 UK Chancellor Gordon Brown called for urgent action to clamp down on 'vulture funds' – financial institutions that buy up the debt of poor countries at knock-down prices and use the Courts to extract payment in full with interest. The activities of 'vulture' funds have come to prominence following the success of Elliott Associates – a New York-based hedge fund – in a legal battle against the government of Peru. Elliott Associates paid US$11 million (£7.5 million) in 1996 on the secondary market to buy US$20 million of Peru's debt and then sued for full repayment plus recapitalised interest (*Guardian*, 6 May 2002: 17). At a UN meeting of world finance leaders, Brown criticised the vulture funds as a distraction from the global campaign for poverty eradication because they were diverting much needed capital resources away from poverty reduction strategies into the coffers of western banks. Heavily indebted countries, caught up in the 'cruel hoax' of the HIPC, are being drawn into courts around the world by these banks as they seek to extract their pound of flesh, but to focus on only these kinds of funds (and not on the wider processes of capitalist exploitation) simply obscures the real and pressing problems at hand. The World Bank also shares the UK chancellor's fear that commercial banks are undermining their own debt 'relief' strategies and the force of their own prescriptions. There are no existing mechanisms whatsoever for tackling the 'vulture capitalists' and the UN meeting concluded with the recommendation that western countries (reflecting their trusteeship and tutelage roles in the South) should provide legal support and technical assistance to allow the countries affected (such as Ethiopia, Peru, Bolivia and Nicaragua) to fight their court battles. The point here is that these cases would never go to court in a world economy that had no place for 'vultures' of this kind. Further, why is it that the notion of trusteeship continues to shape the view of western states towards poverty, poor people and distant others? This image of a vulture seizing upon weak and defenceless victims and intensifying their decay (without resistance) has also been a recurring theme in debates about 'Third World' others.

We have seen how institutions such as the IMF and World Bank engage in a kind of paradigm maintenance (Wade, 1996) fashioning the truth and telling stories about development successes and failures. The success of the East Asian economies in the 1970s and 1980s was seen as a model that could be replicated and should be followed and imitated in Africa because it is 'the most successful model for development that humanity has ever generated', leading to the 'biggest reduction in poverty for the largest number of people that humanity has ever generated' (Clare Short MP, quoted in Dowden, 2002: 3). The Bank's rendition of what it takes to progress is backed up by a multi-million public relations campaign which seeks to popularise and legitimate these explanations (e.g. of why East Asian economies took the Rostovian take-off towards destination 'good life'). Their knowledge will quench the darkness of poverty and poor countries and finally slay the dragon of backwardness. The dominance of modernisation thinking and modernisation geographies is crucial to the very operation of such institutions.

IMAGINING A POST-DEVELOPMENT ERA

A concern with the imagination of a post-development era should not be pilloried but understood for what it can suggest about the powerful nature of this organising principle of social life. The 'nihilism of post-development' (Hart, 2001: 654) has been seen by many as extreme and reactionary. For some, these critiques have run their course and serve only to take us further down a 'cul-de-sac' in development thinking. In addition, there has been a concern that such writings do not confront questions of capitalist development and ignore the 'multiply inflected capitalisms that have gone into the making of globalization' (Hart, 2001: 651). Although the post-development approach focuses on very particular kinds of social movement and the occasionally vague notion of 'alternative development', there is none the less a welcome recentring of 'local' knowledge and practices which seeks to ground global development in the embedded, in particular places and localities. Critics have also suggested that post-development writers present an over-generalised and essentialised view of reality and that they romanticise local traditions, constructing new kinds of 'Thirdworldist' mythology. Perhaps a more important point here, however, concerns the failings of some post-development writings to understand the embeddedness of the local in the global or in playing down ongoing contestation of development 'on the ground'. These disagreements are in great part a consequence of 'contrasting paradigmatic orientations (liberal, Marxist or poststructuralist)' (Escobar, 2000b: 12). Writers of a Marxist persuasion have been particularly critical (see Corbridge, 1995; Pieterse, 1998, 2000; Kiely, 1999).

In many ways, post-development writings pick up on themes raised in the works of anti-colonial writers such as Cabral, Fanon, Freire or Nyerere and draw inspiration from the insights of the dependency scholars and Foucault which cannot be so casually dismissed (Escobar, 2000b). More recently, Escobar's work has taken on a much more geographical concern with 'place-based practices', calling for a 'reassertion of place, non-capitalism and culture'. In response to his critics Escobar notes that it is interesting that many are white male academics in the North, arguing that the post-development movement has been at least more diverse at this level, including men and women from both the North and the South, living and working in both the North and South (Escobar, 2000b: 13). The post-development 'project' seeks, in part, the reclaiming and pluralisation of modernity and its complex genealogies. In this sense we might also explore the possibility of learning to live with development by criticising and changing it rather than by simply rejecting and discarding it. Perhaps 'modernism is discarded too easily' in that at least critical modernism had a concern for examining the causes of material differences with a view to changing them and making the world a better place (Peet with Hartwick, 1999: 12).

For some geographers, a critical modernism and critical developmentalism is possible here (Peet with Hartwick, 1999: 209). This approach is concerned not just to discard development but to think about how to replace it with something better, as a 'universal, liberating activity' (Peet with Hartwick, 1999: 209). This approach tries to go beyond the neglect of the material realities of deprivation in post-development writings and foregrounds practice and action rather than representation. What is particularly useful about this approach is that it seeks to differentiate various types of developmentalism rather than seeing all forms as inherently 'bad' and negative. In some ways, the same is true for development geography in that there have already been critical forms of developmentalist geographies, as we have seen, which can provide the foundations for radicalising the study of development today. These stem from important traditions of opposition to development and an emphasis on the various trajectories of dependent societies (e.g. in Marxism). This would therefore involve a dialogue with these critical traditions of Marxist, feminist and post-structural critiques and would retain a belief 'in the potential, rather than the present practice, of development'. In addition, this kind of development geography would also seek to combine popular discourses of social movements with the liberating ideas of modernism (Peet with Hartwick, 1999: 198). To an extent, this seems useful in that it recognises that ethical, critical and political principles are necessary in helping to form linkages and connections between movements and places and in understanding the connections and similarities between struggles in different but similar contexts. The key question here, however, is the extent to which developmentalism may be seen as a 'mode of progressive thought' which has long contained critical versions and not always been negative (despite its centrality to the ideology and new religion of neoliberalism). Is the development paradigm at its 'last gasp' (Rist, 1997)?

Regardless of the answer to this question, critical approaches to the study of modernism and modernisation theory will continue to remain crucially relevant to the study of (neoliberal) development today. Modernisation saw modern institutional organisations and rational behaviours arriving in fifteenth- and sixteenth-century Europe, seeing a related spreading of social progress and rational action from efficient development institutions (Peet with Hartwick, 1999: 14). In particular the critique of modernisation geography is useful and insightful in interrogating the explanation of regional variations in development in terms of diffusion, 'from the originating cores of modern institutions and rationalised practices' (Peet with Harwick, 1999: 14). These authors are thus cautious of the extremist tendencies of post-development but still see in it one potential way of relearning to view and reassess the realities of communities around the world. Even the post-development approaches can be deconstructed themselves for what they reveal about the way development works implicitly through our assumptions and worldviews.

How then should geographers seek to view and conceptualise the social, political and economic autonomy of marginalised societies? In a way this depends on whether one is hopeful or optimistic that international economic and trade systems and relations will be fundamentally changed and reorganised in the near future. For Rist (1997) the initiatives of those excluded from development (e.g. in social movements) are to be preferred to the 'anyway impossible' transformation of international structures. My own feeling is that it is necessary to go a little further than a reinvented critical modernism by aiming instead to 'shatter the religious structure that protects 'development' ' (Rist, 1997: 245). A good example of this is the reluctance of Peet with Hartwick (1999) to concede the term 'progress'. This mystical, religious structure of protection derives its very authority from developmentalism (however critical) and the notion of organic, naturalised processes of growth. The theorisation of post-development is a pressing task in this regard. It may seem blasphemous to say this, but the critical modernism perspective almost reproduces the arguments of the first generation of postcolonial leaders such as Nehru or Nkrumah who suggested that colonial state machineries could be made to work for the people, if only they were imbued with different purposes and perspectives, or were instead steered by nationalist movements. Perhaps what is necessary therefore is some *distance from development* in order to effectively challenge some of the supposedly given and self-evident ideas of economism. Thus if post-development critiques are about stripping the walls before putting fresh paint on, what point is there in redecorating the edifice with the same colour scheme? To put it another way, if geographies of development were once characterised by a pious Eurocentrism, why are we still worshipping at the altar of western Enlightenment ideals? As Rist argues: 'Whereas "development" offered hope, the rejection of "development" produces new wealth' (Rist, 1997: 248).

All this involves preparing the ground for 'post-development', which should not be confused with anti-development, since to want something different does not mean simply doing the opposite. Admittedly, such approaches raise many more questions about development than they answer, but they do push for new ways of understanding 'that do not reproduce the centrality of Western ways of creating the world' (Escobar, 2002a: 195). My point then is not that we should all sign up to some singular post-development perspective but rather to point up a wider need to imagine a 'post-development' era so as 'to carve out a clearing for thinking other thoughts, seeing other things, writing other languages' (Escobar, 2002a: 199).

Focusing on questions about a 'global sense of place' in development (Massey, 1992) on social relations and on the connections between places (without denying their specificity) is at the centre of such an agenda. Further study of exchange phenomena is also needed so that we do not focus exclusively on the hegemonic idea of the market and begin instead to further understand the mutually constitutive nature of development processes. This also involves interrogating the spatiality of development and a mapping of the apparatus of 'knowledge–power' as well as a different kind of concern for those 'doing the developing' (Ferguson, 1990; Escobar, 2002). The need for the dissemination of other languages is important here since 'development talk' (Sachs, 1992: 1) pervades not only in official declarations but also in the languages of grassroots organisations.

It is also important to understand how development is about the production of stories and narratives (which has come with the turn to discourse), but these will remain just stories and narratives unless there is a greater engagement with the political and the material (Watts, 2000), and greater discussion of alternative stories and narrations of geographical or economic difference. Corbridge (1999) also warns of the dangers of excessive concentration on the discursive aspects of development which focus attention away from the materiality of social problems and the 'very real successes' that we might associate with development since 1950. This is an important point but it ignores a key assertion of post-development writings, namely that the 'successes' of development and their definition are socially constructed and have a particular spatiality.

The cultural politics of post-development thus has to begin with the everyday lives and struggles of real groups of people, such as women (Fagan, 1999). This reminds us that these debates need to be grounded in a concern for the materiality of discursive formations. A common problem of the critiques of post-development writings is that they use the either/or language of 'right' and 'wrong' notions of development. In this sense we must be mindful of the effect of our distance from those whom we write about. Escobar's concerns are indeed inspiring and do not seem to discard, out of hand, the utopian possibility of reimagining other worlds. This work seems to suggest that it is necessary to map out the multiplicity of these journeys of the imagination, rethinking the cultural politics of difference in a more co-operative and collective way. No approach is 'right' or 'wrong' here and what is important for our purposes is the question of what enables and provides the opportunity to take this journey in the first place:

> For me, this is a journey of the imagination, a dream about the utopian possibility of reconceiving and reconstructing the world from the perspective of, and along with, those subaltern groups that continue to enact a cultural politics of difference as they struggle to defend their places, ecologies and cultures.
>
> (Escobar, 2000b: 14)

Figure 10.9 Middle East dartboard
Source: Cartoon by M. Lane

This politics of difference, as it emerges from struggles in particular places, is an important theme in rethinking development and its contested geographies. Despite the postmodern beginnings of such a concern, this should not be seen as incompatible with a concern for material differences and inequalities.

DECOLONISATION AND DEVELOPMENT (GEOGRAPHY): BEYOND TRUSTEESHIP

Themes of cultural difference and identity tend not to be a feature of many undergraduate textbooks on the subject of development. In this way there is evidence of the 'academic socialisation' of geography students into particular models of thinking about the 'less developed world' (Yapa, 2002). In one US textbook on 'development' (Fisher *et al.*, 1995) the authors divided the world into 'more' and 'less' developed realms, tapping into a variety of binaries and divides (developed/underdeveloped, non-problem/problem, knowing-subject/needy-object, industrialised/non-industrialised), with the first term in each binary opposite seen as the primary and privileged one (Yapa, 2002). Some introductory human geography textbook representations of Africa also seem somewhat fixated with notions of 'tribes' and often generalise from highly localised geographies which are 'taken to stand for an entire continent' (Myers, 2001: 523). A further device in many textbooks is to use images of mother–child suffering which highlight helpless victims, thus reinforcing a kind of 'disaster discourse' by showing 'alarming photographs of starving children, starving mothers, desperate refugees, or charred human remains' (Myers, 2001: 527). Some of the images are undignifying to some of the people whom they seek to (mis)represent, but more importantly they have misled many generations of undergraduate geography students around the world. The following section of Lakshman Yapa's critique is well worth quoting at length here:

> To the millions of people who live in Africa, China and India such photographs are a constant reminder of what they are not. Whatever else they may be . . . is banished into oblivion by a universalizing metric that rank orders peoples by the average cash value of the nation's market basket of consumption. They are pitied, 'wretched of the earth', living in the periphery of the world system . . . young American undergraduates . . . are told that Africans and Indians live in less developed countries. . . . Young as they are, they know that they 'rank' higher than millions of those 'other' people from underdeveloped countries. . . . Furthermore, the discourse of the text having created the less developed other, also recreates the undergraduate reader in the image of the more developed self.
>
> (Yapa, 2002: 43)

Surely the use of such images must take some responsibility for producing the patronising ethnocentric attitudes that societies learn to have towards the people of Asia and Africa? There are a number of particularly crucial points that Yapa's intervention makes. First, we have this pictorial reminder of what the poor 'are not'. Second, there is the pervasive influence of 'universalising metrics' which encourage students of development (just as Walt Whitman had done in his anti-communist manifesto of the 1950s) to rank order the peoples of the world, to construct a hierarchy of growth stages of global 'civilisation', viewed as degrees of democratisation or westernisation. Yapa powerfully illustrates how this metric ordering of progress impresses upon 'young American undergraduates' that they must be higher up the global order of things than their Indian or African counterparts rather than seeking to make connections between the ideologies and practices which link the USA, India and Africa. Reading through debates about poverty in America, there is an overriding assumption that such poverty ought not to exist in the richest nation in the world and that it will be possible eventually to eradicate poverty as we know it (Glasmeier, 2002: 161). According to US census data, some 11.7 per cent of the US labour force was born abroad, while immigration is up from 250,000 people a year in the 1950s to some one million a year today (*The Economist*, 1 June 2002). In addition, poverty persists in the United States (and in all other supposedly 'developed' countries), which is today one of the most unequal societies on Earth (George, 2001).

The 'socialisation' of students in these binaries is far from being unique to the US education system however, but has existed across the social sciences for nearly six decades now. *Rethinking Development Geographies* attempts to offer something of a counter-narration, one which problematises all these binaries and several others. The idea that peoples of the South may be considered rich in cultures and economies has been discussed as has the need to highlight subjectivity and agency in meeting the challenges posed by development and its absence. A powerful and enduring feature of many development discourses has been the creation of an 'Other' to stand out against and be contrasted with the 'Anglo-American', 'western' or 'First World'. Thus a key question in future debates about development geography is the extent to which it is possible to 'think beyond' and around the normative perspective of the developed (Crush, 1995). In this respect it is vitally important to 'de-familiarise the familiar', to not accept 'development' as an automatic given, an end product which is self-evident and therefore unworthy of critical attention. This is not to argue that 'language is all there is', as Crush (1995: 5) points out, but rather to build on postcolonial, feminist and post-structuralist critiques of development knowledge and to see what 'development' is and does in new ways, from new vantage points.

What we can learn from this is that 'patronising ethnocentric attitudes' can and must be challenged and deconstructed in the process of rethinking development geographies. The World Regional Approach that Yapa critiques re-creates an image of the undergraduate reader in the reflection of the developed 'western' world with the USA at its summit. The post-structural critique which Yapa (2002) brings to bear on this geo-writing of global development raises the possibility of thinking about social and spatial reality in a variety of new ways. This book has sought to extend this concern and in so doing hopes to build something of an 'antiracist geography' of development (Peake and Kobayashi, 2002), one that seeks to move beyond the colonialist heritage of development geography, to extent the bounds of disciplinary decolonisation and to move away from racist images of an objectified other. This involves critiquing the privileging of whiteness in development geography but also exploring new forms of activism with new social movements as genuine partners in research, centring geographical practices 'in the streets rather than in the academy' (Peake and Kobayashi, 2002: 55). The ways in which place-based identities shape our readings and interpretations of racial difference and value or devalue spaces of cultural diversity are also relevant here. It is thus necessary to contest the way in which development discourses engage in a social construction of these differences, perpetuating them and creating new spaces of exclusion.

Early ideas of development posited a whole set of discourses about 'latecomers'. These stories had it that the club of the world's richest countries can be infinitely expanded, that eventually all latecomers will be able to join up or *catch up* with the party. Postcolonial literatures offer a useful and much needed corrective to this way of thinking, aiming to re-centre development processes around notions of 'time as lived' (Mbembe, 2001) or to ground them in the consciousness of those subjected to development at all levels: local, national and international (Perry and Schenck, 2001). Only very recently, however, has the attention of postcolonial scholars begun to focus on contemporary instances of

BOX 10.2

Postcolonialism, field research and working in developing countries

In some ways 'postcolonialism' shapes the writing of geographies of development most directly in terms of debates about the practice of 'fieldwork' in 'developing countries'. Researchers from 'western countries' are very likely to experience and to be part of many of the inequalities that exist between the researcher and the researched. Jones (2000) asks the important question 'Why is it acceptable to do development research over 'there' (e.g. Africa) but not 'here' (in Britain or the USA)?' and talks about changing vocabularies and common strategies of inclusion across the First World/Third World dichotomy. 'Here' is the First or developed world and 'there' is the other, the Third or developing world. Once again through fieldwork, these definitions of 'here' and 'there' are gradually becoming blurred and less distinct. Fieldwork, more importantly, is not always about 'other'. As Rabinow (1977) argues, research in the field is partly to be understood as the 'comprehension of the self by the detour of the comprehension of the other' (Rabinow, 1977: ix). Dissertations and research projects of various kinds are not only an academic piece of work but also a learning process which can encourage the develop-ment of many practical skills. Rabinow refers to his Ph.D. thesis in anthropology as 'a studied con-densation of a swirl of people, places and feelings' (Rabinow, 1977: 6).

Other binarisms also need to be considered critically – the idea of 'here' versus 'there', 'us' versus 'them' (subjects of research) or of the 'before' and 'after' of fieldwork. Fieldwork can teach us a lot about ourselves and represents partly what Rabinow calls 'the comprehension of the self by the detour of the comprehension of the other'. In addition, the way in which research seeks to gain authority and legitimacy through fieldwork is also important here, involving the construction of the authority to speak for and about the processes and peoples that we are interested in. Through selecting particular methodologies or particular sources in an investigation, decisions have to be made about how best to represent issues and findings in the final research. This question of what constitutes 'respectable' information is a crucial one. Rather than perpetrating this process of 'unequal exchange' there is a need to communicate the outcomes of research to local institutions or peoples. What does 'knowledge' mean in that particular setting? Research involves the establishment and sometimes transgression of social and cultural boundaries but that does not necessarily mean that the whole process is inherently negative. Postcolonialism highlights the importance of these social and cultural differences, but through fieldwork they can be negotiated and reworked in the process of establishing connections with people and other cultures. For interesting examples of South–South fieldwork see the work of Amitav Ghosh *In an Antique Land* (1992) and a discussion of research by African geographers and fieldwork (Cline-Cole, 1999).

Many students also consider working in the international development sector in the future and after graduation. This might be with international agencies such as the EU or UN or with government institutions such as DFID or USAID. Alternatively there are a whole range of NGOs as well as private development consultancies. There are now a number of research and academic opportunities in this field involving work at research institutions of various kinds.

Figure 10.10 We defend and destroy people we don't understand
Source: © Protest Graphics

transnational organisation and protest (Boehmer and Moore-Gilbert, 2002). Much more could be said here about Seattle, Genoa, the Jubilee campaign or the Fair Trade network or about the global movement for reparations from slavery (central to the agenda of the UN conference on racism in Durban, South Africa, in 2001). This does not only involve political action and 'taking to the streets'; transnational resistance has been expressed through new ideas, representations and sentiments, new cultures, customs and habits.

We have seen that resistance to various forms of western domination are not new but can be traced to the struggles against colonialism, the writings of anti-colonial writers such as Frantz Fanon, and the establishment of the non-aligned movement, for example. The political and cultural expressions of transnational resistance thus have had a diversity of historical and geographical forms (Boehmer and Moore-Gilbert, 2002). For some writers on postcolonialism (Ahmad, 1992; Dirlik, 2002) the most pressing ethical and political duty of postcolonial writings today is to mobilise resistance to certain models of globalisation. For others, 'what should be addressed is not the postcolonial condition, but post-Cold War capitalism' (Berger, 2001: 222). Perhaps it is possible to do both simultaneously? There are numerous problems however, as we have seen, with constructing capitalism as the central reference point of development which holds sway everywhere and always will (according to certain stories about globalisation). Communities around the world, despite the much vaunted global reach and penetrative capabilities of capitalism and development, have never allowed their identities to be completely reshaped by these processes. Neither did they simply surrender to development and 'global capitalism'

BOX 10.3

Chapter-related websites

http://www.devnetjobs.org/
Development jobs.

http://www.brettonwoods.org/
Bretton Woods Organisation.

http://www.imf.org/external/np/prspgen/review/2001/index.html/
IMF.

http://geocities.com/ericaquire/g8mobilization.hml/
G-8 mobilisation.

http://www.twnafrica.org/ISSUES/nepad/
Third World Network.

http://www.worldsummit2002.org/texts/AfricanSocialForum.pdf/
African Social Forum.

http://www.g8.gc/ca/2002nsummit/
G-8 Summit.

http://www.egroups.com/group/devjobs/
Development jobs.

http://www.oneworld.net/jobs/
One World Organisation.

http://www.workingabroad.com/
Working abroad.

http://www.devdir.org/
Development jobs directory.

http://www.nt1.ids.ac.uk/eldis/jobs/
Institute of Development Studies.

their indigenous models of economy or nature (Escobar, 2002).

Potter's (2001b) call for geography to study people and places in the 'majority world', to expand their global reach and enlist other geographers in the pursuit of social justice generated an exchange in the journal *Area* involving Adrian Smith (2002), a geographer interested in the economic and social transformations of 'post-Soviet' East-Central Europe and the 'Second World'. What was particularly interesting here was the way in which connections were being made between post-Soviet Europe and different representations of the economic world which move away from 'capitalocentrism'. In this way anti-capitalist activism comes to be seen as much less quixotic and 'more realistic' (Gibson-Graham, 1999: 84). Importantly, this exchange also raises questions about the possibility of transcending simplistic notions of hierarchical scales from local to global and of thinking about the linking and intertwining of places.

This is a key challenge for development geographers, to think about this kind of critical approach to the study of the spatiality of development and this sense of relations over space. Geographers have shown that it is possible and necessary to focus on the transformations of scale that occur through international, national and local development. In this way it becomes important to transform perceptions, representations and images of other cultures and forms of social organisation. In order for this to happen it is necessary to comprehend further the spatial classification of different worlds of development and to understand how this has changed over time. In addition, attention must be shifted to the spatial nature of the associations that exist between First/Second/Third worlds and their various cultures and economies alongside an awareness of the presumed spatial diffusion and spread of capitalist modernity. Can geographers help solve issues of inequality, inequity and social justice 'as well as just identifying and cataloguing their existence' (Golledge, 2002: 11–12)? This will not be easy given that so much of the history of development geography has been characterised by the objective of identifying and cataloguing difference, but social justice has rarely been the outcome of these endeavours.

Glossary

agency This is a concept that relates to the individual and the volition of his or her will and choices. If structures are fixed then agency and agents are terms which refer to the power and capacity of people to operate to some extent independently of the (social) structure.

capitalocentrism This term refers to debates about globalisation and development and is based on the idea of clearing a path for the consideration of the multiplicity of reality (in this case capitalism) and the various struggles around it. Capitalism is often presented as inherently spatial and as naturally stronger than the forms of non-capitalist economy (especially 'Third World' economies) because of its presumed capacity to universalise the market for capital commodities. This increasingly familiar 'script' or story of globalisation clearly lacks nuance and assumes that all other forms of economy will fall away as a result of its universal and overwhelming force in the (largely passive) space of the 'Third World'. This leads us to call into question the naturalness of capitalist identity as the template of all economic activity. What is 'non-capitalism' disappears here before the assumed inevitability of capitalist 'penetration' in the South.

class refers to relationships, in terms of property and employment, for example, that produce systems of social classification and stratification. Karl Marx's work understood the power and wealth of the ruling classes as relational to the marginality and impoverishment of the poor. The consciousness and agency of each group is not always clear however, and there has been a focus on fixed class structures rather than on their interactions with other social relations such as gender.

de-territorialisation This term seeks to focus on the displacement of identities, peoples and meanings that result from globalisation in a postmodern world system. It refers to the way in which territory is seen to lose significance as a result of global-level change and the 'unstoppable' juggernaut of global changes in information, knowledge and technology. It suggests that borders and divides are becoming less important in a global context.

developmentalism This term refers to a view of Third World spaces and their inhabitants as essentialised, homogenised entities. It is also closely associated with an unconditional belief in the concept of progress and the 'makeability' of society. It has complex historical and geographical roots but partly emerged from the organicist and evolutionary thinking of the Enlightenment era in seventeenth- and eighteenth-century Europe.

diaspora This term literally means the scattering of a population such as that of the African Diaspora that resulted from the slave trade. This involves transnational connections that transect the boundaries and borders of national communities. The term points to important geographies of flows and connections and to the openness of culture and identity in the contemporary world.

discourse/discursive These terms refer to written and verbal communications and relate to the variety of social practices through which the world is made meaningful and intelligible to individuals and communities. Knowledge of the world is produced and reproduced through representations and practices that thus come to be naturalised and taken for granted. Discourses vary over time and space and can be institutionalised as they are always, inseparably, bound up with relations of power. Discourses of masculinity and femininity, for example, shape the behaviours, roles and aspirations of men and women.

Enlightenment A period of change and an intellectual fashion associated with the rise of European modernity and liberal theory during the seventeenth and eighteenth centuries.

Enlightenment beliefs posited a singular view of progress and assumed the superiority of western notions of science and reason.

geopolitics The term 'geopolitics' is now familiar to many people bur remains quite difficult to define. It relates to studies of political geography that deal directly with international relations, international conflicts and foreign policies. The term largely refers to the way in which space is important in understanding the constitution of international relations. A critical geopolitics of development may offer some useful ways forward in understanding the history and geography of the 'Third World' as a political idea. As a school of thought it focuses on political meanings and representations and explores the contestation of the geopolitical world.

globalisation A much contested and often poorly defined term that refers to a reordering of time and space, leading to new forms of societal integration across the world. As geographers, this term has especial relevance in that it raises questions about the links between the local and the global and how these links have changed over time and in different places and spaces. Many geographers have studied the globalisation of socio-cultural and economic flows but this remains a complex and ambiguous term.

governance A term that usually refers to the act or process of governing and is therefore sometimes seen as being synonymous with government. In terms of development debates however, the term focuses on the wider range of governmental and non-governmental institutions and actors that shape policy outcomes. The term also refers to the relationships between different actors (ranging from government institutions to non-government organisations (NGOs) and social movements or private companies).

neocolonialism A means of economic and political control articulated through powerful states and capital cities in the 'developed world' over the economies and societies of the 'underdeveloped' world. The dominated states may be formally independent but their economic and political systems remain closely controlled from outside. The origins of these relations lie in imperialism and they are often seen as directly linked to the continual development of 'underdevelopment' in the periphery.

neoliberalism A range of ideas and theories, which have become very widespread in the 1980s and 1990s, that posits the desirability of the market as the central organising principle for social, economic and political life. It is connected to particular liberal economic and political interpretations from the eighteenth and nineteenth centuries. The pro-market position of neoliberalism now closely informs the development theories and practices of the multilateral regulatory institutions such as the World Bank and IMF. The central claim is that free and neutral markets can effectively distribute knowledge and resources socially and politically. This is said to require a minimum of state machinery. Its central point then is that the market mechanism should be allowed to direct the fate of human beings, that economy should dictate its rules to society rather than the other way around.

pharmacopoeia This term refers to the idea of an officially published text or book which contains a list of medicinal remedies and drugs that contain prescribed directions for use.

postcolonialism/postcoloniality These terms refer to a variety of theoretical positions and historical conditions associated primarily (though not exclusively) with the aftermath of colonialism. Postcolonialism seeks to critique western ways of knowing and traditions of thought. It also involves a retracing of the impacts of colonial processes with a particular concern for marginal or subaltern peoples.

post-structuralism This approach seeks to repudiate grand or master narrations of history and geography, focusing on 'local' narratives and stories and drawing attention to the gaps and fissures in these stories and narratives as a better way of understanding the past and the present.

representation This term refers to the ways in which meanings are conveyed or depicted. Geographical representations refer to writings of the world and its economic and political spaces. There are many forms and strategies of representation, as well as a politics to acts of representing others, which are inseparable from the subject being represented.

resistance usually refers to any action, imbued with intent, that aims to challenge, change or retain social, political and economic relations, processes and institutions. It may involve action from individuals or from social groups and movements. Resistance must be seen not as a simple heroic response to some dominating power but as a form of power in its own right and diffused throughout the practices of everyday life.

subject/subjectivity 'The subject' is a term that refers to individual human beings as agents and the ways in which they think their place in the world. There are a variety of theories of subjectivity each imagining the subject in different ways, as bounded, unique, contained and self-knowing, for example. There is no subject prior to knowledge according to some theorists; rather, subjects are produced by power and discourses. The ways in which subjects are used to justify discourses or to produce stable grounds of knowledge is an important theme here.

See also *A Feminist Glossary of Human Geography* edited by Linda McDowell and Joanne P. Sharp (1999) and *The Dictionary of Human Geography* edited by Ron Johnston, Derek Gregory, David Smith and Mike Watts (2001).

Bibliography

Aarts, P. (1999) 'The Middle East: a region without regionalism or the end of exceptionalism?', *Third World Quarterly*, 20 (5), pp. 911–925.

Abbott, J. P. (2001) 'Democracy @internet.asia? The challenges to the emancipatory potential of the net: lessons from China and Malaysia', *Third World Quarterly*, 22 (1), pp. 99–114.

Abdel-Malek, A. (1977) 'Geopolitics and national movements: an essay on the dialectics of imperialism', in Peet, R. (ed.) *Radical Geography*, London: Methuen, pp. 293–308.

Abrahamsen, R. (2001) *Disciplining Democracy: Development Discourse and Good Governance in Africa*, London: Zed Books.

Achebe, C. (1967) *Things Fall Apart*, London: Heinemann.

—— (1975) *Morning yet on creation day: essays*, London: Heinemann.

—— (1987) *Anthills of the Savannah*, London: Heinemann.

—— (1988) *Hopes and impediments: selected essays 1965–1987*, London: Heinemann.

Adams, W. M. (1990) *Green Development*, London: Routledge.

Afzal-Khan, F. and Seshadri-Crooks, K. (eds) (2000) *The Pre-occupation of Postcolonial Studies*, London: Duke University Press.

Agnew, J. (1998) *Geopolitics: Re-visioning World Politics*, London: Routledge.

Agnew, J. and Corbridge, S. (1995) *Mastering Space: Hegemony, Territory and International Political Economy*, London and New York: Routledge.

Ahluwalia, P. (2001) *Politics and Post-colonial Theory: African Inflections*, London: Routledge.

Ahmad, A. (1992) *In Theory: Classes, Nations, Literatures*, London: Verso.

Ainger, K. (2001) 'To open a crack in history', *New Internationalist*, 338, September, Special issue on Global Resistances, available at http://www.oneworld.org/ni/issue338/, pp. 1–8 (3 March 2002).

Ake, C. (1995) 'The new world order: a view from Africa', in Holm, H. H. and Sørenson, G. (eds) *Whose World Order: Uneven Globalization and the End of the Cold War*, Oxford: Westview Press, pp. 19–42.

Allen, T. and Hamnett, C. (eds) (1995) *A Shrinking World? Global Unevenness and Inequality*, Oxford: Oxford University Press in association with the Open University.

Allen, T. and Thomas, A. (2001) *Poverty and Development into the Twenty-first Century*, Oxford: Open University Press.

Almond, G. A. (1970) *Political Development; Essays in Heuristic Theory*, Boston, MA: Little, Brown.

Amin, S. (1974) *Global Accumulation on a World Scale*, New York: Monthly Review Press.

—— (1997a) *Capitalism in the Age of Globalization: The Management of Contemporary Society*, London: Zed Books.

—— (1997b) 'Placing globalization', *Theory, Culture and Society*, 14 (2), pp. 123–137.

—— (2001a) 'Imperialism and globalization', *Monthly Review*, 53, (2), available at http://www.monthlyreview.org/0601amin.html/ (10 March 2002).

—— (2001b) 'Capitalism's global strategy', in Houtart, F. and Pelot, F. (eds) *The Other Davos*, London: Zed Books, pp. 17–24.

—— (2002) 'Africa: living on the fringe', *Monthly Review*, 53 (10), available at http://www.monthlyreview.org/mar2002.html/ (10 March 2002).

Anderson, B. (1998) *The Spectre of Comparisons: Nationalism, Southeast Asia and the World*, London and New York: Verso.

Appadurai, A. (1996) *Modernity at Large: Cultural Dimensions of Globalization*, Minneapolis: University of Minnesota Press.

—— (1999) 'Globalization and the research imagination',

International Social Science Journal, 51 (2), pp. 229–238.
—— (2000) 'Grassroots globalization and the research imagination', *Public Culture*, 12 (1). pp. 1–19.
Apter, D. E. (1987) *Rethinking Development: Modernization, Dependency and Postmodern Politics*, London: Sage.
Armes, R. (1987) *Third World Film-making and the West*, Berkeley: University of California Press.
Arndt, W. (1987) *Economic Development*, Chicago, IL: Chicago University Press.
Arnfred, S. (1995) 'Introduction to issues of methodology and epistemology in postcolonial studies', in Arnfred, S. (ed.) *Issues of Methodology and Epistemology in Postcolonial Studies*, Occasional Paper No. 15/1995, Department of International Development Studies, Roskilde, pp. 1–11.
Arnold, D. (2000) ' "Illusory riches": representations of the tropical world 1840–1950', *Singapore Journal of Tropical Geography*, 21 (1), pp. 6–18.
Arnold, G. (1989) *The Third World Handbook*, London: Cassell.
Asad, T. (1993) *Genealogies of Religion: Discipline and Reasons of Power in Christianity and Islam*, Baltimore, MD: Johns Hopkins University Press.
Ashcroft, B. (1997) 'Globalism, post-colonialism and African studies', in Ahluwalia, P. and Nursey-Bray, P. (eds) *Post-colonialism: Culture and Identity in Africa*, New York: Nova Science Publishers, pp. 131–156.
Ashcroft, B. *et al.* (eds) (1995) *The Post-colonial Studies Reader*, London: Routledge.
Ashcroft, B. *et al.* (eds) (1998) *Key Concepts in Post-colonial Studies*, London: Routledge.
Auty, R. (1979) 'World within worlds', *Area*, 11, pp. 232–235.
Ayoob, M. (1995) *The Third World Security Predicament: State-making, Regional Conflict and the International System*, Boulder, CO: Lynne Reiner.
Bakan, A. (2000) 'After Seattle: the politics of the World Trade Organisation', *International Socialism*, 87, pp. 85–93.
Barnes, T. J. (2000) 'Inventing Anglo-American economic geography, 1889–1960', in Sheppard, E. and Barnes, T. J. (eds) *A Companion to Economic Geography*, Oxford: Blackwell, pp. 11–26.
Bar-Siman-Tov, Y. (1998) 'The United States and Israel since 1948: a special relationship?', *Diplomatic History*, 22 (2), pp. 231–262.
Barbour, K. M. (1961) *The Republic of Sudan: A Regional Geography*, London: University of London Press.
Barwise, J. and White, N. J. (2002) *South-East Asia: A Travellers History*, London: Cassell.
Bauman, Z. (1989) *Modernity and the Holocaust*, Cambridge: Polity Press.
—— (1999) 'The burning of popular fear', *New Internationalist*, 310, pp. 20–23.
Bayart, J. F. (1991) 'Finishing with the idea of the Third World: the concept of the political trajectory' in Manor, J. (ed.) *Rethinking Third World Politics*, London: Longman, pp. 78–99.
—— (1993) *The State in Africa: The Politics of the Belly*, Harlow: Longman.
Bebbington, A. (2000) 'Reencountering development: livelihood transitions and place transformations in the Andes', *Annals of the Association of American Geographers*, 90 (3), pp. 495–520.
Beck, U. (2000) *What is Globalization?*, Cambridge: Polity Press.
Bell, M. (1994) 'Images, myths and alternative geographies of the Third World', in Gregory, D. (ed.) *Human Geography: Society, Space and Social Science*, Basingstoke: Macmillan, pp. 174–199.
Bello, W. (1998) 'East Asia: on the eve of the great transformation?', *Review of International Political Economy*, 5 (3), pp. 424–444.
—— (2002) 'Learning from Doha', paper presented at the 2001 Our World is not for Sale conference, Brussels, 7–9 December 2001. Available at http://www.challengeglobalization,org/news_notices?winter2001/ (10 March 2002).
Bello, W., Cunningham, S. and Poh, L. K. (1998) *A Siamese Tragedy: Development and Disintegration in Modern Thailand*, London: Zed Books.
Benneh, G. (1972) 'Systems of agriculture in Tropical Africa', *Economic Geography*, 48, pp. 244–257.
Bennholdt-Thomsen, V., Faraclas, N. and Von Werlhof, C. (eds) (2001) *There is an Alternative: Subsistence and Worldwide Resistance to Corporate Globalization*, London: Zed Books.
Berger, M. T. (1994) 'The end of the "Third World"?', *Third World Quarterly*, 15 (2), pp. 257–275.
—— (2001) 'The post-cold war predicament: a conclusion', *Third World Quarterly*, 22 (6), pp. 1079–1085.
—— (2001a) 'The rise and demise of national development and the origins of post-Cold War capitalism', *Millennium*: Journal of *International Studies*, 30 (2), pp. 211–235.
Berry, B. J. L. (1973) *The Human Consequences of Urbanisation: Divergent Paths in the Urban Experience of the Twentieth Century*, London: Macmillan.
Bessis, S. (2001) 'The World Bank and women: an "instrumental feminism" ', in Perry, S. and Schenck, C. (eds) *Eye-to-eye: Women Practising Development Across Cultures*, London: Zed Books, pp. 10–24.
Bhaba, H. (1994) *The Location of Culture*, London: Routledge.
Bilgin, P. and Morton. A. D. (2002) 'Historicising representations of "failed states": beyond the cold-war annexation of the social sciences?', *Third World Quarterly*, 23 (1), pp. 55–80.
Bissell, T. (2002) 'Eternal winter: lessons of the Aral Sea disaster', *Harpers Magazine*, 304(1823), April, pp. 41–56.
Black, J. (1990) *Eighteenth Century Europe 1700–1789*, London: Macmillan.
Blaikie, P. (1985) *The Political Economy of Soil Erosion*, London: Zed Books.
Blaut, J. M. (1953) 'A micro-geography', *Malayan Journal of Tropical Geography*, 1 (1), pp. 1–56.

—— (1958) *Chinese Market Gardening in Singapore: A Study in Functional Microgeography*, Ann Arbor: University Microfilms International.

—— (1970) 'Geographic models of imperialism', *Antipode*, 2 (1), pp. 60–85.

—— (1973) 'The theory of development', *Antipode*, 5 (2), pp. 22–26.

—— (1976) 'Where was capitalism born?', *Antipode*, 8 (2), pp. 1–11.

—— (1993) *The Colonizers Model of the World: Geographic Diffusionism and Eurocentric History*, New York: Guilford Press.

Boateng, E. A. (1966) *A Geography of Ghana*, Cambridge: Cambridge University Press.

Boehmer, E. and Moore-Gilbert, B. (2002) 'Introduction to special issue: Postcolonial studies and transnational resistance', *Interventions*, 4 (1), pp. 7–21.

Bond, P. (2000a) *Against Global Apartheid: South Africa meets the World Bank, IMF and International Finance*, London: Pluto Press.

—— (2000b) *Elite Transition: From Apartheid to Neoliberalism in South Africa*, London: Pluto Press.

—— (2001) 'From Seattle to Soweto', *New Internationalist*, 338 (September).

—— (2002a) 'Zimbabwe, South Africa and the power politics of bourgeois democracy', *Monthly Review*, 54 (1), available at http://www.monthlyreview.org/0502bond.html/ (12 April 2002).

—— (2002b) 'NEPAD', *Znet Magazine*, available at http://www.zmag.org/content/ (12 April 2002).

Booth, D. (1985) 'Marxism and development sociology: interpreting the impasse', *World Development*, 13, pp. 761–787.

Braden, K. E. and Shelley, F. M. (2000) *Engaging Geopolitics*, London: Prentice-Hall.

Brass, T. (1995) 'Old conservatism in "New Clothes" ', *Journal of Peasant Studies*, 22, (3), pp. 516–540.

Bratton, M. (1994) 'Economic crisis and political realignment in Zambia', in Widner, J. A. (ed.) *Economic Change and Political Liberalization in Sub-Saharan Africa*, Baltimore, MD, and London: Johns Hopkins University Press.

Brecher, J. and Costello, T. (2002) *Globalization from Below: The Power of Solidarity*, Boston, MA: South End Press. See author's website at http://www.villageorpillage.org/ (12 April 2002).

Brehm, V. (2001) 'Promoting effective North–South partnerships: a comparative study of ten European NGO's', INTRAC Occasional Paper No. 35, Oxford, available at http://www.id21.org/society/ (1 June 2002).

Bretton Woods Project (BWP) (2000) *The World Bank and the State: A Recipe for Change?*, London: N. Hildyard, Bretton Woods Project.

—— (2001a) 'IFI criticism intensifies ahead of meetings', Oxford, available at http://www.brettonwoodsproject.org/ (12 April 2002).

—— (October/November 2001b) *Update*, Oxford: BWP.

—— (2002a) 'World Bank contributes $2 billion to Washington', Bretton Woods *Update*, 28, May–June, p. 6, available at http://www.brettonwoodsproject.org/ (12 April 2002).

—— (2002b) 'G8 "Absolutely zilch" for Africa', July–August, p. 7, available at http://www.brettonwoodsproject.org/ (12 April 2002).

British Broadcasting Corporation (BBC) (2001) 'G8 protestors take to the streets', available at http://www.bbc.co.uk/world/ (19 July 2001).

—— (2002a) 'Israel's battered economy', 21 June, available at http://www.bbc.co.uk/hi/english/business/.

—— (2002b) 'Bush ties aid to reforms', 22 March, available at http://www.bbc.co.uk/hi/english/world/americas/ (25 June 2002).

—— (2002c) 'Blair to call for Africa aid package', 25 June, available at http://www.bbc.co.uk/hi/english/uk_politics/ (25 June 2002).

—— (2002d) 'Blair defends Seville agreement', 24 June, available at http://www.bbc.co.uk/hi/english/uk_politics/ (25 June 2002).

Brookfield, H. C. (1973a) 'The Pacific realm', in Mikesell, M. W. (ed.) *Geographers Abroad*, Chicago: Chicago University Press.

—— (1973b) 'On one geography and a Third World', *Transactions of the Institute of British Geographers*, 68, pp. 1–20.

—— (1975) *Interdependent Development*, London: Methuen.

Brown, L. R. (2001) *State of the World 2001*, London: Earthscan.

Browne, S. (2001) 'Preface: Choices for the poor', in Grinspun, A. (ed.) *Choices for the Poor: Lessons from National Poverty Strategies*, New York: United Nations Development Programme (UNDP), pp. xi–xiii.

Bryceson, D. F. (2002) 'The scramble for Africa: reorienting rural livelihoods', *World Development*, 30 (5), pp. 725–739.

Buchanan, K. M. (1948) Contributor to *Conurbation, A Planning Survey of Birmingham and the Black Country by the West Midland Group*, London: The Architectural Press.

—— (1950a) 'Modern farming in the Vale of Evesham', *Economic Geography*, pp. 235–250.

—— (1950b) 'The "coloured" community in the Union of South Africa', *The Geographical Review*, 40 (3), pp. 397–414.

—— (1953) 'Internal colonization in Nigeria', *The Geographical Review*, 43 (3), 416–418.

—— (1963a) 'The Third World: its emergence and contours', *New Left Review*, 18, pp. 5–23.

—— (1963b) 'Bingo or UNO? Further comments on the affluent and proletarian nations', *New Left Review*, 21, pp. 21–29.

—— (1963c) 'The demolition men: the US and the crisis of neutrality in southeast Asia', *Monthly Review* (February), pp. 575–583.

—— (1964) 'Profiles of the Third World', *Pacific Viewpoint*, 5, pp. 97–126.

—— (1966) 'Mr Angel's obtuse angle: NLR reply to criticisms', *New Left Review.*

—— (1967a) *Out of Asia: Essays on Asian Themes*, London: Wiley.

—— (1967b) *The Southeast Asian World: An Introductory Essay*, London: G. Bell & Sons.

—— (1972) *The Geography of Empire*, London: Spokesman books.

—— (1977a) 'Economic growth and cultural liquidation: the case of the Celtic nations', in Peet, R. (ed.) *Radical Geography: Alternative Viewpoints on Contemporary Social Issues*, London: Methuen, pp. 125–144.

—— (1977b) 'Reflections on a "Dirty word" ', in Peet, R. (ed.) *Radical Geography: Alternative viewpoints on Contemporary Social Issues*, London: Methuen, pp. 363–367.

Buchanan, K. M. and Pugh, J. C. (1955) *Land and People in Nigeria: The Human Geography of Nigeria and its Environmental Background*, London: University of London Press.

Bullard, N., Bello, W. and Malhotra, K. (1998) 'Taming the tigers: the IMF and the Asian crisis', *Third World Quarterly*, 19 (3), pp. 505–555.

Bunting, M. (2002) 'Grassroots gamine', *Guardian*, 7 March, p. 20.

Butlin, R. (2000) 'British geographical societies and the representations of British imperialism and colonialism in the twentieth-century', paper presented at the Historical Geographies of Twentieth-century Britain Conference, University of Sussex.

Cammack, P. (2002) 'Attacking the poor', *New Left Review*, 13 (January/February), pp. 125–134.

Cannon, T. (1975) 'Geography and underdevelopment', *Area*, 7, pp. 212–216.

Center for International Policy (CIP) (2002) 'Colombia Project: peace initiatives in Colombia', available at http://www.ciponline.org/colombia (5 February 2003).

Central Intelligence Agency (CIA) (2000) *Global Trends 2015: A Dialogue about the Future with Non-government Experts*, National Intelligence Council, available at http://www/odci.gov/nic/pubs/2015_files/2015.html (12 April 2002).

—— (2001) 'Research and analysis: the OSS', available at http://www.odci.gov/cia/publications/oss/ (12 April 2002).

Césaire, A. (2000) *Discourse on Colonialism*, London: Monthly Review Press (3rd edn).

Chakrabarty, D. (1998) 'Minority histories, subaltern pasts', *Economical and Political Weekly*, 33 (9), pp. 473–480.

Chaliand, G. (1977) *Revolution in the Third World: Myths and Prospects*, Hassocks, Sussex: Harvester Press.

Chaliand, G. and Rageau, J. P. (1985) *Strategic Atlas: A New and Exciting Survey of the Political Realities of Today's World*, London: Penguin Books.

Chambron, A. C. (1999) 'Bananas: The "Green Gold" of the TNCs', EUROBAN/FLO International.

—— (2000) 'Straightening the bent world of the banana', EUROBAN/FLO International, available at http://www.bananalink.co.uk/ (12 April 2002).

Chanan, M. (1990) 'Introduction: lessons of experience', in Desnoes, E. *Memories of Underdevelopment: Tomás Guitérrez Alea, Director and Inconsolable Memories*, London: Rutgers University Press, pp. 15–22.

Chaturvedi, S. (2000) 'Representing post-colonial India: inclusive/exclusive geopolitical imaginations', in Dodds, K. and Atkinson, D. (eds) *Geopolitical Traditions: A Century of Geopolitical Thought*, London: Routledge, pp. 211–235.

Chaudhry, K. A. (1997) *The Price of Wealth: Economies and Institutions in the Middle East*, New York: Ithaca University Press.

Cheru, F. (2000) 'The local dimensions of global reform', in Pieterse, J. N. (ed.) *Global Futures: Shaping Globalization*, London: Zed Books, pp. 119–132.

—— (2001) 'Overcoming apartheid's legacy: the ascendancy of neoliberalism in South Africa's anti-poverty strategy', *Third World Quarterly*, 22 (4), pp. 505–527.

Chomsky, N. (2001) 'The war in Afghanistan', excerpt from Lakdawala lecture, New Delhi, available at http://www.zmag.org/lakdawalalec.htm/ (12 April 2002).

Chossudovsky, M. (1997) *The Globalisation of Poverty*, London: Zed Books.

CIP (2002) 'Just the facts: US Security Assistance to the Andean region', http://www.ciponline.org/facts/co.html

Clapham, C. (1985) *Third World Politics*, London: Croom Helm.

Cline-Cole, R. (1999) 'Contextualizing the "new internalization" in Anglo(-American) African(ist) human geographies', in Simon, D. and Närman, A. (eds) *Development as Theory and Practice: Current Perspectives on Development and Development Co-operation*, Harlow: Longman.

Cloke, P., Philo, C. and Sadler, D. (1991) *Approaching Human Geography: An Introduction to Contemporary Theoretical Debates*, New York: Guilford Press.

Cobham, A. (2001) 'Capital account liberalisation and poverty', Working Paper No. 70, Queen Elizabeth House, University of Oxford, April, available at http://www.id21.org/society/ (12 April 2002).

Commission on Global Governance (1995) *Our Common Neighbourhood: The Report of the Commission on Global Governance*, Oxford: Oxford University Press.

Connell, J. (1971) 'The geography of development', *Area*, 3 (4), pp. 259–265.

Cooke, B. (2001) 'From colonial administration to development management', IDPM Working Paper No. 63, Institute for Development Policy and Management, Manchester, available at http://www.man.ac.uk/idpm/ (12 April 2002).

Cooke, B. and Kothari, U. (2001) *Participation: The New Tyranny?*, London: Zed Books.

Cooper, F. (1980). 'Africa and the world economy', *African Studies Review*, 24 (2), pp. 1–86.

—— (1997) 'Modernizing bureaucrats, backward Africans and the development concept', in Cooper, F. and Packard, R. (eds) *International Development and the Social Sciences: Essays on the History and Politics of Knowledge*, Berkeley: University of California Press.

Corbridge, S. (1986) *Capitalist World Development: A Critique of Radical Development Geography*, London: Macmillan.
—— (1992) 'Third World development', *Progress in Human Geography*, 16 (4) pp. 584–595.
—— (1995) *Development Studies: A Reader*, London: Edward Arnold.
—— (1998) 'Beneath the pavement only soil: the poverty of post-development,' *Journal of Development Studies*, 34 (6), pp. 138–148.
—— (1999) 'Development, post-development and the global political economy', in Cloke, P., Crang, P. and Goodwin, M. (eds) *Introducing Human Geographies*, London: Arnold, pp. 67–75.
Corbridge, S. and Harriss, J. (2000) *Reinventing India: Liberalization, Hindu Nationalism and Popular Democracy*, London: Polity Press.
Cowen, M. P. and Shenton, R. W. (1996) *Doctrines of Development*, London: Routledge.
Cracknell, B. E. (2000) *Evaluating Development Aid: Issues, Problems and Solutions*, London: Sage.
Credner, W. (1935) *Siam. Das Land der Thai*, Stuttgart: Englehas.
Crow, B. and Thomas, A. (1983) *Third World Atlas*, Milton Keynes and Philadelphia, PA: Open University Press.
Crush, J. (1986) 'Towards a people's historical geography for South Africa', *Journal of Historical Geography*, 12, pp. 2–3.
—— (1994) 'Post-colonialism, decolonization and geography', in Godlewska, A. and Smith, N. (eds) *Geography and Empire*, Oxford: Blackwell, pp. 289–313.
—— (1995) 'Imagining development', in Crush J. (ed.) *Power of Development*, London: Routledge, pp. 1–26.
Dabashi, H. (1993) *Theology of Discontent: The Ideological Foundation of the Islamic Revolution in Iran*, New York and London: New York University Press.
Danaher, K. (2002) *10 Reasons to Abolish the World Bank*, New York: Seven Stories Press.
Darnton, R. (1979) *The Business of Enlightenment: A Publishing History of the Encyclopédie 1775–1800*, Cambridge, MA: Harvard University Press.
Davis, M. (2001) *Late Victorian Holocausts: El Niño, Famines and the Making of the Third World*, London: Verso.
De Blij, H. (1964) *A Geography of Sub-Saharan Africa*, Chicago: Rand McNally.
de Haan, A. (1999) 'Livelihood and poverty: the role of migration. A critical review of the literature', *Journal of Development Studies*, 36 (2), pp. 1–47.
De Janvry, A. (ed.) (2001) *Access to Land, Rural Poverty and Public action*, New York: Oxford Univerity Press.
De Silva, A. (2002) 'The allocation of Canada's bilateral foreign aid', *Canadian Journal of Development Studies*, XXIII (1), pp. 47–67.
De Soto, H. (2000) *El Misterio del Capital: Por qué el capitalismo triunfa en occidente y fracasa en el resto del mundo*, Peru: El Comercio.
—— (2001) 'Capital mystery: why capital triumphs in the West but fails everywhere else', *Geobusiness*, October, pp. 44–47.
De Sousa, A. R. and Porter, P. W. (1976) 'Development geography and radical-liberal dialogue', *Antipode*, 3, pp. 94–97.
Department for International Development (DFID) (2000) *Poverty Elimination and the Empowerment of Women*, London, DFID.
—— (1997) *White Paper on Eliminating World Poverty: A Challenge for the Twenty-first Century*, London: HMSO.
—— (1997a) *Working with British Business*, London: DFID.
—— (1999) *Sustainable Livelihoods Guidance Sheets*, London: DFID.
—— (2000a) *Halving World Poverty by 2015: Economic Growth, Equality and Security*, London: DFID.
—— (2000b) *Eliminating World Poverty: Making Globalisation Work for the Poor*, Cmd 5006, London: HMSO.
Derrida, J. (1976) *Of Grammatology*, Baltimore, MD: Johns Hopkins University Press.
—— (1978) *Writing and Difference*, London: Routledge.
—— (1981) *Dissemination*, trans. Barbara Johnson, London: Athlone Press.
Desai, V. and Potter, R. B. (eds) (2001) *The Companion to Development Studies*, London: Arnold.
Desforges, L. (1998) 'Checking out the planet: global representations/local identities and youth travel', in Skelton, T. and Valentine, G. (eds) *Cool Places: Geographies of Youth Cultures*, London: Routledge, pp. 174–191.
Desnoes, E. (1990) *Memories of Underdevelopment: Tomás Guitérrez Alea (Director)*, London and New Brunswick, NJ: Rutgers University Press.
Dickenson, J., Gould, B., Clarke, C., Mather, S., Prothero, M., Siddle, D., Smith, C. and Thomas-Hope, E. (1996) *A Geography of the Third World*, London: Routledge.
Dirlik, A. (1994) *After the Revolution: Waking to Global Capitalism*, Hanover, MA: Wesleyan University Press.
—— (2002) 'Whither history? Encounters with historicism, postmodernism, postcolonialism', *Futures*, 34 (1), pp. 75–90.
Dixon, C. (1999) *The Thai Economy: Uneven Development and Internationalisation*, London: Routledge.
Dobby, E. H. G. (1950) *South-East Asia*, London: University of London Press.
—— (1955) *Senior Geography for Malayans*, London: Longman.
Dodds, F. (2001) *Earth Summit 2002: A New Deal*, London: Earthscan.
Dodds, K. and Atkinson, D. (eds) (2000) *Geopolitical Traditions: A Century of Geopolitical Thought*, London: Routledge.
Doherty, T. (1993) 'Postmodernism: an introduction', in Doherty, T. *Modernism/Postmodernism*, Hemel Hempstead: Harvester Wheatsheaf, pp. 1–31.
Domosh, M. (1991) 'Towards a feminist historiography of geography', *Transactions of the Institute of British Geographers*, NS 16, pp. 95–104.
Doty, R. L. (1996) 'Repetition and variation: academic discourses on North–South relations', in Doty, R. L. *Imperial Encounters:*

The Politics and Representation in North–South Relations, Minneapolis: University of Minnesota Press, pp. 145–162.

Dowden, R. (2002) 'Clare Short', *Prospect Magazine* (May), available at http://www.prospect-magazine.co.uk/ (31 May 2002).

Driver, F. (1992) 'Geography's empire: histories of geographical knowledge', *Environment and Planning D: Society and Space*, 10, pp. 23–40.

—— (2001) *Geography Militant*, Oxford: Blackwell.

Driver, F. and Yeoh, B. S. A. (2000) 'Constructing the Tropics: Introduction', *Singapore Journal of Tropical Geography*, 21 (1), pp. 1–5.

Duffuor, F. and Bové, J. (2001) *The World is Not for Sale*, London: Verso.

Dussel, E. (1998) 'Beyond Eurocentrism: the world system and the limits of modernity', in Jameson, F. and Miyoshi, M. (eds) *The Cultures of Globalization*, Durham, NC: Duke University Press.

Eade, D. and Ligteringen, E. (eds) (2001) *Debating Development: NGOs and the Future: Essays from Development in Practice*, Oxford: Oxfam.

Earthscan (1997) *The Reality of Aid 1997–1998: An Independent Review of Development Cooperation*, Guildford: Earthscan.

Earthscan (1998) *Internally Displaced People: A Global Survey*, Guildford: Earthscan.

Easterly (2001) *The Elusive Quest for Growth*, New York: MIT Press.

Economist (2001a) 'The Internet's new borders', 11 August, 360 (8234), pp. 9–10.

—— (8 September 2001b) 'The case for brands', p. 9; 'Who's wearing the trousers, pp. 27–31.

—— (17 November 2001c) 'Beyond Doha', p. 11.

—— (19 January 2002a) 'Disarming Sierra Leone: Pax Leone', pp. 52–53.

—— (19 January 2002b) 'Columbia's peace process: The FARC's moment of truth', p. 47.

—— (16 February 2002c) 'Drugs in the Andes: spectres stir in Peru', p. 55.

—— (16 February 2002d) 'Drugs in Bolivia: leaves of discord', p. 56.

—— (1 June 2002e) 'The Americas', pp. 58–60.

Edwards, M. and Hulme, D. (eds) (1992) *Making a Difference: NGOs and Development in a Changing World*, London: Earthscan.

—— (1995) *Non-Governmental Organisations: Performance and Accountability: Beyond the Magic Bullet*, London: Earthscan.

Eliot, S. and Stern, B. (1991) *The Age of Enlightenment: An Anthology of Eighteenth Century Texts*, Volume 1, London: Ward Lock/Open University Press.

Elliott, J. A. (1994) *An Introduction to Sustainable Development*, London: Routledge (2nd edn).

Elliott, L. (2000) 'Poor nation's rights are wronged', *Guardian*, 23 April, p. 25.

Ellis, F. (1998) 'Household strategies and rural livelihood diversification', *Journal of Development Studies*, 35 (1), pp. 1–38.

—— (2000) *Rural Livelihoods and Diversity in Developing Countries*, Oxford: Oxford University Press.

Ellwood, W. (2000) 'Redesigning the global economy', *New Internationalist Magazine*, 327 (September), pp. 2–12.

Elson, D. (1991) *Male Bias in the Development Process*, Manchester: Manchester University Press.

Engardio, P. (2001) 'Smart bombs, so why no smart aid?', *BusinessWeek*, 24 December, p. 58.

Engerman, D. C. (2000) 'Modernization from the other shore: American observers and the costs of Soviet economic development', *American Historical Review*, 105 (2), pp. 383–416.

Englebert, P. (1997) 'The contemporary African state: neither African nor state', *Third World Quarterly*, 18 (4), pp. 767–775.

Escobar, A. (1985) 'Discourse and power in development: Michel Foucault and the relevance of his work to the Third World', *Alternatives*, 10, pp. 377–400.

—— (1988) 'Power and visibility: development and the invention and management of the Third World', *Cultural Anthropology*, 3, pp. 428–441.

—— (1992) 'Reflection on "Development": grassroots approaches and alternative politics in the Third World', *Alternatives*, 10, pp. 377–400.

—— (1995) *Encountering Development; The Making and Unmaking of the Third World*, Princeton, NJ: Princeton University Press.

—— (2000a) 'Place, economy and culture in a post-development era', in *Asia-Pacific Identities: Culture and Identity Formation in the Age of Global Capital*, Lanham, MD: Rowman and Littlefield, pp. 193–217.

—— (2000b) 'Beyond the search for a paradigm? Post-development and beyond', *Development*, 43 (4), pp. 11–14.

—— (2001) 'Culture sits in places: reflections on globalism and subaltern strategies of localization', *Political Geography*, 20 (2), pp. 139–174.

—— (2002) *The Spaces of Neoliberalism: Land, Place and Family in Latin America*, London: Kumarian Press.

Esteva, G. (1992) 'Development', in Sachs, W. (ed.) *The Development Dictionary*, London: Zed Books, pp. 6–25.

—— (2001) 'Mexico: creating your own path at the grass-roots', in Bennholdt-Thomsen, V., Faraclas, N. and Von Werlhof, C. (eds) *There is an Alternative: Subsistence and Worldwide Resistance to Corporate Globalization*, London: Zed Books, pp. 155–166.

Esteva, G. and Prakash, M. S. (1998) *Grassroots Postmodernism: Remaking the Soil of Cultures*, London: Zed Books.

European Union (2001) *Impact of World Commodity Prices on Developing Countries*, Brussels: EU.

Fagan, G. H. (1999) 'Cultural politics and (post)development paradigms', in Munck, R. and O' Hearn, D. (eds) *Critical Development Theory: Contributions to a New Paradigm*, pp. 179–195.

Fanon, F. (1967) *Black Skin, White Masks*, New York: Grove Press.

—— (1968) *The Wretched of the Earth*, Harmondsworth: Penguin.

Farmer, B. H. (1973) 'Geography, area studies and the study of area', *Transactions of the Institute of British Geographers*, 60, pp. 1–15.

—— (1983) 'British geographers overseas, 1933–1983', *Transactions of the Institute of British Geographers*, 8, pp. 70–79.

Fenster, T. (1997) 'Relativism vs. universalism in planning for minority women in Israel', *Israel Social Science Research*, 12 (1), pp. 1–30.

Ferguson, J. (1990) *The AntiPolitics Machine: Development, Depoliticization and Bureaucratic Power in Lesotho*, Cambridge: Cambridge University Press.

—— (1999) *Expectations of Modernity: Myths and Meanings of Urban Life on the Zambian Copperbelt*, Berkeley: University of California Press.

Fieldhouse, D. K. (1999) *The West and the Third World: Trade, Colonialism, Dependence and Development*, London: Blackwell.

Fine, B. (1999) 'The developmental state is dead – long live social capital?', *Development and Change*, 30, pp. 1–19.

—— (2001) *Social Capital versus Social Theory: Political Economy and Social Science at the Turn of the Millennium*, London: Routledge.

Fine, B., Lapavistas, C. and Pincus, J. (eds) (2001) *Development Policy in the Twenty-first Century: Beyond the Post-Washington Consensus*, Routledge Studies in Development Economics, London: Routledge.

Fink, C., Gassert, P. and Junker, D. (1998) *1968: The World Transformed*, Cambridge: Cambridge University Press.

Fisher, C. A. (1964) *South-East Asia*, London: Methuen; New York: Dutton.

Fisher, J. (ed.) (1995) *Geography and Development: A World Regional Approach*, Englewood Cliffs, NJ: Prentice-Hall.

Fitzgerald, W. (1967) *Africa: A Social, Economic and Political Geography of its Major Regions*, London: Methuen (10th edn).

Flint, C. (2002) 'Political geography: globalization, metapolitical geographies and everyday life', *Progress in Human Geography*, 26 (3), pp. 391–400.

Forbes, D. K. (1984) *The Geography of Underdevelopment*, London: Croom Helm.

Forde, C. D. (1934) *Habitat, Economy and Society: A Geographical Interpretation to Ethnology*, London: Methuen.

Foucault, M. (1979) *Discipline and Punish: The Birth of the Prison*, London: Penguin.

—— (1980) *Power/Knowledge: Selected Interviews and Other Writings 1972–1977*, London: Tavistock.

Francis, E. (2000) *Making a Living: Changing Livelihoods in Rural Africa*, London: Routledge.

Frank, A. G. (1969) *Capitalism and Underdevelopment in Latin America: Historical Studies of Chile and Brazil*, New York: Monthly Review Press.

—— (1997) 'The Cold War and me', *Bulletin of Concerned Asian Scholars*, 29 (4), available at http://csf.colorado.edu/bcas/symmpos/syfrank.htm/ (8 January 2001).

Freire, P. (1970) *Pedagogy of the Oppressed*, New York: Herder and Herder.

Friedmann, J. (1966) *Regional Development Policy: A Case Study of Venezuela*, Cambridge, MA: MIT Press.

Frimpong-Ansah, J. H. (1991) 'Development theory and the African experience', in Frimpong-Ansah, J. H. *The Vampire State in Africa: The Political Economy of Decline in Ghana*, London: James Currey, pp. 11–43.

Furtado, C. (1964) *Development and Underdevelopment*, Berkeley: University of California Press.

Gabriel, T. H. (1989) 'Towards a critical theory of Third World films', in Pines, J. and Willemen, P. (eds) *Questions of Third Cinema*, London: BFI, pp. 30–52.

Gandhi, L. (1998) *Postcolonial Theory: A Critical Introduction*, Edinburgh: Edinburgh University Press.

Ganokar, D. P. (ed.) (2001) *Alternative Modernities*, Durham, NC: Duke University Press.

Gariyo, Z. (2001) 'The PRSP process in Uganda', JubileePlus, available at http://www.jubileedebtcampaign.org.uk/ (12 April 2002).

Gay, P. (1973) *The Enlightenment: An Interpretation*, Volume 2: *The Science of Freedom*, London: Wildwood House.

Gelinas, J. B. (1998) *Freedom from Debt*, London: Zed Books.

Gendzier, I. L. (1985) *Managing Political Change: Social Scientists and the Third World*, Boulder, CO, and London: Westview Press.

Geobusiness Magazine (2001) The World Bank, Geneva: Geobusiness.

Geographical Journal (1982) July: 297.

George, S. (2001) 'A short history of neoliberalism: twenty years of elite economics and emerging opportunities for structural change', in Houtart, F. and Polet, F. (eds) *The Other Davos: The Globalization of Resistance to the World Economic System*, London: Zed Books, pp. 7–16.

German, T. and Ewing, D. (2000) *The Reality of Aid 2000: An Independent Review of Poverty Reduction and Development Assistance*, London: Earthscan.

Ghosh, A. (1992) *In an Antique Land*, South Asia Books.

Ghosh, D. (2001) 'Water out of fire: novel women, national fictions and the legacy of Nehruvian developmentalism in India', *Third World Quarterly*, 22 (6), pp. 951–967.

Gibson-Graham, J. K. (1996) *The End of Capitalism (As We Knew It)*, Oxford: Blackwell.

—— (1999) 'Queer(y)ing capitalism in and out of the classroom [1]', *Journal of Geography in Higher Education*, 23 (1), pp. 80–85.

Gibson-Graham, J. K. and Ruccio, D. (2001) 'After development: reimagining economy and class', in Gibson-Graham, J.K., Resnick, S. and Wolff, R. (eds) *Re/presenting Class: Essays in Postmodern Marxism*, Durham, NC: Duke University Press, pp. 158–181.

Giddens, A. (1998) *The Third Way*, Cambridge: Polity Press.

—— (2000) *The Third Way and Its Critics*, Cambridge: Polity Press.

Gilbert, A. (1971) 'Some thoughts on the "new geography" and the study of "development" ', *Area*, 3 (2), pp. 123–128.

—— (1987) 'Research policy and review: from little England to big Englanders: thoughts on the relevance of relevant research', *Environment and Planning A*, 19, pp. 143–151.

Gilbert, E. W. and Steel, R. W. (1945) 'Social geography and its place in colonial studies', *Geographical Journal*, 106, pp. 118–131.

Gilroy, P. (2000) *Between Camps: Nations, Cultures and the Allure of Race*, London: Penguin.

Glasmeier, A. (2002) 'One nation, pulling apart: the basis of persistent poverty in the USA', *Progress in Human Geography*, 26 (2), pp. 155–173.

Gleeson, B. (1999) *Geographies of Disability*, London: Routledge.

Godfrey, (1944) 'Preface', in NID (1944b) *French West Africa*, Volume 1 BR 512, *The Federation*, London: Oxford and Cambridge University Presses, pp. i–iv.

Goldberg, D. T. (1993) *Racist Culture: Philosophy and the Politics of Meaning*, Oxford: Blackwell.

Goldgeie, J. M. and McFaul, M. (1992) 'A tale of two worlds: core and periphery in the post cold-war era', *International Organisation*, 46 (2), pp. 469–470.

Goldsmith, A. A. (2001) 'Donors, dictators and democrats in Africa', *Journal of Modern African Studies*, 39 (3), pp. 411–436.

Golledge, R. G. (2002) 'The nature of geographic knowledge: Presidential address', *Annals of the Association of American Geographers* (AAG), 92 (1), pp. 1–14.

Gonsalves, S. (2002) 'Open-toe sandals and closed minds', Znet Commentary, 19 June, available at http://www.zmag.org/ (12 April 2002).

González de la Rocha, M. (2001) 'Angola', in Grinspun, A. (ed.) *Choices for the Poor: Lessons from National Poverty Strategies*, New York: United Nations Development Programme (UNDP), pp. 205–216.

Gordon, D. and Spickler, P. (eds) (1999) *The International Glossary on Poverty*, London: Zed Books.

Gould, B. (1998) 'Viewpoint: New Labour, new international development policy?', *Third World Planning Review*, 20 (1), pp. iii–vi.

Gould, P. (1960) 'The development of transportation patterns in Ghana', Northwestern Studies in Geography, No. 5, Evanston, IL.

—— (1970) 'Tanzania 1920–1963: the spatial impress of the modernization process', *World Politics*, 22, pp. 149–170.

Gourou, P. (1960) *Les Paysans du Delta Tonkinois*, Paris: Mouton.

—— (1931) *Le Tonkin*, Paris: Exposition Coloniale Internationale.

—— (1940) *La Terre et L'homme en extrême-Orient*, Paris: Mouton.

—— (1947) *Les Pays Tropicaux*, Paris: Mouton.

—— (1953) *The Tropical World*, trans. by B. D. Laborde, London, New York and Toronto: Longmans, Green and Co.

—— (1966) *The Tropical World: Its Social and Economic Conditions and its Future Status*, London: Longmans (4th edn).

Gowan, P. (1999) *The Global Gamble: Washington's Faustian Bid for World Dominance*, London: Verso.

—— (2001) 'Neoliberal cosmopolitanism', *New Left Review*, September–October, 11, pp. 79–93.

Gramsci, A. (1971) *Selections from the Prison Notes of Antonio Gramsci*, edited by Q. Hoare and G. Nowell-Smith, London: Lawrence.

Grinspun, A. (2001) 'Introduction: Stimulating policy change', in Grinspun, A. (ed.) *Choices for the Poor: Lessons from National Poverty Strategies*, New York: United Nations Development Programme (UNDP).

Grove, A. T. (1970) *Africa South of the Sahara*, Oxford: Oxford University Press (2nd edn).

Guardian (1 June 1988) 'A poor show', Media supplement, pp. 2–3.

—— (3 September 2001) 'Be fair to the poor: box off fair trade', Duncan Green and Matthew Griffiths.

—— (7 March 2002) 'Roy goes to prison and agonises over fine or serving longer term', Luke Harding, p. 1.

—— (6 May 2002) 'Stop debt vultures demands Brown', Larry Elliott, p. 17.

Guha, R. and Spivak, G. (1988) *Selected Subaltern Studies*, New Delhi, India: Oxford University Press.

Gupta, A. (1998) *Postcolonial Developments: Agriculture in the Making of Modern India*, Durham, NC: Duke University Press.

Haas, W. H. (1952) *The Rural Land Classification Program of Puerto Rico*, NorthWestern University, Studies in Geography, No. 1, p. v.

Hadjor, K. B. (1993) *Dictionary of Third World Terms*, London: Penguin.

Haffajee, F. (2001) 'From Seattle to Soweto', *New Internationalist*, 338, September, available at http://www.oneworld.org/ni/issue338/ (12 April 2002).

Haggard, S. M. (1998) 'Business, politics and policy in East and Southeast Asia', in Rowen, H. S. (ed.) *Behind East Asian Growth: The Political and Social Foundations of Prosperity*, pp. 78–104, London: Macmillan.

Hall, S. (1991) 'The local and the global; globalization and ethnicity', in A. King (ed.) *Culture, Globalization and the World System*, London: Macmillan.

—— (1992) 'The West and the Rest: discourse and power', in Hall, S. and Gieben, B. (eds) *Formations of Modernity*, Cambridge: Polity Press, pp. 276–320.

Hall, S. and Gieben, B. (eds) (1992) *Formations of Modernity*, Cambridge: Polity Press.

Halliday, F. (1979) *Iran: Dictatorship and Development*, London: Penguin (2nd edn).

—— (1988) 'The Iranian Revolution: uneven development and religious populism', in Halliday, F. and Alavi, H. (eds) *State and Ideology in the Middle East and Pakistan*, London: Macmillan, pp. 31–63.

—— (1995) *Islam and the Myth of Confrontation: Religion and Politics in the Middle East*, London: I B Tauris.

—— (2000) *Nation and Religion in the Middle East*, London: Saqi Books.

Halliday, F. and Alavi, H. (eds) (1988) *State and Ideology in the Middle East and Pakistan*, London: Macmillan.

Hampson, N. (1968) *The Enlightenment*, London: Penguin.

Hance, W. A. (1964) *The Geography of Modern Africa*, New York: Columbia University Press.

Hanlon, J. (1996) *Peace Without Profit: How the IMF Blocks Rebuilding in Mozambique*, Oxford: Irish Mozambique, Solidarity & the International African Institute in association with James Currey.

Hardoy, J. E., Mitlin, D. and Satterthwaite, D. (2001) *Environmental Problems in an Urbanizing World: Finding Solutions for Cities in Africa, Asia and Latin America*, London: Earthscan.

Hardt, M. and Negri, A. (2000) *Empire*, Cambridge, MA: Harvard University Press.

Hari, J. (2002) 'Now the protestors box clever', *New Statesman*, 1 April, available at http://www.newstatesman.co.uk/ (11 January 2002).

Harrison-Church, R. J. (1947) 'The case for colonial geography', *Geographical Journal*, 109, pp. 15–25.

—— (1971) *Africa and the Islands*, New York: Columbia Press.

Harriss, J. and de Renzio, P. (1997) 'Policy arena: "missing link" or "analytically missing"? The concept of social capital – an introductory bibliographic essay', *Journal of International Development*, 9 (7), pp. 919–937.

Hart, G. (2001) 'Development critiques in the 1990s: culs-de-sac and promising paths', *Progress in Human Geography*, 25 (4), pp. 649–658.

Harvey, D. (1972) 'Social justice and spatial systems', in Peet, R. (ed.) *Geographical Perspectives on American Poverty*, Antipode Monograph in Social Geography, No.1, Worcester, MA.

Haskell, T. L. (1985a) 'Capitalism and the origins of the humanitarian sensibility, part 1', *American Historical Review*, 90, pp. 339–361.

—— (1985b) 'Capitalism and the origins of the humanitarian sensibility, part 2', *American Historical Review*, 90, pp. 547–566.

Held, D. and McGrew, A. (1993) 'Globalisation and the liberal democratic state', *Government and Opposition*, 28 (2), pp. 262–263.

Heper, M. and Israeli, R. (eds) (1984) *Islam and Politics in the Modern Middle East*, London: Croom Helm.

Hettne, B. (1995) *Development Theory and the Three Worlds*, London: Longman.

Hewitt, A. (2001) 'Beyond poverty: the new UK policy on international development and globalisation', *Third World Quarterly*, 22 (2), pp. 291–296.

Hirschmann, A. O. (1958) *The Strategy of Economic Development*, New Haven, CT: Yale University Press.

Hodder, B. (1965) 'Some comments on the origins of traditional markets in Africa south of the Sahara', *Transactions of the Institute of British Geographers*, 26, pp. 97–105.

Holland, M. (1998) 'World Bank book (shh)', *The Nation*, 226 (10), pp. 4–5.

Holloway, S. and Hubbard, P. (2001) *People and Place: The Extraordinary Geographies of Everyday Life*, London: Pearson.

Hoogevelt, A. (1997) *Globalisation and Postcolonialism*, London: Macmillan.

—— (2001) *Globalisation and the Postcolonial World: The New Political Economy of Development*, London: Palgrave (2nd edn).

Hooson, D. (1998) 'Obituary: Keith Buchanan', *Asia Pacific Viewpoint*, 39 (1), pp. 1–28.

—— (2000) 'Oskar H.K. Spate, 1911–2000', *Geographical Review*, 90 (4), pp. 629–635.

Houtart, F. (2001) 'Preface: From Davos to Porto Alegre', in Houtart, F. and Polet, F. *The Other Davos: The Globalization of Resistance to the World Economic System*, London: Zed Books, pp. vi–viii.

Hughes, A. (2001) 'Global commodity networks, ethical trade and governmentality: organizing business responsibility in the Kenyan cut flower industry', *Transactions of the Institute of British Geographers*, 26, pp. 390–406.

Hunter, J. (1967) 'Seasonal hunger in a part of the West Africa savanna', *Transactions of the Institute of British Geographers*, 41, pp. 167–185.

Ignatieff, M. (1998) 'The stories we tell: television and humanitarian aid', in Moore, J. (ed.) *Hard Choices: Moral Dilemmas in Humanitarian Intervention*, Lanham, MD: Rowman & Littlefield, pp. 287–302.

Independent (2 March 2002) 'Mugabe says Blair is biased and can "Go to Hell" ', Kathy Marks, p. 5.

Index on Censorship (1999) *Word Power*, Vol. 22, No. 2, March–April 1999, Issue 187, available at http://www.indexoncensorship.org/ (12 April 2002).

Ingstad, B. and Whyte, S. R. (eds) (1995) *Disability and Culture*, Berkeley: University of California Press.

International Broadcasting Trust (IBT) (1998) *The Bank, the President and the Pearl of Africa*, London: IBT/Channel Four Television.

Jackson, R. H. (1990) *Quasi-states: Sovereignty, International Relations and the Third World*, Cambridge: Cambridge University Press.

Jacobs, J. (1996) *Edge of Empire: Postcolonialism and the City*, London: Routledge.

Jalée, P. (1969) *The Third World in World Economy*, New York: Monthly Review Press.

James, J. (1999) *Globalization, Information Technology and Development*, Basingstoke: Macmillan.

Jansen, K. (2001) 'Thailand: the making of a miracle?', *Development and Change*, 32, pp. 343–370.

Jarosz, L. (1992) 'Constructing the dark continent: metaphor as geographic representation of Africa', *Geografiska Annaler* (B), 74 (2), pp. 105–115.

Jawad, H. A. (ed.) (1994) *The Middle East in the New World Order*, London: Macmillan.

Jerve, A. M. (2001) 'Lebanon and Palestine', in Grinspun, A. (ed.) *Choices for the Poor: Lessons from National Poverty*

Jessop, B. and Brown, C. M. (1990) *Karl Marx's Social and Political Thought: Critical Assessments*, London: Routledge.

Strategies, New York: United Nations Development Programme (UNDP), pp. 303–328.
Johnson, C. (1982) *MITI and the Japanese Miracle: The Growth of Industrial Policy, 1925–1975*, Stanford, CA: Stanford University Press.
Johnston, R. J. (1983) *Geography and Geographers: Anglo-American Human Geography Since 1945*, London: Arnold.
—— (1997) *Geography and Geographers: Anglo-American Human Geography Since 1945*, London: Arnold, and New York: Oxford University Press (5th edn).
—— (1999) 'Classics in human geography revisited: commentary on Llwyd L (Keith Buchanan) (1968)', *Progress in Human Geography*, 23 (2), pp. 253–266.
Johnston, R., Gregory, D. and Smith, D. M. (eds) (1994) *The Dictionary of Human Geography*, Oxford: Blackwell (3rd edn).
Johnston, R. J., Gregory, D., Pratt, G. and Watts, M. (eds) (2000) *The Dictionary of Human Geography*, Oxford: Blackwell (4th edn).
Jolly, R. (1999) 'New composite indices for development co-operation', *Development*, 42 (3), pp. 36–42.
Jones, C. and Murphy, E. C. (2002) *Israel: Challenges to Identity, Democracy and the State*, London: Routledge.
Jones, D. (1957) *Malcolm Jarvis Proudfoot Memorial Volume*, Evanston: Northwestern University, pp. 1–3.
Jones, P. S. (2000) 'Why is it alright to do development "over there" but not "here"? Changing vocabularies and common strategies across the "First" and "Third Worlds" ', *Area*, 32, pp. 237–241.
Jubilee Debt Campaign (2002) 'Blair's "heal Africa tour" doomed without full debt cancellation: new report reveals development funding black hole', available at http://www.jubileedebtcampaign.org.uk/ (23 March 2002).
Kabeer, N. (1994) *Reversed Realities: Gender Hierarchies in Development Thought*, London: Verso.
Karaosmanoğlu, A. L. (1984) 'Islam and its implications for the international system', in Heper, M. and Israeli, R. (eds) *Islam and Politics in the Modern Middle East*, London: Croom Helm, pp. 103–118.
Keddie, N. R. (1988) 'Ideology, society and the state in post-colonial Muslim societies', in Halliday, F. and Alavi, H. (eds) *State and Ideology in the Middle East and Pakistan*, London: Macmillan, pp. 9–30.
Keen, S. (2001) *Debunking Economics: The Naked Emperor of the Social Sciences*, London: Pluto.
Kelley, D. G. (1999) 'A poetics of anticolonialism', *Monthly Review*, 51 (10), available at http://www.monthlyreview.org/1199kell.html/ (17 April 2002).
Kelly, P. F. (1997) 'Globalization, power and the politics of scale in the Philippines', *Geoforum*, 28 (2), pp. 151–171.
—— (2000) *Landscapes of Globalisation: Human Geographies of Economic Change in the Philippines*, London: Routledge.
Kelsall, T. (2002) 'Go figure(?): Why has postcolonial theory had so little impact on African Political Studies?', unpublished paper.
Khor, M. (ed.) (2001) *Rethinking Globalisation; Critical Issues and Policy Choices*, London: Zed Books.
Kiely, D. (2000) 'Globalization: from domination to resistance', *Third World Quarterly*, 21 (6), pp. 1059–1070.
Kiely, R. (1999) 'The last refuge of the Noble Savage? A critical assessment of post-development theory', *European Journal of Development Research*, 11 (1), pp. 30–55.
Kiely, R. and Marfleet, P. (eds) (1998) *Globalisation and the Third World*, London: Routledge.
Klein, N. (2000) *No Logo: Taking Aim at the Brand Bullies*, London: Picador.
Kolb, A. (1942) *Die Phillipinen*, Leipzig: Quelle & Meyer.
Koonings, K. and Kruijt, D. (eds) (2001) *Political Armies: The Military and Nation Building in the Age of Democracy*, London: Zed Books.
Kothari, R. (1988) *Rethinking Development: In Search of Human Alternatives*, Delhi: Ajanta.
Kothari, U. (1996) 'Development studies and post-colonial theory', institute for Development Policy and Management (IDPM) Discussion Paper Series No. 47, University of Manchester.
Lacoste, Y. (1973) 'An illustration of geographical warfare: bombing of the dikes on the Red River, North Vietnam', *Antipode*, 5 (2), pp. 1–13.
Lacquer, T. W. (1989) 'Bodies, details and humanitarian narrative', in Hunt, L. (ed.) *The New Cultural History*, Berkeley: University of California Press, pp. 176–204.
Laïdi, Z. (ed.) (1988) 'Introduction: What use is the Soviet Union?, in *The Third World and the Soviet Union*, London: Zed Books, pp. 1–23.
Langevin, M. S. and Rosset, P. (1997) 'Land reform from below: the landless workers movement in Brazil', available at http://www.mstbrazil.org/ (31 July 2002).
Lee, J. M. (1967) *Colonial Development and Good Government: A Study of the Ideas Expressed by the British Official Classes in Planning Decolonization 1939–1964*, Oxford: Clarendon Press.
Lester, A. (1999) 'Historical geographies of imperialism', in Graham, B. and Nash, C. (eds) *Modern Historical Geographies*, Harlow: Longman, pp. 100–120.
—— (2002) 'Obtaining the "due observance of justice": the geographies of colonial humanitarianism', *Environment and Planning D: Society and Space*, 20, pp. 277–293.
Lewis, M. W. and Wigen, M. W. (1997) *The Myth of Continents: A Critique of Metageography*, Berkeley: University of California Press.
Leys, C. (1996) *The Rise and Fall of Development Theory*, London: James Currey.
Liddell, I. and Donovan, P. (2000) 'Unpeeling the banana trade', Fairtrade Foundation, London, available at http://www.fairtrade.org.uk/unpeeling/ (15 April 2002).
Lister, M. (1997) *The European Union and the South: Relations with Developing Countries*, London: Routledge.
Livingston, S. (1996) 'Suffering in silence: media coverage of war and famine in the Sudan', in Rotberg, R. I. and Weiss, T. G. (eds) *From Massacres to Genocide: The Media, Public Policy and Humanitarian Crises*, Washington, DC: Brookings, pp. 68–89.

Livingstone, D. N. (1992) *The Geographical Tradition: Episodes in the History of a Contested Enterprise*, Oxford: Blackwell.

—— (2000) 'Tropical hermeneutics: fragments for a historical narrative – an afterword', *Singapore Journal of Tropical Geography*, 21 (1), pp. 92–98.

Llwyd, L. (Keith Buchanan) (1968) 'A preliminary contribution to the geographical analysis of a Pooh-scape', *IBG Newsletter*, 6, pp. 54–63.

Logan, M. I. (1972) 'The development process in the less developed countries', *The Australian Geographer*, 12 (2), pp. 146–179.

Long, C. M. (2001) *Participation of the Poor in Development Initiatives: Taking Their Rightful Place*, London: Earthscan.

Loomba, A. (1998) *Colonialism/postcolonialism*, London: Routledge.

Louie, M. C. Y. (2001) *Sweatshop Warriors: Immigrant Women Workers Take on the Global Economy*, Cambridge, MA: South End Press.

Luciani, G. (1990) 'Allocation vs. production states: a theoretical framework', in Luciani, G. (ed.) *The Arab State*, Berkeley: University of California Press.

Luciani, G. and Salamé, G. (eds) (1988) *The Politics of Arab Integration*, London: Croom Helm.

Luke, T. L. (1993) 'Discourses of disintegration, texts of transformation: re-reading realism in the New World Order', *Alternatives*, 18, pp. 229–258.

Luke, T. W. (1991) 'The discourse of development: a genealogy of "developing nations" and the discipline of modernity', *Current Perspectives in Social Theory*, 11, pp. 271–293.

Lummis, C. D. (2002) 'Political theory: why it seems universal, but isn't really', *Futures*, 34 (1), pp. 63–73.

Mabogunje, A. (1962) *Yoruba Towns*, Ibadan.

—— (1968) *Urbanization in Nigeria*, New York.

—— (2002) 'Comments on Professor Keith Buchanan', online, e-mail Marcus@geog.leeds.ac.uk (21 April 2002).

McCaughey, R. A. (1984) *International Studies as Academic Enterprise: A Chapter in the Enclosure of American Learning*, New York: Columbia University Press.

McClintock, A. (1992) 'The angel of progress: pitfalls of the term post-colonialism', *Social Text*, 31/32, pp. 84–98.

McDowell, L. and Sharp, J. (eds) (1999) *A Feminist Glossary of Human Geography*, London: Arnold.

MacEwan, A. (2002) 'Economic debacle in Argentina: the IMF strikes again', *Foreign Policy in Focus*, available at http://www.fpif.org/ (2 January 2002).

McEwan, C. (1998) 'Cutting power lines within the palace. Countering paternity and eurocentrism in the geographical tradition', *Transactions of the Institute of British Geographers*, NS 23 (3), pp. 371–384.

—— (2001) 'Postcolonialism', in Desai, V. and Potter, R. B. (eds) *The Companion to Development Studies*, London: Arnold, pp. 127–131.

McGee, T. G. (1974) 'In praise of tradition: towards a geography of anti-Development', *Antipode*, 6 (3), pp. 30–47.

McGrew, A. (2000) 'Sustainable globalization? The global politics of development and exclusion in the new world order', in Allen, T. and Thomas, A. (eds) *Poverty and Development into the Twenty-first Century*, Oxford: Open University Press, pp. 345–364.

Macrae, J. (2001) *Aiding Recovery? The Crisis of Aid in Chronic Political Emergencies*, London: Zed Books.

—— (2002) 'Aid in chronic political emergencies', *Opinions*, 6, Overseas Development Institute, available at http://www.odi.org.uk/opinions/ (19 April 2002).

Machel, G. (1996) 'UN study on the impact of armed conflict on children', available at http://www.unicef.org/graca/women/html (23 April 1998).

Main, L. (2001) 'The global information infrastructure: empowerment or imperialism?', *Third World Quarterly*, 22 (1), pp. 83–97.

Mamdani, M. (1996) *Citizen and Subject*, Princeton, NJ: Princeton University Press.

Mander, J. (2001) 'Who benefits most?', *Resurgence*, 208, pp. 12–15.

Manzo, K. (1991) 'Modernist discourse and the crisis of development theory', *Studies in Comparative International Development*, 26 (2), pp. 3–36.

—— (1995) 'Black consciousness and the quest for a counter-modernist development', in Crush, J. (ed.) *Power of Development*, London: Routledge.

—— (1999) 'The "new" developmentalism: political liberalism and the Washington Consensus', in Slater, D. and Taylor, P. J. (eds) *The American Century: Consensus and Coercion in the Projection of American Power*, Oxford: Blackwell, pp. 98–114.

Marcuse, P. (2000) 'The language of globalization', *Monthly Review*, 52 (3), pp. 1–4, available at http://www.monthlyreview.org/700marc.html/ (24 June 2002).

Marcuse, P. and Kempen, R. (2000) *Globalising Cities: A New Spatial Order?*, Oxford: Blackwell.

Marglin, F. and Marglin, S. (1990) *Dominating Knowledge: Development, Culture and Resistance*, Oxford: Clarendon Press.

Marshall, D. D. (2001) 'The New World group of dependency scholars: reflections on a Caribbean avant-garde movement', in Potter, R. B. and Desai, V. (eds) *The Companion to Development Studies*, London: Arnold, pp. 102–106.

Martinez, E. B. (2001) 'Telling the story of our America', *Monthly Review*, 53 (2), available at http://www.monthlyreview.org/0601martinez.html/ (24 June 2002).

Masina, P. (ed.) (2001) *Rethinking Development in East Asia: From Illusory Miracle to Economic Crisis*, Copenhagen: NIAS.

Mason, M. (1997) *Development and Disorder: A History of the Third World Since 1945*, Hanover, NH: University of New England Press.

Massey, D. (1992) 'A global sense of place', in Gray, A. and McGuigan, J. (eds) *Studying Culture*, London: Edward Arnold, pp. 232–240.

—— (1994) *Space, Place and Gender*, Cambridge: Polity Press.

Massey, D. and Jess, P. (eds) (1995) *A Place in the World?: Places, Cultures and Globalization*, Milton Keynes: Open University Press.

Matsu, K. (2002) 'A genuine progress indicator for Australia', *Adbusters Magazine*, August, p. 59, available at http://www.abusters/campaigns/ (24 June 2002).

Maxwell, S. (2002) 'More aid? – Yes – and use it to reshape aid architecture', *ODI opinions*, 3, available at http://www.odi.org.uk/opinions/ (24 June 2002).

Maxwell, S. and Riddell, R. (1998) 'Conditionality or contract: perspectives on partnership for development', *Journal of International Development*, 10 (2), pp. 257–268.

Mazrui, A. (1993) 'Language and the quest for liberation in Africa: the legacy of Frantz Fanon', *Third World Quarterly*, 14 (2), pp. 351–363.

—— (1999) 'Globalization and cross-cultural values: the politics of identity and judgement, *Arab Studies Quarterly*, 21 (3), pp. 96–109.

Mbembe, A. (1992) 'Provisional notes on the postcolony', *Africa*, 62 (1), pp. 1–37.

—— (2001) *On the Postcolony*, Berkeley: University of California Press.

Mercer, C., Mohan, G. and Power, M. (2003) 'Towards a critical political geography of African development,' *Geoforum*, forthcoming.

Milne, S. (2002) 'Colonialism and the New World Order', *Guardian*, 7 March, p. 20.

Mitchell, T. (1995) 'The object of development: America's Egypt', in Crush, J. (ed.) *Power of Development*, London: Routledge, pp. 129–157.

Mohan, G. and Mohan, J. (2002) 'Placing social capital', *Progress in Human Geography*, 26 (2), pp. 191–210.

Mohan, G. and Stokke, K. (2000) 'Participatory development and empowerment: the dangers of localism', *Third World Quarterly*, 21 (2), pp. 247–268.

Mohanty, C. (1988) 'Under Western eyes: feminist scholarship and colonial discourses', *Feminist Review*, 30, pp. 61–88.

—— (1991) 'Cartographies of struggle: Third World women and the politics of feminism', in Mohanty, C., Parker, A. and Russo, A. (eds) *Third World Women and the Politics of Feminism*, London: Routledge.

Molyneux, M. (2002) 'Gender and the silences of social capital: lessons from Latin America', *Development and Change*, 33 (2), pp. 167–188.

Monbiot, G. (2001) 'Tinkering with poverty', *Guardian*, 20 November, p. 17.

Moore-Gilbert, B. (1997) *Postcolonial Theory: Contexts, Practices, Politics*, London: Verso.

Moran, W. (2000) 'Classics in human geography revisited. Exceptionalism in the antipodes', *Progress in Human Geography*, 24 (3), pp. 429–438.

Morgan, W. T. (1969). 'Peasant agriculture in tropical Africa', in Thomas, M. and Whittington, G. (eds) *Environment and Land Use in Africa*, London: John Wiley, pp. 89–105.

—— (1973) *East Africa*, Geographies for Advanced Study Series, London: Longman.

Morrell, R. (ed.) (2001) *Changing Men in Southern Africa*, London: Zed Books.

Morrissey, O. (2002) 'Introductory brief: ODI opinions on effective expansion of aid', *ODI opinions*, 1, available at http://www.odi.org.uk/opinions/ (24 June 2002).

Moser, C. (1993) *Gender Planning and Development: Theory, Practice and Training*, London: Routledge.

Moser, C. and Clark, F. C. (eds) (2001) *Victims, Perpetrators or Actors? Gender, Armed Conflict and Political Violence*, London: Zed Books.

Moser, C. and Norton, A. (2001) *To Claim our Rights: Livelihood Security, Human Rights and Sustainable Development*, London: Overseas Development Institute.

Mountjoy, A. B. (1976) 'Worlds without end', *Third World Quarterly*, 2 (4), pp. 753–757.

Mountjoy, A. B. and Embleton, C. (1965) *Africa: A Geographical Study*, London: Hutchinson.

Movimento dos Trabalhadores Rurais Sem Terra (MST) (2000) 'The MST's manifesto to the Brazilian people', available at http://www.mstbrazil.org/ (31 July 2002).

—— (2002) 'MST social projects', available at http://www.mstbrazil.org/ (31 July 2002).

Mudimbe, V. (1988) *The Invention of Africa: Gnosis, Philosophy, and the Order of Knowledge*, Bloomington: Indiana University Press, and London: James Currey.

Multilateral Investment Guarantee Agency (MIGA) (2001) *Investment Promotion Toolkit: A Comprehensive Guide*, Washington: World Bank, Vols 1, 3 and 9.

Munck, R. (1999) 'Deconstructing development discourse: of impasses, alternatives and politics', in Munck, R. and O Hearn, D. (eds) *Critical Development Theory: Contributions to a New Paradigm*, London: Zed Books, pp. 198–210.

Muni, S. D. (1979) 'The Third World: concept and controversy', *The Third World Quarterly*, 16 (4), pp. 691–701.

Murray, C. (2001) 'Rural livelihoods', in Desai, V. and Potter, R. B. (eds) *The Development Studies Reader*, London: Arnold.

Myers, G. (2001) 'Introductory human geography textbook representations of Africa', *Professional Geographer*, 53 (4), pp. 522–532.

Myrdal, G. (1957) *Economic Theory and Underdeveloped Areas*, London: Duckworth.

—— (1968) *The Asian Drama: An Inquiry into the Poverty of Nations*, Harmondsworth: Penguin.

Naipaul, S. (1997) *North of South: An African Journey*, London: Penguin.

Naipaul, V. S. (1985) 'A thousand million invisible men', *The Spectator*, 254, pp. 9–11.

—— (1986) *The Enigma of Arrival: A Novel in Five Sections*, Harmondsworth: Viking Press.

Narayan, D. (2000) *Voices of the Poor: Can Anyone Hear Us?*, Oxford: World Bank/Oxford University Press.

Narayan, D., Chambers, R., Shah, M. K. and Petesch, P. (2000) *Voices of the Poor – Crying Out for Change*, Oxford: Oxford University Press for the World Bank.

Narmada Bachao Andolan (NBA) (1991) 'The History of suffering and of the resistance of the people of the Narmada Valley', India: NBA.

Naschold, F. (2002) 'Aid and the millennium development goals', *ODI Opinions*, 4, available at http://www.odi.org.uk/opinions/ (31 July 2002).

Nash, C. (2002) 'Cultural geography: postcolonial cultural geographies', *Progress in Human Geography*, 26 (2), pp. 219–230.

Nashel, J. (2000) 'The road to Vietnam: modernization theory in fact and fiction', in Appy, C. G. (ed.) *Cold War Constructions: The Political Culture of United States Imperialism 1945–1966*, Amherst: University of Massachusetts Press.

Naval Intelligence Division (NID) (1942) *French Equatorial Africa and Cameroons*, BR 515, London: Oxford and Cambridge University Presses.

—— (1944a) *The Belgian Congo*, BR 522, London: Oxford and Cambridge University Presses.

—— (1944b) *French West Africa*, Volume I: *The Federation*, BR 512, London: Oxford and Cambridge University Presses.

—— (1944c) *French West Africa*, Volume II: *The Colonies*, BR 512a, London: Cambridge and Oxford University Presses.

Ndione, E. S. (1994) *Réinventer le présent. Quelques jalons pour l'action*, Dakar: ENDA GRAF Sahel.

Newman, D. (2002) 'Citizenship, identity and location: the changing discourse of Israeli geopolitics', in Dodds, K. and Atkinson, D. (eds) *Geopolitical Traditions: A Century of Geopolitical Thought*, London: Routledge, pp. 302–331.

Ngũgi wa Thiong'o (1986) *Decolonising the mind: The Politics of Language in African Literature*, London: James Currey.

Niblock, T. (1994) 'A framework for renewal in the Middle East?', in Jawad, H. A. (ed.) *The Middle East in the New World Order*, London: Macmillan, pp. 1–12.

Nkrumah, K. (1961) *I Speak of Freedom*, London: Panaf Books.

Nyamnjoh, F. (2001) 'Expectations of modernity in Africa or a future in a rearview mirror', *Journal of Southern African Studies*, 27, pp. 363–369.

O'Brien, R., Goetz, A. M., Scholte, J. A. and Williams, M. (2000) *Contesting Global Governance: Multilateral Economic Institutions and Global Social Movements*, Cambridge: Cambridge University Press.

Observer (24 February 2002) 'Third world finance dries up', Business supplement, page 2.

O'Connor, A. (1976) 'Third World or One World', *Area*, 8, pp. 269–271.

—— (2001) 'Poverty in global terms', in Desai, V. and Potter, R. B. (eds) *The Development Studies Reader*, London: Arnold, pp. 37–41.

O'Keefe, P. and Wisner, B. (1977) 'African drought: the state of the game', in Richards, P. (ed.) *African Environment*, London: Routledge, pp. 31–39.

Ó Tuathail, G. (1996) *Critical Geopolitics: The Politics of Writing Global Space*, Minneapolis: University of Minnesota Press.

—— (1999) 'Borderless worlds? Problematising discourses of deterritorialisation', *Geopolitics*, 4 (2), pp. 139–154.

Ó Tuathail, G. and Agnew, J. (1992) 'Geopolitics and discourse: practical geopolitical reasoning in American foreign policy', *Political Geography*, 11 (2), pp. 190–204.

Olsen, G. R. (2001) 'European public opinion and aid to Africa: is there a link?', *Journal of Modern African Studies*, 39 (4), pp. 645–674.

Ong, A. (1999) *Flexible Citizenship: The Cultural Logics of Transnationality*, Durham, NC: Duke University Press.

Organization for Economic Cooperation and Development (OECD) (1996) *Shaping the Twenty-first Century: The Contribution of Development Cooperation*, Paris: OECD.

Oxfam (1993) *Africa: Make or Break. Action for Recovery*, Oxford: Oxfam.

—— (1999) *Africa's Forgotten Crises: People in Peril*, Oxford: Oxfam.

—— (2002) *Europe's Double Standards: How the EU Should Reform Its Trade Policies with the Developing World*, Oxford: Oxfam.

Pacione, M. (ed.) (1999) *Applied Geography*, London: Routledge.

Palast, G. (2001) 'The globalizer who came in from the cold', available at http://www.portoalegre2002.org/, (13 November 2001).

Palestine Solidarity Campaign (PSC) (2002) 'Economic development under fire', PSC Archives, 9 April, http://www.aquascript.com/psc/ (29 June 2002).

Palma, G. (1981) 'Dependency: a formal theory of underdevelopment or a methodology for the analysis of concrete situations of underdevelopment?', in Streeton, P. and Jolly, R. (eds) *Recent Issues in World Development*, London: Pergamon Press.

Panos (2001) *Food for All: Can Hunger be Halved?*, London: Panos Institute.

Paolini, A. (1997) 'The place of Africa in discourse about the postcolonial, the global and the modern', *New Formations*, 31, pp. 83–106.

Parnwell, M. and Rigg, J. (2001) 'Global dissatisfactions: globalisation, resistance and compliance in Southeast Asia', *Singapore Journal of Tropical Geography*, 22 (3), pp. 205–11.

Pasuk, P. and Baker, C. (2000) *Thailand's Crisis*, Singapore: Institute of South East Asian Studies, National University of Singapore.

Patel, R. (2001) 'Knowledge, power, banking', *Znet Magazine*, available at http://www.zmag.org/ (7 March 2002).

Pearce, J. (2000) *Development, NGOs and Civil Society: Selected Essays from Development in Practice*, Oxford: Oxfam.

Peake, L. and Kobayashi, A. (2002) 'Policies and practices for an antiracist geography at the millennium', *Professional Geographer*, 541, pp. 50–61.

Peck, J. and Tickell, A. (2002) 'Neoliberalizing space', *Antipode*, pp. 380–404.

Peet, R. (1977a) 'Introduction', in Peet, R. (ed.) *Radical Geography: Alternative Viewpoints on Contemporary Social Issues*, London: Methuen, pp. 1–4.

—— (1977b) 'The development of radical geography in the United States', in Peet, R. (ed.) *Radical Geography: Alternative Viewpoints on Contemporary Social Issues*, London: Methuen, pp. 6–30.

—— (1977c) 'Inequality and poverty: a Marxist-geographic theory', in Peet, R. (ed.) *Radical Geography: Alternative Viewpoints on Contemporary Social Issues*, London: Methuen, pp. 112–124.

—— (2002) 'Ideology, discourse and the geography of hegemony: from socialist to neoliberal development in post-apartheid South Africa', *Antipode*, pp. 54–84.

Peet, R. with Hartwick, E. (1999) *Theories of Development*, London: Guilford Press.

Peet, R. and Watts, M. (1993) 'Introduction: development theory and environment in the age of market triumphalism', *Economic Geography*, 69 (3), pp. 227–253.

Perry, S. and Schenck, C. (eds) (2001) *Eye-to-eye: Women Practising Development Across Cultures*, London: Zed Books.

Petras, J. (2000) 'The Third Way: myth and reality', *Monthly Review*, 51 (10), available at http://www.monthlyreview.org/300petras.html/ (31 July 2002).

—— (2001) 'The geopolitics of Plan Colombia', *Monthly Review*, 53 (1), available at http://www.monthlyreview.org/may2001.html/ (11 July 2002).

—— (2002) 'US offensive in Latin America: coups, retreats and radicalization', *Monthly Review*, 54 (1), available at http://www.monthlyreview.org/0502petras.html/ (21 July 2002).

Petras, J. and Veltmeyer, H. (2002) 'Age of reverse aid: neo-liberalism as catalyst of regression', *Development and Change*, 33 (2), pp. 281–293.

Philo, C. (1992) 'Foucault's geography', *Environment and Planning D: Society and Space*, 10, pp. 137–161.

Philpott, S. (2000) *Rethinking Indonesia: Postcolonial Theory, Authoritarianism and Identity*, Basingstoke: Macmillan.

Pieterse, J. N. (1991) 'Dilemmas of development discourse: the crisis of developmentalism and the comparative method', *Development and Change*, 22, pp. 5–29.

—— (1998) 'My paradigm or yours? Alternative development, post-development, reflexive development', *Development and Change*, 29 (2), pp. 343–373.

—— (2000) 'After post-development', *Third World Quarterly*, 21 (2), pp. 171–191.

—— (2001a) *Globalization and Social Movements*, London: Macmillan.

—— (2001b) *Development Theory: Deconstructions/ Reconstructions*, London: Sage.

—— (2001c) 'Changing definitions', *Third Text*, 53, pp. 91–92.

Pieterse, J. N. and Parekh, B. (ed.) (1995) *Decolonization of the Imagination: Culture, Knowledge and Power*, New York: St Martins Press.

Pieterse, J. N., Hamel, P. and Roseneil, S. (eds) (2000) *Global Futures: Shaping Globalization*, London: Zed Books.

Pincus, J. (2001) 'The post-Washington consensus and lending operations in agriculture: new rhetoric and old operational realities', in Fine, B., Lapavitsas, C. and Pincus, J. (eds) *Development Policy in the Twenty-first Century*, London: Routledge, pp. 182–218.

Pinto, M. de C. (1999) *Political Islam and the United States*, Reading, MA: Ithaca University Press.

Pletsch, C. (1981) 'The three worlds or the division of social scientific labour 1950–1975', *Comparative Studies in Society and History*, 23, pp. 565–590.

Pollock, N. C. (1968) *Africa: A Systematic Regional Geography*, Volume 9, London: University of London Press.

Porter, R. (1990) *The Enlightenment*, London: Macmillan.

Potter, R. B. (1993) 'Little England and little geography: reflections on Third World teaching and research', *Area*, 25, pp. 291–294.

—— (2001a) 'Correspondence: what ever happened to development geography?', *Geographical Journal*, 168, pp. 188–189.

—— (2001b) 'Geography and development: "core and periphery"?', *Area*, 33 (4), pp. 422–427.

—— (2001c) 'Progress, development and change' (editorial), *Progress in Development Studies*, 1, pp. 1–3.

—— (2002) 'Making progress in development studies', *Progress in Development Studies*, 2 (1), pp. 1–3.

Potter, R. B. and Unwin, T. (1988) 'Developing area research in British geography 1982–1987', *Area*, 20 (2), pp. 121–126.

Potter, R. B., Binns, T., Elliott, J. A. and Smith, D. (1999) *Geographies of Development*, London: Longman.

Power, M. (2000) 'The short cut to international development: representing Africa in New Britain', *Area*, 32, pp. 91–100.

—— (2001) 'Alternative geographies of global development and inequality', in Daniels, P., Bradshaw, M., Shaw, D. and Sidaway, J. D. (eds) *Human Geography: Issues for the Twenty-first Century*, London: Pearson, pp. 274–302.

—— (2002) 'Patrimonialism and petro-diamond capitalism: peace, (geo)politics and the economics of war in Angola', *Review of African Political Economy*, 90, pp. 489–502.

Power, M. and Sidaway, J. D. (2004) 'The degeneration of tropical geography', *Annals of the Association of American Geographers*, forthcoming.

Preston, P. W. (1996) *Development Theory: An Introduction*, Oxford: Blackwell.

Priestey, M. (1999) *Disability Politics and Community Care*, London: Jessica Kingsley.

Prothero, R. M. (1957) *Migrant Labour from Sakoto Province*, Nigeria, Kaduna: Government Printer.

Prothero, R. M. (ed.) (1969a) *A Geography of Africa: Regional Essays on Fundamental Characteristics, Issues and Problems*, London: Routledge.

—— (1969b) 'Understanding Africa', in Prothero, R. M. (ed.) *A Geography of Africa*, London: Routledge, pp. 1–20.

Proudfoot, M. J. (1952) 'Preface', in Haas, W. H. *The Rural Land Classification Program of Puerto Rico*, Northwestern University Studies in Geography, No. 1, p. v.

Pugh, J. C. and Perry, A. E. (1960) *A Short Geography of West Africa*, London: University of London Press.

Rabinow, P. (1977) *Reflections on Fieldwork in Morocco*, Berkeley: University of California Press.

Radcliffe, S. (1994) '(Representing) post-colonial women: authority, difference and feminisms', *Area*, 26, pp. 25–32.

—— (1999) 'Re-thinking development', in Cloke, P. Crang, P. and Goodwin, M. (eds) *Introducing Human Geographies*, London: Arnold, pp. 84–91.

Rahnema, M. (1991) 'Global poverty: a pauperizing myth', *Interculture*, 24 (2), pp. 4–51.

—— (1992) 'Poverty', in Sachs, W. (ed.) *The Development Dictionary: A Guide to Knowledge as Power*, London: Zed Books.

—— (1997a) 'Signposts for post-development: in search of a different language and new paradigms', *ReVision*, 19 (4), pp. 4–12.

—— (1997b) 'Towards post-development: searching for signposts, a new language and new paradigms', in Rahnema, M. and Bawtree, V. (eds) *The Post-Development Reader*, London: Zed Books, pp. 377–403.

Rahmena, M. and Bawtree, V. (eds) (1997) *The Post-Development Reader*, London, Zed Books.

Rajan, R. S. (1997) 'The Third World academic in other places: on the postcolonial intellectual revisited', *Critical Inquiry*, 23 (3), pp. 600–619.

Ramutsindela, M. (1999) 'African boundaries and their interpreters', *Geopolitics*, 4 (2), pp. 180–198.

Ravenhill, J. (1985) *Collective Clientelism: The Lomé Conventions and North–South Relations*, New York: Columbia University Press.

Redclift, M. (1987) *Sustainable Development: Exploring the Contradictions*, London: Routledge.

Reddy, S. G. and Hogge, T. W. (2002) 'How *not* to count the poor', unpublished paper, Department of Economics, University of Columbia, available at http://www.socialanalysis.org/ (26 February 2002).

Redfield, P. (2000) *Space in the Tropics: From Convicts to Rockets in French Guiana*, Berkeley, Los Angeles and London: University of California Press.

Regan, C. and Walsh, F. (1977) 'Dependence and underdevelopment: the case of mineral resources and the Irish Republic', in Peet, R. (ed.) *Radical Geography: Alternative Viewpoints on Contemporary Social Issues*, London: Methuen, pp. 315–338.

Reporting the World (2002) 'Representation and responsibility in covering stories about Africa and African people: seminar summary', available at http://www.reportingtheworld.org/ (26 February 2002), pp. 1–5.

Results Educational Fund (2001) 'The World Bank's proposed changes to its information disclosure policy: a critical review', a report by REF, unpublished paper, Washington DC, available at http://www.resultsinfo.org/docs/ (7 August 2001).

Riddell, B. (1970) *The Spatial Dynamics of Modernization in Sierra Leone*, Evanston, IL: Northwestern University Press.

Riech, B. (1996) 'The United States and Israel: the nature of a special relationship', in Lesch, D. W. (ed.) *The Middle East and the United States: A Historical and Political Reassessment*, Boulder, CO: Westview Press, pp. 233–248.

Rieff, D. (1990) *Los Angeles: Capital of the Third World*, New York: Simon & Schuster.

Rigg, J. (ed.) (1995) *Counting the Cost: Economic Growth and Environmental Change in Thailand*, Singapore: Institute of South East Asian studies.

—— (1997) *Southeast Asia: The Human Landscape of Modernization and Development*, London: Routledge.

—— (2001) 'The Asian crisis', in Desai, V. and Potter, R. B. (eds) *The Development Studies Reader*, London: Arnold.

Rikowski, G. (2001a) 'Transfiguration: globalisation, the WTO and the national faces of the GATS', *Information for Social Change*, 14, available at http://libr.org/ISC/articles/ (26 February 2002).

—— (2001b) *The Battle in Seattle: Its Significance for Education*, London: Tufnell Press.

Rist, T. (1997) *History of Development: From Western Origins to Global Faith*, London: Routledge.

Robequain, C. (1931) *L'Indochine Française*, Paris: Presses universitaires de France.

—— (1944) *The Economic Development of French Indo-China*, London: Oxford University Press.

—— (1958) *Madagascar et les bases dispersées de l'Union française*, Paris: Presses universitaires de France.

Robinson, J. (2002) 'Global and world cities: a view from off the map', *International Journal of Urban and Regional Research*, 26(3), pp. 531–554.

Robinson, P. (1999) 'The CNN effect: can the news media drive foreign policy?', *Review of International Studies*, 25 (2), pp. 301–309.

Rocha, J. (2002) 'On the frontier', *Guardian*, 26 June, p. 11.

Rose, G. (1993) *Feminism and Geography: The Limits to Geographical Knowledge*, Cambridge: Polity Press.

Rosenblatt, L. (1996) 'The media and the refugee', in Rotberg, R. I. and Weiss, T. G. (eds) *From Massacres to Genocide: The Media, Public Policy and Humanitarian Crises*, Washington, DC: Brookings, pp. 122–141.

Rostow, W. W. (1960) *The Stages of Economic Growth: A Non-communist Manifesto*, Cambridge: Cambridge University Press.

Routledge, P. (1997) 'A spatiality of resistances: theory and practice in Nepal's revolution of 1990', in Pile, S. and Keith, M. (eds) *Geographies of Resistance*, London: Routledge.

—— (1998) 'Introduction: anti-geo-politics', in Ó Tuathail, G. (ed.) *The Geopolitics Reader*, London: Routledge, pp. 245–255.

—— (1999) 'Survival and resistance', in Cloke, P. Crang, P. and Goodwin, M. (eds) *Introducing Human Geographies*, London: Arnold, pp. 77–83.

Rowbotham, M. (2000) *Goodbye America: Globalisation, Debt and the Dollar Empire*, London: John Carpenter.

Roy, A. N. (1999) *The Third World in the Age of Globalism: Requiem or New Agenda?*, London: Zed Books.

Ryan, C. (1999) *Action on the Front Lines*, WorldWatch Report 12/6, Washington, DC: WorldWatch Institute.

Sachs, W. (1992) 'Introduction', in Sachs, W. (ed.) *The Development Dictionary: A Guide to Knowledge as Power*, London: Zed Books.

Saddy, F. (1982) 'OPEC capital surplus funds and Third World indebtedness: the recycling strategy reconsidered', *Third World Quarterly*, 4 (4).

Said, E. (1978) *Orientalism*, London: Penguin.

—— (1993) *Culture and Imperialism*, New York: Alfred A. Knopf.

—— (2000) *Orientalism*, New York: Vintage Books (2nd edn).

—— (2001) 'Islam and the West are inadequate banners', *Observer*, 16 September, available at http://www.edwardsaid.org/ (16 February 2002).

—— (2002) 'Thoughts about America', *Al-Ahram*, weekly online, 28 February to 6 March, Issue 575, available at http://www.edwardsaid.org/ (21 February 2002).

Sammy, W. (2002) 'NEPAD – new partnership for Africa's development. Is NEPAD the answer to Africa's problems?, *TWN Africa Secretariat*, Available at http://www.twnafrica.org/ (24 February 2002).

Samoff, J. and Stromquist, N. P. (2001) 'Managing knowledge and storing wisdom? New forms of foreign aid?', *Development and Change*, 32, pp. 631–656.

Santiso, C. (2001) 'Conditionality and the reform of multilateral development finance: the role of the Group of Eight', Johns Hopkins University, G8 Governance No. 7, Available at http://www.g7.utoronto.ca/ (26 February 2002).

Santos, M. (1974) 'Geography, Marxism and underdevelopment', *Antipode*, 6 pp. 1–9.

Sautter, G. and Pelissier, P. (1962) 'Pour un atlas des terroir Africains', *L'Homme*, 1 (1), pp. 56–72.

Schaffer, N. M. (1965) 'The competitive position of the port of Durban', Evanston, IL: Northwestern University Studies in Geography, No. 8.

Schafer-Gross, L. (1995) *The International World of Electronic Media*, London: McGraw-Hill.

Schech, S. and Haggis, J. (2000) *Culture and Development: A Critical Introduction*, Oxford: Blackwell.

Schenck, C. (2001) 'Developing subjects', in Perry, S. and Schenck, C. (eds) *Eye-to-eye: Women Practising Development Across Cultures*, London: Zed Books, pp. 235–258.

Schmitz, G. (1995) 'Democratization and desmystification: deconstructing "governance" as development paradigm', in Moore, D. and Schmitz, G.(eds) *Debating Development Discourse*, London: Macmillan, pp. 54–90.

Schuurman, F. J. (1993) *Beyond the Impasse: New Directions in Development Theory*, London: Zed Books.

—— (2001) 'Gobalization and development studies: introducing the Challenges', in Schuurman, F. J. (ed.) *Globalization and Development: Challenges for the 21st Century*, London: Sage, pp. 3–16.

Schwarz, W. (2001) 'The future is orange: food sovereignty is boosted by fairtrade support', *Guardian*, 19 December, p. 9.

Scott, D. (1999) *Refashioning Futures: Criticism after Postcoloniality*, Princeton, NJ: Princeton University Press.

Scott, E. (1972) 'The spatial structure of rural northern Nigeria', *Economic Geography*, 48, pp. 316–32.

Sen, A. (1983) 'Poor, relatively speaking', Oxford Economic Papers, No. 35, Dublin: Economic and Social Research Institute.

—— (1984) *Resources, Values and Development*, Oxford: Blackwell.

—— (1985) *Commodities and Capabilities*, Amsterdam: North Holland.

—— (2001) *Development as Freedom*, Oxford: Oxford University Press (2nd edn).

Shapiro, J. (2001) *Mao's War against Nature: Politics and the Environment in Revolutionary China*, Cambridge: Cambridge University Press.

Sharp, J. (1988) 'Two worlds in one country: "First World" and "Third World", in South Africa', in Boonzaier, E. and Sharp, J. (eds) *South African Keywords: The Uses and Abuses of Concepts*, Cape Town: David Phillip Publishers.

Sharp, J. E., Routledge, P, Philo, C. and Raddison, R. (2000) *Entanglements of Power*, London: Routledge.

Shatkin, G. (1998) 'Fourth world cities in the global economy: the case of Phnom Penh', *IJURR*, 22, pp. 378–393.

Shaw, J. D. (2001) *The United Nations World Food Programme and the Development of Food Aid*, Basingstoke: Palgrave.

Shaw, M. (1996) *Civil Society and Media in Global Crises: Representing Distant Violence*, London: Pinter.

Shiva, V. (1989) *Staying Alive: Women, Ecology and Development*, London: Zed Books.

—— (2001) 'Globalization and poverty', in Bennholdt-Thomsen, V., Faraclas, N. and Von Werlhof, C. (eds) *There is an Alternative: Subsistence and Worldwide Resistance to Corporate Globalization*, London: Zed Books, pp. 57–66.

—— (2002) 'The poor can buy Barbie Dolls', *Renaissance Universal* magazine, available at http://www.ru.org/10-4Shiva.htm/, pp. 1–6 (26 February 2002).

Shohat, E. and Stam, R. (1994) *Unthinking Eurocentrism: Multiculturalism and the Media*, London: Routledge.

Sidaway, J. D. (1994) 'Geopolitics, geography and "terrorism" in the Middle East', *Environment and Planning D: Society and Space*, 12, pp. 357–372.

—— (1997) 'The remaking of the western "geographical tradition"; some missing links', *Area*, 29 (1), pp. 72–80.

—— (2000) 'Postcolonial geographies: an exploratory essay', *Progress in Human Geography*, 24 (2), pp. 591–612.

—— (2001) 'Post-development', in Desai, V. and Potter, R. B. (eds) *The Development Studies Reader*, London: Arnold, pp. 16–20.

—— (2002) *Imagined Regional Communities: Integration and Sovereignty in the Global South*, London and New York: Routledge.

Sidaway, J. D. and Pryke, M. (2000) 'The strange geographies of "emerging markets" ', *Transactions of the Institute of British Geographers*, RGS/IBG, NS 25, pp. 187–201.

Sidnes, A. K. and Brown, M. M. (2001) 'Foreword', in Grinspun A. (ed.) *Choices for the Poor: Lessons from National Poverty Strategies*, New York: UNDP.

Simon, D. (2001) 'Dilemmas of development and the environment in a globalising world: theory, policy and praxis', Inaugural lecture given at Royal Holloway, University of London, 25 October. Copy available from author.

—— (2002) Book review of Thompson, N. and Thompson, S. (2000) *The Baobab and the Mango Tree: Lessons about Development – African and Asian Contrasts*, London and New York: Zed Books, *Third World Quarterly*, forthcoming.

Simone, A. M. (2001) 'Straddling the divides: remaking associational life in the informal African city', *International Journal of Urban and Regional Research*, 25 (1), pp. 102–117.

—— (2002) 'Africa and intersecting politics' (editorial), *Geoforum*, 33 (3), pp. 271–411.

Sklair, L. and Robbins, P. T. (2002) 'Global capitalism and major corporations from the Third World', *Third World Quarterly*, 23 (1), pp. 81–100.

Slater, D. (1973) 'Geography and underdevelopment I', *Antipode*, 5, pp. 21–53.

—— (1976) 'Anglo-Saxon geography and the study of underdevelopment', *Antipode*, 8 (3), pp. 88–93.

—— (1977) 'Geography and underdevelopment II', *Antipode*, 9 (3), pp. 1–31.

—— (1992) 'On the borders of social theory: learning from other regions', *Environment and Planning D*, 10, pp. 307–327.

—— (1993) 'The geopolitical imagination and the enframing of development theory', *Transactions of the Institute of British Geographers*, 18 (4), pp. 419–37.

—— (1998) 'Post-colonial questions for global times', *Review of International Political Economy*, 5 (4), pp. 647–678.

—— (2002) 'Other domains of democratic theory: space, power, and the politics of democratization', *Environment and Planning D: Society and Space*, 20, pp. 255–276.

Slater, D. and Bell, M. (2002) 'Aid and the geopolitics of the post-colonial: critical reflections on New Labour's overseas development strategy', *Development and Change*, 33 (2), pp. 335–360.

Smillie, I. (1996) 'Mixed messages: public opinion and development assistance in the 1990s', in Foy, C. and Helmich, H. (eds) *Public Support for International Development*, Paris: OECD, pp. 27–54.

Smith, A. (2002) 'Trans-locals, critical area studies and geography's Others, or why "development" should not be geography's organizing framework: a response to Potter', *Area*, 34 (2), pp. 210–213.

Smith, N. (1997) 'The Satanic geographies of globalization: uneven development in the 1990s', *Public Culture*, 10 (1), pp. 168–180.

Smith, N. and Godlewska, A. (1994) 'Introduction: Critical histories of geography', in Smith, N. and Godlewska, A. (eds) *Geography and Empire*, Oxford: Blackwell, pp. 1–13.

Sogge, D. (2002) *Give and Take: What's the Matter with Foreign Aid?*, London: Zed Books.

Soja, E. W. (1968) *The Geography of Modernization in Kenya: A Spatial Analysis of Social, Economic and Political Change*, New York: Syracuse University Press.

Solanas, F. and Gettino, O. (1976) 'Towards a Third Cinema', in Nichols, B. (ed.) *Movies and Methods*, Vol. 1, Berkeley: University of California Press, pp. 56–78.

South Commission (1990) *The Challenge to the South*, Geneva, Switzerland: South Centre.

Spate, O. H. K. (1954) *India and Pakistan*, London: Methuen.

Spurr, D. (1993) *The Rhetoric of Empire: Colonial Discourse in Journalism, Travel Writing and Imperial Administration*, Durham, NC: Duke University Press.

Spybey, T. (1992) *Social Change, Development and Dependency: Modernity, Colonialism and the Development of the West*, London: Polity Press.

Stallings, B. (1995) 'The new international context of development', in Stallings, B. (ed.) *Global Change, Regional Response*, Cambridge: Cambridge University Press.

Stamp, L. D. (1934) *A Short Geography of the World*, Paris: Longmans.

—— (1938) 'Land utilisation and erosion in Nigeria', *Geographical Review*, 28, pp. 32–45.

—— (1948) *Asia: A Regional and Economic Geography*, London: Methuen.

—— (1953) *Africa: A Study in Tropical Development*, New York and London: John Wiley.

Steel, R. W. (1961) 'A review of IBG publications 1946–60', *Transactions of the Institute of British Geographers*, 29, pp. 129–147.

—— (1964) 'Geographers and the Tropics', in Steel, R. W. and Prothero, R. M. (eds) *Geographers and the Tropics*, Liverpool: University of Liverpool, pp. 1–30.

Steel, R. W. and Prothero, R. M. (eds) (1964) *Geographers and the Tropics: Liverpool Essays*, London: Longman.

Stiglitz, J. E. (2002) *Globalization and its Discontents*, London: Penguin.

Stoddart, D. (1996) 'Letter to the Editor', *The Geographical Journal*, 162, pp. 354–355.

Stöhr, W. B. and Taylor, D. R. F. (1981) *Development from Above or Below? The Dialectics of Regional Planning in Developing Countries*, Chichester: John Wiley.

Stone, E. (1999) *Disability and Development*, Leeds: University of Leeds, Disability Press.

Students for Global Justice (SFGJ) (2000) 'On WEF cocktail party', available at http://www.studentsforglobaljustice.org/ (19 June 2002).

Sukarno, I. (1955) 'Modern History Sourcebook: President Sukarno of Indonesia: Speech at the Opening of the Bandung Conference, April 18 1955', available at http://www.fordham.edu/halsall/mod/1955sukarno-bandong.html/ (7 April 2002).

Sutcliffe, B. (2001) *100 Ways of Seeing an Unequal World*, London: Zed Books.

Sweetman, C. (ed.) (2000) *Gender in the Twenty-first Century*, Oxford: Oxfam.

Sylvester, C. (1999) 'Development and postcolonial studies: disparate tales of the "Third World",' *Third World Quarterly*, 20 (4), pp. 703–721.

Talbot, C. (2002) 'Blair's neocolonialist vision for Africa', available of http://www.wsws.org/articles/2002/feb2002/ (26 February 2002).

Teivainen, T. (2002) *Enter Economism, Exit Politics: Experts, Economic Policy and the Damage to Democracy*, London: Zed Books.

Telatin, M. (2001) 'Sierra Leone: the IMF's planned route from conflict to poverty', Jubilee Plus, available at http://www.jubileedebtcampaign.org.uk/ (26 February 2002).

Thérien, J. P. and Lloyd, C. (2000) 'Development assistance on the brink', *Third World Quarterly*, 21 (1), pp. 21–38.

Thiong'o, N. wa (1986) *Decolonising the Mind*, London: Heinemann.

Third World Network (TWN) (2001) 'Introduction: about TWN', available at http://www.twnside.org.sg/twintro.htm/ (24 September 2001).

Third World and Environment Broadcasting Project (3WE) (2000) 'Losing, perspective: global affairs on British terrestrial television 1989–1999', available at, http://www.ibt.org.uk/research/ (6 February 2003).

Thomas, A. (2000) 'Meanings and views of development', in Allen, T. and Thomas, A. *Poverty and Development into the Twenty-first Century*, Oxford: Oxford University Press, pp. 23–48.

Thompson, N. and Thompson, S. (2000) *The Baobab and the Mango Tree: Lessons about Development – African and Asian Contrasts*, London and New York: Zed Books.

Thrift, N. (2000) 'It's the little things', in Dodds, K. and Atkinson, D. (eds) *Geopolitical Traditions: A Century of Geopolitical Thought*, London and New York: Routledge, pp. 380–387.

—— (2002) 'The future of geography', *Geoforum*, 33, pp. 291–298.

Thrift, N. and Walling, D. (2000) 'Geography in the United Kingdom 1996–2000', *The Geographical Journal*, 166, pp. 96–124.

Tipps, D. (1973) 'Modernization theory and the comparative study of societies: a critical perspective', *Comparative Studies in Society and History*, 15(2), pp. 199–226.

Tipton, F. B. (1998) *The Rise of Asia: Economics, Society and Politics in Contemporary Asia*, Honolulu: University of Hawaii Press.

Townsend, J. (2000) 'Whose ideas count: how can NGOs Challenge global development fashions?', University of Durham, Department of Geography, unpublished paper.

Truman, H. (1949) Public Papers of the President, 20 January, United States Government Printing Office, Washington, DC.

United Nations (UN) (2002) 'Trusteeship Council', copy available at http://www.un.org/Overview/Organs/ (18 July 2002).

United Nations Children's Fund (UNICEF) (2000) *The Progress of Nations 2000*, available at http://www.unicef.org/ (22 February 2002).

United Nations Conference on Trade and Development (UNCTAD) (2000) *FDI Determinants and TNC Strategies: The Case of Brazil*, New York and Geneva: UN.

—— (2000–2001) *Africa Competitiveness Report, 2000–2001*, Geneva: UNCTAD.

—— (2001a) *The World of Investment Promotion at a Glance: A Survey of Investment Promotion Practices*, UN Advisory Series No. (17), Geneva: UN.

—— (2001b) *World Investment Report: Promoting Linkages*, New York and Geneva: UN.

United Nations Conference on Trade and Development and International Chamber of Commerce (UNCTAD-ICC) (2001) *An Investment Guide to Mozambique: Opportunities and Conditions*, New York and Geneva: UN.

United Nations Development Programme (UNDP) (1991) *Human Development Report*, New York: UN.

—— (1998) *Mapping Living Conditions in Lebanon: An Analysis of the Housing and Population Database*, Beirut: Ministry of Social Affairs and the UNDP.

—— (2000) *Human Development Report*, New York: UNDP.

—— (2001) *African Development Indicators*, New York: UNDP.

United Nations Industrial Development Organization (UNIDO) (2002) *Quality, Technology, Investment: Improving Productivity and Competitiveness*, UNIDO Technology and Investment branch, Vienna, available at http://www.unido.org/ (15 March 2002)

United Nations Statistics Division (1999) 'Integrated and coordinated implementation and follow-up of major United Nations conferences and summits: a critical review of the development of indicators in the context of conference follow-up', paper presented to the thirtieth session of the Statistical Commission, 1–5 March, New York, available at http://www.gopher.un.org/70/00/esc/docs/1999/ (10 March 2002).

United States Agency for International Development (USAID) (2002a) 'Program highlights', available at http://www.usaid.gov/pubs/ (15 July 2002).

—— (2002b) 'A history of foreign assistance', available at http://www.usaid.gov/about/ (15 July 2002).

University Grants Committee (1961) *Report of the Committee on Oriental, Slavonic, East European and African Studies* (The Hayter Report).

Unwin, T. (ed.) (1994) *Atlas of World Development*, Chichester: John Wiley.

Van Ausdal, S. (2001) 'Development and discourse among the Maya of Southern Belize', *Development and Change*, 32, pp. 577–606.

Van der Gaag, N. (ed.) (2002) 'Reducing poverty: is the World Bank's strategy working?', London: Panos Institite/ Medianet.

Van Riessen, M. (1999) *EU 'Global Player': The North–South Policy of the European Union*, Utrecht: International Books.

Van Valkenberg, S. (1925) 'Java: the economic geography of a tropical island', *Geographical Review*, 15, pp. 563–583.

Varley, W. J. and White, H. P. (1958) *The Geography of Ghana*, London: Longman.

Vía Campesina (2002) 'What is the *Vía Campesina*?', available at http://www.viacampesina.org/ (31 July 2002).

Vidal, J. (2002) 'Britons sink into ignorance as TV turns to trivia in third world', *Guardian*, 10 July, p. 7.

Vitalis, R. (1996) 'The New Deal in Egypt: the rise of Anglo-American commercial competition in World War II and the fall of neocolonialism', *Diplomatic History*, 20 (2), pp. 211–240.

—— (2002) 'Black gold, white crude: an essay on American exceptionalism, hierarchy, and hegemony in the Gulf', *Diplomatic History*, 26 (2), pp. 185–213.

Vitalis, R. and Heydemann, S. (2000) 'War, Keynesianism and colonialism: explaining state–market relations in the post-war Middle East', in Heydemann, S. (ed.) *War and Society in the Middle East*, Berkeley: Universtity of California Press.

Voluntary Services Overseas (VSO) (2001) *The Live Aid Legacy*, London: VSO.

Wade, R. (1996) 'Japan, the World Bank, and the art of paradigm maintenance: the East Asian miracle in political perspective', *New Left Review*, 217, pp. 3–33.

—— (2001) 'Winners and losers', *Economist* (April), pp. 79–82.

Wallerstein, I. (1990) 'Culture as the ideological battleground in the modern world system', in Featherstone, M. (ed.) *Global Culture*, London: Sage.

—— (1994) 'Development: lodestar or illusion?', in Sklair, L. (ed.) *Capitalism and Development*, London: Routledge, pp. 3–20.

—— (1997) 'The unintended consequences of cold war area studies', in Chomsky, N. *et al.* (eds) *The Cold War and the University: Toward an Intellectual History of the Postwar Years*, New York: New York Press, pp. 199–200.

—— (1998) 'The rise and future demise of world-systems analysis', *Review*, 21 (1), pp. 103–112.

Ward, R. G. (2001) 'Obituary: Oskar Hermann Khristian Spate (1911–2000)', *Australian Geographical Studies*, July, 39 (2), pp. 253–255.

Watnick, M. (1952) 'The Appeal of Communism to the peoples of underdeveloped areas', *Economic Development and Cultural Change*, 1, pp. 22–37.

Watters, R. (1998) 'The geographer as radical humanist: an appreciation of Keith Buchanan', *Asia Pacific Viewpoint*, 39 (1), pp. 1–28.

Watts, M. J. (1983) *Silent Violence: Food, Famine and Peasantry in Northern Nigeria*, Berkeley: California University Press.

—— (1993a) 'Development I: Power, knowledge, discursive practice', *Progress in Human Geography*, 17 (2), pp. 257–272.

—— (1993b) 'The geography of post-colonial Africa; space, place and development in sub-Saharan Africa (1960–1993)', *Singapore Journal of Tropical Geography*, 14 (2), pp. 173–190.

—— (2000) 'Poverty and the politics of alternatives at the end of the millennium', in Pieterse, J. N. (ed.) *Global Futures: Shaping Globalization*, London: Zed Books, pp. 133–147.

—— (2001) '1968 and all that. . .', *Progress in Human Geography*, 25 (2), pp. 157–188.

Weisskopf, W. A. (1964) 'Economic growth and human well-being', *Quarterly Review of Economics and Business*, 4, September, Part 2.

Wellington, J. H. (1955) *Southern Africa: A Geographical Study*, Vol. 1, *Physical Geography*, Cambridge: Cambridge University Press.

Werbner, R. (2002) 'Introduction: Postcolonial subjectivities: the personal, the political and the moral', in Werbner, R. (ed.) *Postcolonial Subjectivities in Africa*, London: Zed Books, pp. 1–21.

White, H. (1999) 'Global poverty reduction: are we heading in the right direction?', *Journal of International Development*, 11, pp. 503–519.

—— (2001) 'Pro-poor growth in a globalised economy', *Journal for International Development*, 13 (4), pp. 549–570.

White, P. (1999) 'The role of UN specialised agencies in complex emergencies: a case study of FAO', *Third World Quarterly*, 20 (1), pp. 223–238.

Wilks, A. (2001) *A Tower of Babel on the Internet?: The World Bank's Development Gateway*, London: Bretton Woods Project.

Williams, G. (1981) 'The World Bank and the peasant problem', in Heyer J., Roberts, P. and Williams, G. (eds) *Rural Development in Tropical Africa*, London: Macmillan.

Williams, M. (1993) 'Re-articulating the Third World coalition: the role of the environmental agenda', *Third World Quarterly*, 13 (1), pp. 22–30.

Williams, R. (1976) *Keywords: A Vocabulary of Culture and Society*, London: Fontana.

—— (1983) *Toward 2000*, London: Penguin.

Wills, J. (2002) 'Political economy III: neoliberal chickens, Seattle and geography', *Progress in Human Geography*, 26 (1), pp. 90–100.

Winner, L. (2001) 'Questioning the unquestioned', *Resurgence*, 208, pp. 6–8.

Wise, M. J. (2001) 'Textbooks that moved generations: becoming a geographer around the Second World War', *Progress in Human Geography*, 25 (1), pp. 111–121.

Wise, M. J. and Johnston, R. J. (1999) Obituary, *Geographical Journal*, 165.

Wisner, B. (1969) 'Editors Note', *Antipode*, 1 (1), pp. 1–3.

—— (1976) 'An overview of drought in Kenya', Working Paper No. 30, Boulder, CO: Institute for Behavioral Science.

—— (1977) 'Man-made famine in western Kenya', in O' Keefe, P. and Wisner, B. (eds) *Land Use and Development*, London: Blackwell, pp. 194–215.

Wolfe-Phillips, L. (1987) 'Why Third World – origins, definitions and usage', *Third World Quarterly*, 9 (4), pp. 1311–1319.

Wolfensohn, J. (2002), 'Time for the UN to act to sustain global development', *The Straits Times*, 2 September, p. 13.

Wood, A. and Welch, C. (1998) *Policing the Policemen: The Case for an Independent Evaluation Mechanism for the*

IMF, London: Bretton Woods Project, Friends of the Earth US.
Wood, C. (1998) *Development Arrested: Race, Power and Blues in the Mississippi Delta*, London: Verso.
Woodridge, S. W. (1950) 'Reflections on regional geography in teaching and research', *Transactions of the Institute of British Geographers*, 16, pp. 1–11.
Woods, N. (2000a) *The Political Economy of Globalisation*, Basingstoke: Macmillan.
—— (2000b) 'The challenge of good governance for the IMF and World Bank themselves', *World Development*, 28 (5), pp. 823–841.
World Bank (1995) *Priorities and Strategies for Education*, Washington, DC: World Bank.
—— (1997) *The State in a Changing World*, World Development Report, Washington, DC: World Bank.
—— (1998) *Assessing Aid: What Works, What Doesn't and Why*, Washington, DC: World Bank.
—— (1999) *Knowledge for Development*, World Development Report, Washington, DC: Oxford University Press for the World Bank.
—— (2000a) *Assessing Globalisation*, Washington, DC: World Bank.
—— (2000b) *Can Africa Claim the Twenty-first Century?*, Washington, DC: World Bank.
—— (2001a) *World Development Indicators*, Washington, DC: World Bank.
—— (2001b) *Attacking Poverty*, World Development Report, Washington, DC: Oxford University Press for the World Bank.
—— (2001c) 'New World Bank Report urges broader approach to reducing poverty', News Release 2001/042/S, available at http://www.worldbank.org/news/ (15 July 2002).
—— (2001d) 'World Bank and IMF will not hold annual meetings', News Release 2002/084/S, available at http://www.worldbank.org/news/ (15 July 2002).
—— (2002) *Building Institutions for Markets*, World Development Report, Washington, DC: World Bank.
World Commission on Environment and Development (WCED) (1987) *Our Common Future*, Oxford: Oxford University Press.
World Social Forum (WSF) (2001) 'Postscript: Porto Alegre call for mobilization, January 2001 WSF', in Houtart F. and Polet F. *The Other Davos: The Globalization of Resistance to the World Economic System*, London: Zed Books, pp. 122–125.
—— (2002) 'Porto Alegre II: Call of social movements', available at http://www.viacampesina.org/ (31 July 2002).
Worseley, P. (1979) 'How many worlds?', *Third World Quarterly*, 1 (2), pp. 100–108.
Wright R. (1995) *The Color Curtain: A Report on the Bandung Conference*, Mississippi: University of Mississippi.
Yapa, L. (1996) 'What causes poverty? A postmodern view', *Annals of the Association of American Geographers*, 86, pp. 707–728.
—— (2002) 'How the discipline of geography exacerbates poverty in the Third World', *Futures*, 34, pp. 33–46.
Yeung, H. (1999) 'Third World multinationals revisited: a research critique and future agenda', *Third World Quarterly*, 15, pp. 287–317.
Zoubir, Y. H. (2002) 'Libya in US foreign policy: from rogue state to good fellow?', *Third World Quarterly*, 23 (1), pp. 31–53.
Zuza, L. B. (2002) 'Vital interests and budget deficits: US foreign aid after September 11', *Middle East Insight*, March–April, pp. 25–29.

Index

Page numbers in *italics* denote references to Figures.

Aarts, P. 39, 179
Abdel-Malek, A. 52
Abrahamsen, R. 77, 80, 81, 84, 86, 114
Achebe, Chinua 123–4, 129
Adebayo, Adedeji 223
adivasi activists *206*, 209
Afghanistan 1, 37–8, 80, 117
Africa: alternative geography of 66–7; anti-colonialism 83, 110; Bandung conference 103; Blair's neocolonialism 133–4; British public perception of 36; Buchanan 55; Césaire 129; CIA report 219; colonial artificial territorial boundaries 98; dependency approach 81; development aid 35, 37; Eurocentrism 102; European view of history *50*, *51*; forgotten crises *172*; free trade 156; globalisation 6, 146, 151, 154–5; Hegel 76; images of 232; imagined geographies 6; IMF conditionality *164*, *165*; Institute of British Geography 58; media representations 170; military geographers 49; modernisation theory 59; modernity 138, 146, 151; nationalism 53, 55, 83, 108; New Economic Partnership for African Development 104, 133, 135–6, 223; new states 200; non-farm income 180; popular perceptions of 10; postcolonialism 108, 122–3, 124–5, 129–30, 137, 140; Prothero 58; reduction of US aid 114; role of the state 227; share of world trade 146; state labelling 97–8; stereotypes 139; structural adjustment 154; textbooks 231; 'three worlds' concept 105, 111; underdevelopment debates 64; Western handbooks 48–9; Western hegemony 84
African National Congress (ANC) 210–12
Africanist discourse 46
Afro-pessimism 136
agency 119, 126, 236
Agnew, J. 202
agriculture 156, 157, 180–1, 203–4
Ahluwalia, P. 136, 137, 139, 146, 152
Ahmad, Aijaz 106
Ainger, K. 204, 206, 209, 214, 215
Alea, Tomás Gutiérrez 127
Algeria 116
alienation 130
Allen, T. 6, 11, 20, 27, 77
Allende, Salvador 104
Americocentrism 7, 27
Amin, Idi 60
Amin, Samir 42, 64, 81, 107, 163; Bandung Era 103; globalisation 146; World Bank 182; World Forum for Alternatives 208
ANC *see* African National Congress
Anderson, B. 53
Angola: anti-imperial struggles 53; corruption 162; good governance agenda 166; humanitarian crisis *172*; peace demonstrations *82*; postcolonialism 121; Poverty Reduction Strategy Papers 177; socialism 81; US intervention 115; Western sponsoring of anti-socialist guerrilla movements 80
Ankomah, Baffour 170
Annan, Kofi 15, 36, 223
anti-capitalism protests 18, 25–6, 42, 89, *195*, 235; *see also* anti-globalisation
anti-colonialism 55, 64, 110, 116; Césaire 129; Marxism 83; public opinion 114; resistance 197, 234; *see also* decolonisation
anti-communism 31–2, 78, 80, 104, 114, 132
anti-development 49, 83, 89, 90, 230; localism 207; McGee 61, 66
anti-geopolitics 202–3

anti-globalisation 18, 199; misleading nature of term 42; protests 26, 143, 155, 156, *195*, 209; *see also* anti-capitalism protests
Antipode (journal) 60, 80, 111
apartheid 55, 97, 111, 113, 210, 211
Appadurai, A. 163, 187, 198
Apter, D. E. 1, 4, 27–8, 79, 89, 90, 91
Arap Moi, Daniel 223
area studies 58–9, 106, 110
Argentina: crisis 197; IMF policies 42; neoliberal 'success stories' 166; postcolonialism 121; resistance 152, 196, 197, 217; 'Third Way' 127
arms trade 108, 133
Arnold, G. 114
Ashcroft, B. 123, 137, 139
Asia: anti-colonialism 110; Bandung conference 103; development aid 35, 37; financial crisis 42, 153, 165, 207, 214; globalisation 6; images of 232; imagined geographies 6; IMF conditionality *164, 165*; macroeconomic instability 152; nationalisms 53; new states 200; popular perceptions of 10; postcolonialism 140; postwar independent countries 31; role of the state 227; 'three worlds' concept 105, 111; tiger economies 107, 160; Western pre-war research 47; *see also* Southeast Asia
Asian Development Bank 38
Australia 35, 102
Ayoob, Mohammed 224

Bakan, A. 147
Baker, S. J. 52
banana trade 213
Bandaranaike, Sirimavo 104
Bandung conference 49, 102, 103, 104, 116, 129
Bangladesh 158, 177
Banzer, Hugo 33
Barbour, Michael 57
Barnes, T. J. 66
basic needs approach 66
Bauman, Z. 25
Bayart, Jean-Francis 69, 224
Bebbington, A. 198
Belgian Congo 48
Belize 89
Bell, M. 108, 111
Bello, Walden 207
Berger, M. T. 105, 140–1, 166, 168, 198, 227, 234
Bessis, S. 227
Bhaba, H. 27, 119
bilateral aid 32, 34, 35, 228
Bilgin, P. 224, 225
biotechnology corporations 157
Bishop, Maurice 104
Black, J. 73
Blair, Tony *3*, 84, 117, 133–4, 223
Blaut, Jim 55, 56, 57, 61, 67, 68
Boehmer, E. 141–2, 197, 217
Bokassa, J. B. 60
Bolivia: debt reduction 23; drug industry 33; Poverty Reduction Strategy Papers 175; US aid 34; vulture funds 228
Bond, Patrick 135, 136, 211, 212
Bonds Boycott 211, 212
Botswana 121
Bourdet, Claude 102
Braden, K. E. 96, 98, 99
brands 13
Brazil 9, 99; banana trade 213; cinema 127; DARG research projects 64; dependency approach 81; financial crisis 165; *Movimento dos Trabalhadores Rurais Sem Terra* 203–4, 209, 216, 222; postcolonialism 121; resistance movements 203–4, 216, 217; reverse cultural currents 109; Western defence of dictatorships 80
Brazilian geographers 60
Bretton Woods institutions 24, 84; *see also* International Monetary Fund; World Bank
Bretton Woods Project (BWP) 200, 220, 223, 228
bribarisation 160, 163
Britain *see* Great Britain
Brookfield, H. C. 56, 59, 61
Brown, Gordon 228
Browne, S. 176
Brundtland World Commission on Environment and Development (WCED) 13
Buchanan, Keith 49, 54–7, 220–1; critique of Rostow 78–9; fieldwork 67; imperialist power relations 62; radical geography 61, 66, 68, 69
Buchanan, R. O. 49
Burkina Faso 222
Burma (Myanmar): Bandung conference 103; independence 31; military geographers 49; perceived importance of 99; Western pre-war research 47
Bush, George W. *2*, 15, 33, 38, 223, 226
Butlin, R. 47, 57, 58
BWP *see* Bretton Woods Project

Cabral, A. 229
cacerolazos 197
Cambodia 177
Canada: bilateral aid programme 32; development aid 35, 223; development geographies 7; World Trade Organisation 147
Canadian International Development Agency (CIDA) 32, 173, 183, 188
capital liberalisation 152, 163
capitalism 2, 28, 43, 76, 234, 235; Asian financial crisis 153; Cold War 11, 97; colonisation 100; critiques of 56, 60; dependency theory 81, 82; economic rationality 87; foreign aid 39; globalisation 143, 145, 146, 149, 152–4; hegemonic 149; inequalities 9, 78; Islam 111; 'little d' development 16; Marxism 62, 155; modernisation theory 78, 79; neocolonialism 102; post-development 229; postcolonialism 121; Poverty Reduction Strategy Papers 23; resistance to 28, 150, 209; 'Third Way' 117; *see also* anti-capitalism protests
capitalocentrism 149, 150, 235, 236
Cardoso, Fernando Henrique 203

Caribbean: colonisation 100; dependency approach 81, 90; development aid 35; popular perceptions of 10
CBOs *see* community-based organisations
Central America 6
Central Intelligence Agency (CIA) 12–14, 39, 59, 149, 219
Césaire, Aimé 107, 129
Ceylon (Sri Lanka) 1, 99; Bandung conference 103; colonisation 100; independence 31; military geographers 49; socialism 104; tropical geography 52
Chad 114
Chakrabarty, D. 124
Chaliand, G. 111
Chambron, Anne-Claire 213
'champagne glass' of income distribution 9–10, *10*, 192, 221
Chaturvedi, S. 76, 205
Chaudhry, K. A. 227
Chile: dependency theory 81, 83; socialism 104; US intervention 115; Western defence of dictatorships 80
China 9, 48, 141, 228; Buchanan 55, 57; CIA report 219; communism 31; cultural revolution 68, 107; dam construction 199; DARG research projects 64; development aid 35; images of 232; perceived importance of 99; relatively little attention given to 65; socialism 18, 82; 'three worlds' concept 102, 104; tropical geography 53
Chirac, Jacques 15
Chomsky, N. 37, 209
Chossudovsky, M. 13
CIA *see* Central Intelligence Agency
CIDA *see* Canadian International Development Agency
cinema 109, 127–8
cities 65, 109, 139, 140, 151; *see also* urban areas
citizenship 19, 124, 141, 166
civil society 98, 117, 135, 216, 228; globalisation 148; map of development discourse *62*; Poverty Reduction Strategy Papers 175, 176, 177; social movements 196
'civilisation' concept 48, 53, 75
class 56, 149, 162, 170, 236; livelihood discourses 181, 182; social movements 196; women 125
'CNN effect' 36, 170
Coetzee, J. M. 123
Cold War 11, 16, 80; Afghanistan 38; anti-geopolitics 202; end of 106, 114; foreign aid 31, 32, 64, 132; geopolitics 41, 43, 59, 97, 107, 224; India 30; Marshall Plan 31; modernisation theory 79; Non-Aligned Movement 103–4; 'three worlds' conceptualisation 96, 105
collectivism 107
Colombia: banana trade 213; dependency approach 81; Escobar 85–6; good governance agenda 166; US Plan Colombia 33–4
colonialism 19, 28–31, 86, 100, 227; artificial territorial boundaries 98; Bandung conference 103; 'colonial geography' 48; critiques of 129; dates of independence *17*; dependency theory 81, 82, 83, 100; Gourou 53; independence movements 10, 11; internal 141; modernisation theory 56, 61; Orientalism 46; radical geography 68; Southeast Asia 140; Third World definition 116; tropical geography 58; 'Tropics' concept 45; trusteeship 131; *see also* anti-colonialism; imperialism; neocolonialism; postcolonialism
Columbus, Christopher 95
Commonwealth 31, 60, 128
communism: anti-communism 11, 31–2, 78, 80, 104, 114, 132; Cold War 11, 97; imagined geographies 57; modernisation theory as counter to 77; 'three worlds' concept 102; *see also* Marxism; socialism
communities 4, 7
community-based organisations (CBOs) 183
competition 222
Conable, Barber 24
conditionality *164*, *165*, 176, 223, 224, 226
conflicts 12, 98, 201
Congo, Democratic Republic of *172*
Connell, J. 56
consumerism 13, 108, 149
consumption 13, 221; developed countries 67; Rostow 78, 79, 81, 82
control 15–16
Cooke, B. 131, 132, 188, 201
Coppock, J. T. 49
Corbridge, S. 27, 66, 205, 230
core-periphery relations 31, 40, 62–3; dependency approaches 81, 82, 83; modernisation theory 79, 80; postcolonialism 119
corporate liberalism 25
corruption 60, 107, 162; CIA report 219; Mozambique 160; 'the West' 101
Costa Rica 213
Cousins, Jim 162
Cowen, M. P. 28, 73, 130–1
critical geopolitics 95, 96–7
Crow, B. 108
Crush, Jonathan 4, 14, 64, 138, 232
Cuba: cinema 109, 127, *128*; dependency theory influence 83; development aid 35; as 'rogue' state 225; socialism 18, 81, 82; US intervention 115
cultural studies 110
culture: colonialism 130; globalisation 146; imperial 120; 'Third World' 109

Dabashi, H. 90, 101, 141
Dakar declaration 216, 223
DANIDA 135
DARG *see* Developing Areas Research Group
Darwinism 47, 77
data collection 21
Davies, Mike 57
De Soto, Hernando 42, 43
De Sousa, A. R. 56
debt 22–5, 175; Argentina 197; cancellation 25, 133–4, 176, 216; crisis 23, 35, 150, 168; private capital flows 35; South Africa 211; vulture funds 228; *see also* Heavily Indebted Poor Countries; Jubilee 2000 campaign; Jubilee Debt Campaign
decolonisation 11, 16, 30, 59, 77, 129; cinema 127; of development thinking 25, 29, 41, 67; of geography 47, 49, 232; geopolitics 97; incompleteness 137; India 79, 224;

Middle East 141; paradigm shifts 46; postcolonialism 120, 121; postwar 101; Subaltern Studies Group 124; Third World cities 139, 140; trusteeship 131; *see also* anti-colonialism
deconstruction 16, 90, 174
deliberalisation 145, 222
democracy 84, 114, 152, 166, 205, 215
democratisation 89, 169, 225
Denmark 135
Department for International Development (DFID) 132, 133, 134, 179–81, 182, 188; development dissemination 192; poverty 173, 190; research opportunities 233
dependency approaches 23, 56, 60, 72, 81–3, 90; capitalism 155; colonialism 81, 82, 83, 100; critiques of 82–3, 88–9; hegemony *91*; neocolonialism 122; post-development 88–9, 229; resistance 194; underdevelopment 117
depoliticisation 81, 87, 89
deregulation 9
Derrida, Jacques 16, 72, 93, 102
Desforges, L. 5
deterritorialisation 6, 152, 163, 196, 236
Developing Areas Research Group (DARG) 64
developing countries 11, 18; agricultural exports 156; CIA report 12–13; data collection 21; fieldwork 67; 'knowledge gaps' 186; media representations 170, 171; Western intervention 87; *see also* Third World
development 16–19, 232–5; 'big D/little d' distinction 16, 28; colonialism 28–30; critical developmentalism 229; decolonisation 11; deconstruction of 16; definitions 1–2, 4; deliberalisation of 222; disability issues 92–3; discourse of 14; dissemination of 19, 169, 172, 183, 191, 192; Enlightenment 74, 75; global moral imperative 172–3; Hegelian principle of 76; indicators 20–2; international financial institutions 161, 162; map of development discourse 62–3, *62–3*; meanings of *94*; modernisation theory 59–60, 77–81; post-development critique 26–8; radical geography 60–6, 68, 69–70; 'religious' conception of 7–8; spatiality 5, 41, 100, 230, 235; uneven 9, 107–8, 170; World Bank 'development knowledge' 184–8; *see also* anti-development; developmentalism; post-development; sustainable development; underdevelopment
developmentalism 28, 29; critical 229; critiques of 74; decolonisation 224; definition 236; demise of 196; evolutionary thinking 77; nationalist 77; progress 230; state-centred 202; Third World cities 65, 109
DFID *see* Department for International Development
diaspora 121–2, 236
difference 18, 89, 111, 195, 233; cultural politics of 231; Orientalism 46; postcolonialism 123
digital divide 184
Dimbleby, Jonathan 171
Dirlik, A. 27, 94, 123
disability 92–3, 227
discourse 115, 116, 169, 170, 224; definition 14, 236; post-development 85, 86, 87, 90; postcolonialism 119
diversity 18, 137, 215
Doherty, T. 74, 76
Domosh, M. 66
Donovan, P. 213
Doty, Roxanne L. 66, 86, 107, 116
Driver, F. 45, 46–7, 52
Drop the Debt campaign 133

'the East' 95–6
East Asia 87, 160, 207, 228
East Indies 100
East Timor 153, 170
Easterly, William 165
Eastern Europe 106, 191, 235; IMF conditionality *164, 165*; official aid 37; postcolonialism 140; social assistance 177; socialism 105, 202–3; US aid missions 114
Economic Commission for Latin America (ECLA) 82
economic determinism 82
economic geography 72
economic growth 60, 136, 191, 192, 221; Middle East 179; neoliberalism 9
economics 143, 191; neoclassical *62, 63*, 64, 72, 155; neoliberal 9, 94, 237; new institutional *63*
Ecuador 214
education 22
Edwards, M. 201
Egypt: Bandung conference 104; development aid 39; development texts 91; historical particularity 58; reverse cultural currents 109
El Salvador 115
elites 98–9, 102, 115
Elliott, L. 176
Ellis, F. 180
Ellwood, W. 143, 155
empowerment 132, 194, 199–202, 228
end of geography 6, 152
Engardio, P. 31
Enlightenment 28, 72–7, 83, 89, 230, 236–7; Africa 58; colonial geography 48; developmentalism 236; European colonialism 100; influence on neoliberalism 73, 94; legacies of 71, 90, 119; trade 155; trusteeship 131; universalism 90
environment 13, 108, 111; globalisation effect on 145, 148, 166; millennium development goals 22; resistance movements 203, 207
environmental determinism 58
Equatorial Guinea 49, 60
Eritrea *172*
Escobar, Arturo 22, 52, 115, 116; critique of developmentalism 74; 'domains of objects' 169; knowledge 93; neoliberalisation 145; participation 201; place 151; popular struggles 196; post-development 85–6, 87, 88–9, 102, 229, 230, 231; subjectivities 186
Esteva, Gustavo 71, 86, 91, 204, 224
Ethiopia 36, 58, *172*, 228
ethnic politics 98
ethnicity 60, 125; *see also* race
ethnocentrism 61, 115, 123, 232; *see also* Eurocentrism
EU *see* European Union
Eurocentrism 27, 64, 96, 102, 166, 230; Africanist discourse 46; Enlightenment thinking 71; globalisation 154; imagined

geographies 7; modernisation theory 90; modernity 72; postcolonial critique 119, 120, 123; postmodernism 91, 93; racism distinction 115; socialism 107; status of geography 49; systems of ordering the world 118
Europe 14, 67, 100; capitalist expansion 62; colonialism 29, 45, 68, 100–1; Enlightenment 72–7; free trade 133; nationalism ideals 114–15; Orientalism 46; perceived importance of 99; pre-war research 47–8; 'three worlds' concept 102; *see also* Eastern Europe
European Union (EU): development aid 34, 35–6, 223; research opportunities 233; World Trade Organisation 112, 147, 155–6
evolutionary theories 76–7
exclusion 7, 151, 175, 227, 230; neoliberalism 222; new spaces of 233; social 190, 191

FAA *see* Foreign Assistance Act
fair trade 147, 149, 153, 213, 234
Fair-Trade Network 135, 234
famine 36, 64, 111, 171
Fanon, Frantz 102, 119, 129, 229, 234; alienation 130; colonialism 100–1; national culture 127
FAO *see* Food and Agriculture Organisation
Farmer, B. H. 48, 50, 52, 57
FDI *see* foreign direct investment
feminism 69, 108, 111, 199–200, 229, 232; postcolonialism 125; postmodernism 93; radical geography 66; *see also* gender
FENOP *see* National Federation of Farmworkers Organizations
Ferguson, James 75, 86, 87, 88, 121
Fieldhouse, D. K. 102
fieldwork 55, 57, 67, 139, 140, 233
FIJI 35, 49
Fine, Ben 158, 160, 161, 162, 182
Fink, C. 107
First World: colonial ideas about intervention 131–2; culture 109; Fanon 101; feminism 199–200; South Africa 113; 'Thirdworldisation' 11; 'three worlds' conceptualisation 96, 102, 105, 106
First World War 100
'Firstworldisation' 11
Fisher, C. A. 48, 49, 58–9
Fisher, W. B. 49
Floyd, Barry 57
Focus on the Global South 145–6, 177, 207, 216
Food and Agriculture Organisation (FAO) 21, 52
Forbes, D. K. 60
foreign aid 15, 18, 22, 25, 32–7; conditionality 223, 224, 226; critics of 32; geopolitics 39–40, 41, 97; net flows *32*; postwar 31; US security interests 226; Zambia 88
Foreign Assistance Act (FAA) 31, 41
foreign direct investment (FDI) 35, 39, 152, 158, 159–60
foreign investment 42, 150, 152, 158, 159–60
Foucault, Michel 15, 16, 72, 85, 98, 197, 229
fourth world 111
France: development aid 35; globalisation 150; imperialism 137; Movement of the Unemployed 222; overseas research 47, 52; three 'Estates' 74, 102
Frank, André Gunder 60, 78, 79, 81, 82
free trade 19, 162, 163, 207, 222; globalisation 143, 148, 153, 155–7; resistance to 203, 209, 216; Smith 73; US promotion of decolonisation 101; World Trade Organisation 147; *see also* liberalisation; neoliberalism
Freire, P. 229
French Equatorial Guinea 49
Friedman, Milton 9

G-8 states 25, 26, 84, 207, 220, 223
G-77 states 103, 104, 155
Gandhi, L. 138
Ganokar, D. P. 66
Gariyo, Zie 176
Gates, Bill 185, 192
GATT *see* General Agreement on Tariffs and Trade
Gay, P. 73, 74
GDN *see* Global Development Network
Gelinas, J. B. 23
gender 64, 170, 212–14; identity 141, 196; modernisation theory 80; postcolonialism 123, 125; social movements 196, 215; South Africa 212; World Bank 227; *see also* feminism; women
General Agreement on Tariffs and Trade (GATT) 147
geopolitics 18, 25, 37–40, 41, 98, 99; Cold War 41, 43, 59, 97, 107, 224; critical 95, 96–7; definition 237; from below 202, 214; influence on World Bank 166; Plan Colombia 33–4; of race 12; Russian national development 105; *see also* politics
George, S. 158, 219, 222
Germany 35, 47, 52
Gettino, O. 127
Ghana 10–11, 59–60, 104, 128
Ghosh, Amitav 233
Ghosh, D. 128–9
Gibson-Graham, J. K. 149, 235
Gilbert, Alan 56, 65
Gilbert, E. W. 48
Gilroy, P. 11–12
Gleeson, B. 92
Global Development Network (GDN) 185
Global Exchange 188
global village concept 163
globalisation 5–6, 19, 83, 112–14, 143–68, 188; capitalocentrism 236; CIA report 219; critiques of 139; cultural identity 130; definition 237; deterritorialisation 196, 236; from below 187, 198, 202; modernisation paradigm 94; monoculture 208; post-development 89, 90; postcolonialism 123, 137, 139; resistance to 26, 42, 194, 195, 234; sub-disciplinary interaction 65; Zambia 88; *see also* anti-globalisation
glocalisation 154
GNP *see* Gross National Product
Goldsmith, A. A. 39
Golledge, R. G. 235
Gonsalves, S. 26

'good governance' 114, 131–2, 166–8, 169, 176, 198, 224
Gould, Peter 59–60, 61, 180
Gourou, Pierre 46, 47, 52–3
governance: definition 237; everyday practices 196; global 148; 'good governance' 114, 131–2, 166–8, 169, 176, 198, 224; international institutions 150; macroeconomic 145, 166; resistance 195; *see also* state
grand narratives 72, 84, 237
grassroots organisations (GROs) 183, 209, 210, 215, 217, 230; *see also* social movements
Great Britain: Buchanan 55; Commonwealth 31; datasets 21; development aid 31, 35; good governance 132; imperialism 137; India 76, 124, 224; Institute of British Geography 58, 64; overseas research 47, 48, 52, 58; partnerships 132; postcolonialism 133–4; principal ports *45*; television 171; urban life 109; *see also* Department for International Development; United Kingdom
Grenada 104
Grinspun, A. 174, 175, 177
GROs *see* grassroots organisations
Gross National Product (GNP) 3, 8–9, 34, 221
Group of 8 (G-8) 25, 26, 84, 207, 220, 223
Group of 77 (G-77) 103, 104, 155
Guatemala 80
Guevara, Che *112*
Guha, Ranajit 124
Guinea Bissau 53
Gupta, Akhil 124, 224

Haas, William H. 53
Hadjor, K. B. 114
Haffajee, F. 210, 211, 212
Hall, Stuart 99–100
Halliday, F. 39, 141
Hance, William A. 53
Hanlon, Joe 160
Hardt, M. 108–9, 207
Harriss, J. 205
Hart, G. 16, 28, 161, 228, 229
Hartwick, E. 22, 27, 29, 68, 78, 94, 229–30
Harvey, David 57, 61
Hatoum, Mona 127
Hayter, William 58
HDI *see* Human Development Index
healthcare 22
Heavily Indebted Poor Countries (HIPC) 23, 27, 104, 154; debt relief 133, 175, 176; NEPAD 136; vulture funds 228
Hegel, G. W. F. 76
hegemony 84, *91*, 101, 114, 224; Africanist discourse 46; capitalism 149; cultural 122; geopolitics from below 202; globalisation 164; United States 225
Held, D. 151
Hettne, B. 71, 89
Hewitt, A. 179–80
HIPC *see* Heavily Indebted Poor Countries
Hirschmann, A. O. 78
history 76, 126, 224
Holland, M. 158
Holloway, S. 197
Hong Kong 61, 153
Hoogevelt, A. 26, 154
Hooson, David 57
households 199
Houtart, F. 194, 202
Hubbard, P. 197
Hulme, D. 201
Human Development Index (HDI) 2–3, 191
human geography 64, 174
human rights 89
humanism 72
humanitarian assistance 138, *172*, 173
humanitarian crises 34, 36, 170, *172*
Huntingdon 180
hybridity: postcolonialism 119, 120, 123, 130, 139; resulting from colonialism 110

IBG *see* Institute of British Geography
ICC *see* International Chamber of Commerce
IDA *see* International Development Association
identity 109, 198, 218; critical geopolitics 97; cultural 130, 194; ethnic 196; gender 141, 196; multiple 154; postcolonialism 120, 123–4, 126, 130, 137, 139; social movements 196; travel relationship 5; *see also* national identity
identity politics 196
ideology 71, 100, 200
IDPM *see* Institute for Development Policy and Management
IFAD *see* International Fund for Agricultural Development
IFC *see* International Finance Corporation
IFG *see* International Forum on Globalization
IFIs *see* International Financial Institutions
Ignatieff, M. 36, 115
ILO *see* International Labour Office
imagined geographies 6–7, 57, 174, 220
IMF *see* International Monetary Fund
immigration 116, 232
imperialism 29–30, 43, 48, 100, 109, 116; British 124; core–periphery relations 31; globalisation 139; multiple identities 154; neocolonialism 237; NEPAD 136; postcolonialism 120, 123, 138; radical geography 61, 62; reason 76; Said 137; trusteeship 131; United States 39, 111; *see also* colonialism; neocolonialism
import-substitution industrialisation (ISI) 82
incomes: 'champagne glass' of distribution 9–10, *10*, 192, 221; diversification 180, 181; inequalities 144; mapping of *9*
independence movements 10, 11, 98–9, 106, 197–8
India 9, 48, 99, 141, 228; *adivasi* activists *206*, 209; agriculture 156, 157; banana trade 213; Bandung conference 103, 104; CIA report 219; citizenship 166; colonialism 76, 100, 224; corruption 162; dam projects 125–6, 205; DARG research projects 64; decolonisation 79; Enlightenment project 76; images of 232; independence 31, 198; military geographers 49; *Narmada Bachao Andolan* 126, 205–6, 209; Nehru 30; NGOs 183, 184; popular fictions 128–9; postcolonialism 124, 129; relatively little attention given to

65; resistance movements 205–6, *206*, 209, 214, 217; reverse cultural currents 109; Subaltern Studies Group 124; tropical geography 53
indigenous cultures 120
indigenous knowledge 187, 192, 194, 219
indigenous peoples 47, 124, *125*, 168, 205, 214, 215
Indonesia: Asian crisis 153; Bandung conference 103, 104; IMF assistance 171; media representations 170
industrialisation 64, 73, 76, 77; import-substitution 82; postcolonialism 123
inequalities 1, 111, 137, 170, 222; capitalism 78; 'champagne glass' representation 9–10, *10*, 192, 221; free trade 153; globalisation 151, 161, 166; households 199; income 144; livelihood discourses 181; macroeconomic 177; media representation 6; Middle east 39; postcolonialism 121; Thailand 207; Third World labelling 96
infant mortality 22
Institute of British Geography (IBG) 58, 64
Institute for Development Policy and Management (IDPM) 131
International Chamber of Commerce (ICC) 158, 159
International Development Association (IDA) 24
International Finance Corporation (IFC) 24
International Financial Institutions (IFIs) 27, 39, 160, 163, 168; economistic view of development 219; good governance 132; market fundamentalism 180; NEPAD 136; notion of failure 225; poverty reduction 228; PRSPs 175–6; reform of 220; resistance 214, 215, 216; social capital 148, 161, 162; steady state model 222; unaccountability 150; US political agenda 166; *see also* International Monetary Fund; World Bank; World Trade Organisation
International Forum on Globalization (IFG) 166
International Fund for Agricultural Development (IFAD) 89
International Labour Office 82, 92
International Monetary Fund (IMF) 24, *43*, 145; Argentina 42, 197; Asian financial crisis 207, 214; conditionality *164*, *165*; currency regulation 152; EU block 35; failure of poverty alleviation 165; Indonesia 170–1; influence on national governments 23; issues ignored by 225; market-based loans 22; neoliberalism 163; NGO funding 117; paradigm maintenance 228; protests against 25, *150*, 206, 207, 209, 214; PRSPs 175, 176; terrorism 37; US domination 128; Western influence 84; WTO cooperation 156
internationalisation 64
internationalism 99, 107
Internet 185
Investment Promotion Agencies (IPAs) 158, 159, 160
Iran 4, 38, 109, 141; Afghanistan conflict 38; dependency theory influence 81; Islamic revolution 101, 117; as 'rogue' state 225; war on terrorism 39; Western defence of dictatorships 80
Iraq 39, 117, 225
Ireland, Republic of 35, 61, 90
ISI *see* import-substitution industrialisation
Islam 39, 101, 110–11, 141; diversity 227; fundamentalism 116; Third Way 117
Israel: contested political space 189; internal colonialism 141; occupation of Palestine 178–9; radical geography 111; US support 41, 178; war on terrorism 39; 'the West' 141

Jackson, R. H. 107
Jalée, P. 109
Jamaica 104, 127–8, 213
James, Preston E. 53
Japan: development aid 35; development discourse 87; perceived importance of 99; 'three worlds' concept 102; 'the West' 141; World Trade Organisation 147
Jarosz, L. 36
Java 47
JDC *see* Jubilee Debt Campaign
Jerve, A. M. 178, 190
Johannesburg Summit (2002) 1, *2*, 13, 211
Johnson, B. L. 49
Johnson, R. 14
Jolly, R. 108
Jones, P. S. 233
Jordan 178
Jospin, Lionel 150
Jubilee 2000 campaign 25, 135, 176
Jubilee Debt Campaign (JDC) 134, 234
Jubilee Plus 134
Jubilee Research 133
Jubilee South 27, 133, 216
Jubilee South Africa 211
justice: economic 195, 217; social 61, 225, 235

Kanbur, Ravi 162
Keen, S. 64, 158
Kelley, D. G. 129
Kelly, P. F. 163
Kennedy, J. F. 31, 41
Kenya 64, 114, 130
Keynesianism *62*
Khomeini, Ayatollah 101, 117
Kiely, R. 18, 27
knowledge 72, 90, 170; Enlightenment 73, 74, 75; globalisation 145; indigenous 187, 192, 194, 219; post-development 84, 85; power relationship 169; unequal power relations 135; World Bank 'development knowledge' 184–8
Kobayashi, A. 232, 233
Kohler, Horst 24
Kosovo 170
Kothari, R. 110, 119, 122, 136–7, 140, 201
Kureshi, Hanif 127

Lacquer, T. W. 138
Laïdi, Z. 83
laissez-faire 62, 81, 82
language 19, 93, 169, 172, 188, 230; deconstruction 16; exclusion 175; postcolonialism 125, 129–30, 140; 'toxic' keywords 116; World Bank 189
Lao People's Democratic Republic 153, 177, 199
Latin America 95, 221; anti-colonialism 110; colonisation 100; dependency theory 81, 82; development aid 35, 37; failure of

Western policies 42; financial crises 165; globalisation 6; imagined geographies 6; IMF conditionality *164*, *165*; neoliberalisation 145; popular perceptions of 10; postcolonialism 140; social movements 216; 'three worlds' concept 105, 111; US 'Alliance for Progress' 31; World Bank 24
LDCs *see* 'less developed countries'
League of Nations 48, *187*
Learmonth, A. T. R. 49, 57
Lebanon 117, 178
Lee, John 224
legitimacy: colonial states 30; state 98, 152
Lesotho 87
'less developed countries' (LDCs) 10, 61, 62, 158
Lester, A. 29
Leys, C. 82
liberalisation 23, 225; capital 152, 163; CIA report 219; Middle East 179; Mozambique 159; trade 155, 156, 157, 158, 177; *see also* free trade; neoliberalism; privatisation
liberation movements 53, 141, 202
Liberia 114, 115
Libya 225–6
Liddell, I. 213
Lister, M. 35
literature 126–8, 129
livelihoods discourse 19, 172, 179–82, 183, 189
Livingstone, D. N. 46
Llwyd, L. 45
loans 22
the local 128, 137, 151, 154, 163, 217, 229
localisation 154, 163
Logan, M. I. 56
Loomba, A. 119
Luke, T. W. 97
Lummis, C. D. 9, 67, 166

Mabogunje, Akin 55
McCullagh, Ron 170
McDowell, L. 119
MacEwan, A. 42, 197
McEwan, C. 66, 119, 120, 140
McGee, Terry 56, 61–2, 66
McGrew, A. 104, 151
Machel, Samora *131*
Malaya 53
Malaysia 1, 31, 53, 153
Malloch Brown, Mark 177
Mamdani, M. 124
Mandela, Nelson 211
Manley, Michael 104, 128
Manuel, Trevor 15
Mao Tse Tung 68, 107
Marcuse, P. 145
marginalised groups 2, 67, 194, 227
market triumphalism 94
markets: map of development discourse *62*; neoliberalism 9, 237; World Development Report 153
Marquez, Gabriel Garcia 123
Marshall, D. D. 90
Marshall, George 31
Marshall Plan 31
Marx, Karl 62, 76, 78, 131, 155; class 236; dependency theory 81; modernity 77
Marxism 56, 62, 64, 66, 72, 77; dependency approaches 82, 83; national liberation movements 141; neocolonialism 122; political economy 155; post-development 229; social movements 108; *see also* communism; neo-Marxism
masculinity 200, 212, 214
Massey, Doreen 57, 122, 151, 230
Mauritania 176
Maxwell, S. 132
Mazrui, Ali 130
Mbeki, Thabo 136, 223
Mbembe, A. 126, 233
MDGs *see* millennium development goals
meaning 16
media 170–1; 'dumbing down' 171; globalisation 6; humanitarian crises 36; representation of anti-capitalist protesters 26; representations of the Third World 6, 115, 170, 171; stereotypical images 86, 189
Meer, Fatima 211
Melanesia 49
mental maps 6–7
Mexico 99, 166; resistance movements 204–5, 214, 217; reverse cultural currents 109; student protests 107; Zapatistas 97, 203, 204–5, *205*, 208, 214, 215
'micro-geography' approach 55
Middle East 4, 53, 87, 109, 227; CIA report 219; dependency approach 81; development aid 35, 39; Islamic ideology 101; Israel/Palestinian conflict 178–9; lack of geographical interest in 58, 64; military geographers 49; oil reserves 41; politics of development 90; postcolonialism 140, 141; PSI/PRSP processes 178; rogue states 225–6; US involvement 203
MIGA *see* Multilateral Insurance Guarantee Agency
migration 122, 123, 137, 144, 166
military geographers 49, 52
millennium development goals (MDGs) 19, 22, 36, 133, 172, 179, 223
Milne, S. 134
Ministry of Overseas Development 31
missionaries 130
Mitchell, T. 91
modernisation theory 18, 72, 77–81, 83, 105, 229–30; affluence 70; critiques of 56, 59, 61–2, 67–8, 81–2, 230; dominance of 228; Eurocentrism 90; hegemony *91*; linear stage theories 66; map of development discourse *63*; New International Economic Order 162–3; replaced by globalisation paradigm 94; Rostow 78–9; United Nations 84; Zambia 88
modernism 72, 75, 76, 83; critical 229, 230; shortcomings of 27
modernity 14, 59, 62, 72, 154; Africa 138, 146, 151; critiques of 56, 84; Enlightenment 71, 74, 75; Marx 77; post-development 90; Western 101; Zambia 88

Mohan, G. 135, 183
Mohan, J. 183
Mohanty, C. 200
Monterrey consensus 15
Moore, Michael 15, 156
Moore-Gilbert, B. 129, 141–2, 197, 217
Morgan, W. B. 52
Morrell, R. 212
Morton, A. D. 224, 225
Movement of the Unemployed (France) 222
Movimento dos Trabalhadores Rurais Sem Terra (MST) 203–4, 209, 216, 222
Mozambique *104*, *124*, 198; foreign investment 158, 159–60; perceived importance of 99; postcolonial cinema 127; Poverty Reduction Strategy Papers 176; socialism 81, 82; women's march *198*
MST *see Movimento dos Trabalhadores Rurais Sem Terra*
Mudimbe, V. 136
Mugabe, Robert 133, 134
multiculturalism 120
multilateral aid 34, 228
Multilateral Insurance Guarantee Agency (MIGA) 24, 158, 159
multinational corporations 64; *see also* transnational corporations
Munck, R. 27, 71, 83, 84, 90, 123, 169, 196
Murray, C. 180, 181–2
Myers, G. 231
Myrdal, Gunnar 78

NAFTA *see* North American Free Trade Agreement
Naipaul, Shiva 109
Naipaul, V. S. 114, 115, 123
NAM *see* Non-Aligned Movement
Narmada Bachao Andolan (NBA) 126, 205–6, 209
Nash, C. 142
Nashel, J. 80
Nasser, G. A. 103, 104
nation building 67, 89
nation-state 148
national belonging 19, 99, 123
National Federation of Farmworkers Organizations (FENOP) 222
national identity 77, 90, 99, 109, 125
nationalism 53, 77, 115, 198; Africa 53, 55, 83, 108; decolonisation 30; India 129; politics 98; postcolonialism 123–4
nationhood 19, 120
Naval Intelligence Division (NID) 48
NBA *see Narmada Bachao Andolan*
Ndione, E. S. 169
Ndungane, Njongonkulu 211
Negri, A. 108–9, 207
Négritude 129
Nehru, Jawaharlal 30, 103, 104, 117, 205, 230
neo-Marxism 66, 82
neo-populism *62*
neoclassical economics *62*, *63*, 64, 72, 155
neocolonialism 30, 42, 55, 107, 110; Blair 133–4; capitalism 102; critiques 109; definition 237; Marxism 122; Mazrui 130; Middle East 141, 227; partnerships 132; shared history of 114; subjectivities 126; Thirdworldism 200; 'three worlds' concept 99; trusteeship 131
neoliberalisation 145, 162, 189
neoliberalism 9, 10, 19, 158–60, 169, 222; alternatives to 193; colonisation of social sciences 161, 182; crises of 218; definition 237; denaturalisation of 222; developmentalism 229; DFID 180; domination of 130, 170; export markets 166; globalisation 26, 42, 143, 145, 149, 150, 152–3, 165; hegemony *91*; increase in numbers of poor people 191; international financial institutions 23, 43, 162, 163; map of development discourse *63*; Mexico 204; Middle East 179; modernisation theory comparison 94; Mozambique 159; NEPAD 219, 223; partnership concept 132; poverty 171, 175, 191; resistance to 28, 168, 170, 194, 195, 202, 209, 215; Russia 37; Sinatra Doctrine 224; single economic blueprint 207; Smith influence 73, 94; South Africa 210, 212; sub-disciplinary interaction 65; sustainable development incompatibility 13; 'Third Way' 117; USAID 188; World Bank 24, 175, 186, 200, 219; World Trade Organisation 147; *see also* free trade; liberalisation
NEPAD *see* New Economic Partnership for African Development
Netherlands 47
New Age romanticism 27
New Economic Partnership for African Development (NEPAD) 104, 133, 135–6, 223
New Economics Foundation 134
'new geographers' 56; *see also* radical geography
new institutional economics *63*
New International Economic Order (NIEO) 104, 162
New Labour 132, 134, 179–80
New Left Review 57
newly industrialising countries (NICs) 35, 221
Newman, D. 141
NGOs *see* non-governmental organisations
Nguema, Marcias 60
Ngũgĩ wa Thiong'o 123, 129–30
Ngwane, Trevor 211, 212, 223
Nicaragua 80, 115, 228
NICs *see* newly industrialising countries
NID *see* Naval Intelligence Division
NIEO *see* New International Economic Order
Nigeria: Buchanan 55, 57; civil war 60; DARG research projects 64; military aid 156
Nixon, Richard 40
Nkrumah, Kwame 11, 104, 117, 230
'noble savage' concept 100
Non-Aligned Movement (NAM) 49, 103–4, 129, 234
non-governmental organisations (NGOs) 5, 117, 145, 183–4; development dissemination 192; empowerment 201–2; partnerships 134–5; Poverty Reduction Strategy Papers 177; protests against WTO 155, 156; research opportunities 233; Uganda 176

'the North' 82, 95, 96, 116
North America 7, 29, 67, 100
North American Free Trade Agreement (NAFTA) 152, 168, 216
North Korea 18, 225
Nyamnjoh, F. 121, 126
Nyerere, Julius 104, 136, 229

Ó Tuathail, G. 152, 163, 202
OA *see* official aid
O'Connor, A. 174
ODA *see* overseas development assistance
OECD *see* Organisation for Economic Co-operation and Development
official aid (OA) 37
oil 41, 227
O'Keefe, P. 64
Olsen, G. R. 35
Oman 39
Ong, A. 67
OPEC *see* Organisation of Petroleum Exporting Countries
organic model of development 78, 81
Organisation for Economic Co-operation and Development (OECD): development aid 34, 35, 41; European development 177; global protests 25; International Monetary Fund 24; poverty reduction 173; *Shaping the Twenty-first Century* report 36, 132
Organisation of Petroleum Exporting Countries (OPEC) 64, 150
Orientalism 45, 46, 52, 106, 120
'the other' 46, 97, 101, 116, 141, 232
'Other Davos' 208, 222
overseas development assistance (ODA) 25, 31, 35, 37
Oxfam *43*, 147

Pacific Islands Trust territory 131
Pakistan: Afghanistan conflict 38; Bandung conference 103; independence 31; military aid 156; military geographers 48, 49; US lifting of sanctions 166
Palau 131
Palestine 178–9, 189
Panitchpakdi, Supatchai 147
Panos Institute 156
Paolini, A. 111, 120, 121; Africanist discourse 46; globalisation 149, 150, 151, 154
Papua New Guinea 35, 49
Parekh, B. 154, 179
Parikrama *226*
Parnwell, M. 194, 198, 207
participation 142, 194, 200–1; livelihood discourses 182; NGOs 183, 184; World Bank 189
partnership 19, 123, 131, 132–5, 137, 175
Patel, R. 75, 189, 190, 191
Patkar, Medha 125–6, 206, *206*, 209
patriarchy 157, 168
Peake, L. 232, 233
Peck, J. 26, 145
Peet, Richard 22, 27, 29, 60–1, 64, 180; concerns of radical geography 111; critical developmentalism 229; market triumphalism 94; modernisation theory 68, 78; post-development 230; South Africa 210; spatial theory 59
perceptions 6, 10, 108
Pergau dam project 31, 132
periphery *see* core–periphery relations
Péron, Juan 117, 127
Perry, S. 128, 199
Peru 33, 228
Petras, J. 33, 117, 134
pharmacopoeia 42, 171, 237
Philippines 47, 153
Philpott, S. 55
Pieterse, J. N. 72, 121, 122, 132; capitalism 146, 154; Middle East 179; post-development 83, 88, 89; postcolonialism 217; poverty eradication 172
Pincus, J. 162
place: definition of 5; globalisation 151; postcolonialism 123
Plan Colombia 33–4
Pletsch, C. 102, 105, 106, 107, 116
Pogge, T. W. 20
political ecology 64, 72
political economy *62*, *63*, 155
political geography 57, 96
politics: cultural 231; democratisation 225; depoliticisation 87; foreign aid 22, 32; instabilities 12, 144, 219; Islamic ideology 101; nationalist 98; NGOs 201; of representation 116; resistance movements 202–10, 214–18, 222, 234; 'the South' 98; 'Third Way' 117, 132, 134; Third World 96; *see also* Cold War; geopolitics; governance; state
'the poor' 4, 7, 11, 43, 86, 139
Porter, P. W. 56
Portugal *29*, 35
positivism 46, 68
post-development 18, 23, 26–8, 72, 83–9, 228–31; anti-capitalism 25, 150; critiques of 88–9, 91, 229; Eurocentrism 90, 91; picture of the world 115–16; postcolonialism common concerns 120; postmodernism 91–3; poverty eradication 172; representation of Third World 102; resistance 217, 218; spatiality 41
post-structuralism 72, 85, 93, 125, 229, 232; definition 237; postcolonialism 119, 120
'post-Washington consensus' 148, 160, 165
postcolonialism 18–19, 110, 114, 119–42, 232, 233–4; Africa 108; Bandung project 104; critiques of 119–20, 121, 139, 140–2; definition 237; Enlightenment 74; globalisation 146, 148, 154, 163; good governance 224; India 30; internationalism 107; nationalism 98; poverty 189; radical geography 62, 66, 69; resistance 217, 218, 234; Said 49; solidarity 210; sub-disciplinary interaction 65; Zambia 88
postcoloniality 119, 130
postmodernism 27, 72, 91–3, 231; anti-capitalist protests 18; discursive production 15; postcolonialism 119, 120, 122
postmodernity 6
Potter, R. B. 27, 64, 65, 68–9, 78, 234

poverty 1, 4, 10, 111, 171–5, 188–92; acronyms *189*; alternative ways of addressing 169; 'blocked and generalised' 221; critiques of modernisation theory 56; developed countries 67; discourse 86; environmental issues 13; free trade 156, 157; global trade 150; globalisation 144, 145, 166; as handicap 86; Human Development Index 2–3; IFI-scripted strategies 228; livelihoods discourse 179–82, 183; material causes 171; millennium development goals 19, 22; misrepresentation of 172; monetary measures of 8–9; Monterrey consensus 15; post-development 83–4; radical geography 111; resistance movements 209, 215; social capital 182–3; statistics 20, 22; technification of 116; United States 7, 232; World Bank 7–8, 24, 42, 170, 175–7; *see also* Poverty Reduction Strategy Papers
Poverty Reduction and Growth Facility (PRGF) 175
Poverty Reduction Strategy Papers (PRSPs) 23, 27, 163, 175–8, 189, 216; lack of transparency 227–8; Mozambique 159; NEPAD 136
Poverty Strategies Initiative (PSI) 177–8
power 41, 90, 170, 196–7, 202, 230; aid programmes 31; critical geopolitics 97; development studies 122; empowerment 201; Enlightenment knowledge 74; Foucault 15, 197; globalisation 145; inequalities in knowledge production 135; knowledge relationship 169; neoliberalism 26; post-development 84, 85, 86, 87; postcolonialism 142; radical geography 62; resistance 195; spatiality relationship 99; state 98, 99; technologies of domination 15–16
Power, M. 180
Prakash, M. S. 91
Prescott, Victor 57
PRGF *see* Poverty Reduction and Growth Facility
privatisation 23, 42, 168; Argentina 197; Mozambique 159; South Africa 211; USAID promotion of 79; World Bank 160, 163; Zambia 88; *see also* deregulation; liberalisation
progress 2, 20, 137, 222, 230; developmentalism 29, 236; Enlightenment 28, 71, 72, 73, 75, 237; hegemony *91*
Prothero, Mansell 57–8, 65
Proudfoot, Malcolm Jarvis 53
PRSPs *see* Poverty Reduction Strategy Papers
PSI *see* Poverty Strategies Initiative
public opinion 35, 36, 114
public resources 144
public sphere *63*
Pugh, J. C. 54–5

Qaddafi, Muammar 112, 136
Qatar 156
Quebec 222

Rabinow, P. 233
race 11–12, 47, 66, 108, 116, 123; *see also* ethnicity
racism 12, 115, 116, *140*, 232; Bandung conference 103; Césaire 129; colonialism 100–1, 130; developed countries 67; Enlightenment reason 76; internal 114; media representations of Africa 170; neoliberal globalisation 168; *see also* apartheid
Radcliffe, S. 41, 119
radical geography 46, 49, 57, 60–6, 69–70, 72; Buchanan 61, 66, 68, 69; development as 'liberation' 111, 220–1; *see also* 'new geographers'
Rahnema, Majid 4, 5, 10, 27, 200–1
rationalism 73
Ravenhill, J. 36
Reagan, Ronald 117
realism 72
reason 73, 75, 76
Reddy, S. G. 20
Redfield, P. 66
redistribution of wealth 210–11
regional geography 47, 49, 50–2, 56, 58
Reporting the World 170–1
representation 116, 123, 237
research 233
resistance 19, 194–9, 202–10, 214–18, 222; capitalism 28, 150, 209; definition 237; free trade 203; globalisation 26, 42, 194, 195, 234; neoliberalism 28, 168, 170, 194, 195, 202, 209, 215; post-development 27; postcolonialism 119, 128; South Africa 210–12, 214; terrains of 195; Western discourse 87; women 196, *198*, 199–200, *206*, 209, 217; World Social Forum 143–4
revolutions 101, 107, 111–12, 117
Rhodesia (Zimbabwe) 53, 57, 58, 134
Ricardo, David 155
Riddell, R. 132
Rieff, David 109
Rigg, J. 194, 198, 207
Rikowski, G. 147
Rio Earth Summit (1992) 13
Rist, T. 26, 84, 86, 157, 199, 217; Bandung conference 103; crisis of development 173; dependency approaches 81, 83; development activities 8; exclusion 230; semantic conjuring 188
Robequain, Charles 47, 48
Robinson, Jenny 65–6, 109, 139–40, 151
Rocha, J. 203
Rogerson, Andrew 166
rogue states 225–7
romanticism 27, 89, 91, 229
Rose, G. 66
Rosenblatt, L. 36
Rostow, Walt Whitman 59, 78–9, 80, 81, 82, 83, 232
Routledge, P. 194, 195, 196, 202
Rowbotham, M. 22, 23, 26, 150, 153, 163
Roy, Arundhati 125–6
rural areas 180–1, 190
Russia: Afghanistan 38; development aid 37; financial crisis 165; perceived importance of 99; poverty 67; *see also* Soviet Union, former
Rwanda 137, 170

Sachs, W. 10, 11
Said, E. 49, 106, 107, 119, 120, 129; anti-Americanism 38; imperialism 137; Islam 227; multiple identities 154; Orientalism 45–6

Sammy, W. 223
Samoff, J. 186, 187
sanctions 225–6
Sangavi, Sanjay 209
Santiso, C. 223
SAPs *see* Structural Adjustment Programmes
SAPSN *see* Southern African People's Solidarity Network
Sartre, Jean-Paul 108
SatireWire 15
Saudi Arabia 35, 39, 79, 227
Sauvy, Alfred 102, 106
Schaffer, Manfred 53, 54
Schenck, C. 128, 199
Schmitz, G. 41
Schuurman, F. J. 2, 27, 83, 137, 148, 149
science 75
Scott, D. 49, 104
Scottish Enlightenment 73
Seattle protests 25
Second World 18, 96, 99, 102, 106
Second World War 31, 48, 77, 80, 101
Sen, Amartya 191
Senghor, Léopold Sédar 129
September 11th attacks 1, 12, 192, 225
sexism 168
SFGJ *see* Students for Global Justice
Sharp, J. 83, 112, 113, 119
Sharp, J. E. 26
Shelley, F. M. 96, 98, 99
Shenton, R. W. 28, 73, 130–1
Shiva, Vandana 143, 156–7, 208
Shohat, E. 95, 107, 109–10, 114; cinema 127; discourse 115; Eurocentrism 102
Short, Clare 132, 133, 171, 179, 211, 228
Sidaway, J. D. 85, 140
Sierra Leone *172*
Simon, D. 91, 93, 128, 135
Simone, Abdou Maliq 139
Sinatra Doctrine 79, 224
Singapore 55–7, 153
Slater, David 56, 80, 90, 122, 195; dependency approaches 194; globalisation 148; resistance 198
slavery 29, 122, 123, *138*, 234
Smillie, I. 36
Smith, Adam 73, 94, 155
Smith, Adrian 235
Smith, Neil 94
social capital 148, 161, 162, 172, 182–3, 189
social change 76
social class *see* class
social engineering 89
social exclusion 190, 191
social history 64
social justice 61, 225, 235
social movements 19, 89, 108, 145, 227; critical developmentalism 229; partners in research 232–3; resistance 194–6, 198, 202–10, 214–18; South Africa 210, 211–12; World Social Forum 143–4, 168, 209, 211, 214–15, 216, 222; *see also* grassroots organisations
social reproduction 162
social responsibility 13
social sciences 75, 76, 106, 116, 161, 182
social semantics 102, 107
socialism 56, 81, 82, 102, 105; 'Bandung regimes' 104; collapse of 99; Eastern Europe 105, 202–3; Islam 111; map of development discourse *62*, 63, *63*; mythical images of 111; revolutions 46, 107, 111–12; Second World 18; state role 227; 'three worlds' concept 106; trusteeship 131; *see also* communism; Marxism
Sogge, D. 21, 41, 179, 182; development aid 34, 35, 36; geopolitics 37; NGOs 183
Solanas, F. 127
solidarity 26, 194, 204, 209–10, 215; anti-capitalist protests 25; coalitions 19; Dakar manifesto 216; post-development 27; South Africa 211, 212
Somalia 114, 131
'the South' 6, 10, 11, 18, 95, 96; exploitation 82; non-capitalism 149; politics 98; 'politics of representation' 116; *see also* Third World
South Africa 1, 187; apartheid 55, 97, 111, 113, 210, 211; banana trade 213; Buchanan 55, 57; DARG research projects 64; First World/Third World classification 112, 113; NEPAD 136; 'new wave' of geography 64; Non-Aligned Movement 104; radical geography 111; resistance 210–12, 214, 217; 'the West' 141
South Commission 136
South Korea 35, 153, 166, 222
Southeast Asia: Buchanan 55–7, 68; colonialism 31, 140; financial crisis 214; growth 160; military geographers 49; private capital flows 35; tropical geography 53; Western pre-war research 47
Southern African People's Solidarity Network (SAPSN) 193
sovereignty 99, 107, 152, 200
Soviet Union, former (USSR) 57, 104, 105–6; Cold War 80, 97, 105; collapse of 114; development aid 32; Middle East 178; poverty 192; *see also* Russia
space: critical geopolitics 97; disability issues 92
Spain 35
Spate, O. H. K. 49
spatialisations of development 21
spatiality 4–5, 41, 99, 100, 230, 231, 235
Spivak, G. 119
Spurr, D. 115
Sri Lanka (Ceylon) 1, 99; Bandung conference 103; colonisation 100; independence 31; military geographers 49; socialism 104; tropical geography 52
Stam, R. 95, 107, 109–10, 114; cinema 127; discourse 115; Eurocentrism 102
Stamp, Lawrence Dudley 47, 53, 54, 55
state 5, 200, 224; Africa 97–8; capitalist 56; colonial 30; geopolitics 97–8; intervention 23; legitimacy 152; map of development discourse *62*; neoliberalism 166; NGOs relationship 201; post-development perspective 87; role 152–3, 227; *see also* nation-state; rogue states

statistics 20, 21–2
Steel, R. W. 48, 58
stereotypes 111, 118; Africa 139; media images 86, 189; Third World 96, 170
Stiglitz, Joe 160, 162, 163
stigmatisation 115
Stoddart, David 64
Stokke, K. 135
Stone, E. 93
Stromquist, N. P. 186, 187
Structural Adjustment Programmes (SAPs) 23, 154, 159, 175, 180, 188, 191, 216
structuralism 60, 122
structuration theory 72
Students for Global Justice (SFGJ) 143
Subaltern Studies Group 124
'the subject' 84
subjectivity 89–90, 186, 237; post-development 84; postcolonialism 125, 126, 139, 218
Sudan 39, 57, 225
Suharto, President 153, 170
Sukarno, President 103, 104, 117
Summers, Lawrence 37
sustainable development 13, 20, 72, 89, 174; DFID 180; Johannesburg Summit 1, *2*; millennium development goals 22
sweatshops *7*
Sylvester, C. 77, 79, 81–2, 120, 121, 136
Syria 225

Taaffe, Edward J. 53, 60
Taiwan 35, 153
Talbot, C. 133
Taliban 37
Tanzania 61, 81, 82, 104
technical assistance 21
technologies of domination 15–16, 98
technology 145, 184, 185
Teivainen, T. 169
Telatin, M. 23
television 171
territoriality 99
territory: colonial artificial territorial boundaries 98; critical geopolitics 96, 97; definition 99; globalisation effect on 6; *see also* deterritorialisation
terrorism: politics of representation 116; September 11th attacks 1, 12, 192, 225; war on 18, 33, 37, 38–9, 223, 226, 227
Thailand 128, 140; Asian crisis 153; neoliberal 'success stories' 166; resistance movements 206–7, 214, 217; Western pre-war research 47
Thatcher, Margaret 117
Thiongo, Ngugi Wa 123, 129–30
'Third Way' 117, 127, 132, 134
Third World 10, 18, 109–116; blame 43; cinema 127; cities 65, 109, 140, 151; colonialism 101; debt 22–3; developmentalism 28; elites 98–9; exploitation 82; Fanon 101; feminism 199–200; 'Firstworldisation' 11; foreign aid 22; geopolitics 96, 97, 98, 99; globalisation 150–1; imagined geographies 220; media representations 6, 115, 170, 171; modernisation theory 79, 81; nation building 67; nationalism 77; post-development 85, 86, 87; postcolonialism 123; revolutionary pressures 53, 57; security-building 224; stereotypes 96, 111, 118, 170; student travellers 5, 111; 'three worlds' conceptualisation 96, 102–8, 109, 110–11; Western modernity 101; women 111, 112, 125; *see also* developing countries; 'the South'
Third World Network (TWN) 112–13, 166, 223
'Thirdworldisation' 11
Thirdworldism 18, 82, 98, 109, 110, 137; Islamic countries 117; linked resistances 116, 122, 210; neocolonialism 200; Non-Aligned Movement 103, 104; self-determination 215
Thomas, A. 2, 6, 11, 77, 201; balance sheet concept 20; post-development 27; student perceptions 108
Thompson, N. 128
Thompson, S. 128
'three worlds' concept 18, 93, 96, 99, 102–8, 116; critiques of 84, 109, 111; decentring of 117; Islam 110–11; Williams 221; *see also* First World; Second World; Third World
Thrift, N. 64
Tickell, A. 26, 145
tiger economies 107, 160
Tito, Marshal 117
Tobin, James 150
Togo 48–9
Townsend, J. 179, 183
trade 100, 133; banana trade 213; globalisation 143, 144, 150, 155–7; New International Economic Order 162–3; *see also* free trade; World Trade Organisation
transition economies 67
transnational corporations (TNCs) 5, 158, 209; banana trade 213; brands 13; foreign direct investment 159; globalisation 145, 146, 148, 165; Poverty Reduction Strategy Papers 23; resistance against 215, 216; state sovereignty 152; *see also* multinational corporations
transport 59–60
travel 5
trickle-down effects 60, 77, 78, 80, 211
tropical geography 46, 52–3, 58, 66, 69; Buchanan contrast 55, 56, 57; legacy of 68
'the Tropics' 45, 46–7, 59, 66, 67, 100; radical geography 61, 68; regional geography 49, 52–3
Truman, Harry 11, 30, 31, 49, 59, 71, 86, 106
trusteeship 74, 123, 130–1, 137, 228
Tunisia 39, 117
Turkey: development aid 35; geographical handbooks 48; non-colonial history 109; Thirdworldism 117; war on terrorism 39
Tutu, Desmond 211
TWN *see* Third World Network

Uganda 1, 60, 158; debt reduction 23; perceived importance of 99; Poverty Reduction Strategy Papers 176; trade 147
UN *see* United Nations

UNCTAD *see* United Nations Conference on Trade and Development
underdevelopment 11, 30, 52, 58, 68; Africa 64; capitalist expansion 62; dependency approaches 60, 82, 83, 117; geopolitics 97; India 124; modernisation theory 77, 79; neocolonialism 237; post-development perspective 86, 87
UNDP *see* United Nations Development Programme
UNESCO *see* United Nations Educational, Scientific and Cultural Organisation
UNICEF *see* United Nations Children's Fund
United Arab Emirates 39
United Kingdom 136, 174; *see also* Great Britain
United Nations (UN) 1, *3*, 7, *14*, 77, 101, *184*; Afghanistan conflict 38; Charter 131; conference on racism 234; development aid 32–4; 'Development Decades' 32; digital divide 184; disability issues 92; Food and Agriculture Organisation 21; Human Development Index 2; modernisation 84; monetary criteria 8; Monterrey consensus 15; New International Economic Order 162; NGO funding 117; poverty 191, 192; progress definitions 20; research opportunities 233; Rio Earth Summit 13; Statistics Division 21; Trusteeship Council 131; US geopolitical concerns 166
United Nations Children's Fund (UNICEF) 16, *21*, 92, 172–3
United Nations Conference on Trade and Development (UNCTAD) 112, 158, 159, 160, 190
United Nations Development Programme (UNDP): Afghanistan crisis 38; development definition 2; disability issues 92; global inequalities 9–10; multilateral aid 34; poverty 174, 177–8, 189, 191; Programme of Assistance to the Palestinian People 178; sustainable livelihoods *181*; trusteeship 131
United Nations Educational, Scientific and Cultural Organisation (UNESCO) 21, 52, 92
United States: anti-Americanism 38; anti-communism 31–2, 80, 132; anti-geopolitics 202; areas of influence 116; Bretton Woods institutions 24; CIA 12–14, 39, 59, 149, 219; Cold War 97, 106; datasets 21; decolonisation 101; dependency approach 81; development aid 31, 32, 34, 40, *40*, 223, 226; farm subsidies *157*; Foreign Assistance Act 31, 41; geopolitical concerns 166; global empire 11; IMF domination 128; imperialism 39, 111; interventions 115; Israel relationship 41, 178; living standards 1; Marshall Plan 31; Middle East 178, 203; modernisation theory 68, 79, 80, 94; NAFTA 152; NEPAD 136; Office of Strategic Services 59; overseas research 53–4, 59; perceived importance of 99; Plan Colombia 33–4; poverty 7, 174, 232; radical geography 61; reduction of aid 34, 114; rogue states 225–6; security assistance *40*; 'three worlds' concept 102; trade barriers 162; transnational corporations 158; urban life 109; Vietnam War 104; war against Afghanistan 37–8; World Bank 220; World Trade Organisation 112, 147, 155–6
United States Agency for International Development (USAID) 31, 35, 41, 79, 91; co-optation of NGO agendas 135; neoliberalism 188; poverty reduction 173; research opportunities 233; trusteeship 131
universities 52
Unwin, T. 64
urban areas 80, 139, 151, 190; *see also* cities
urban geography 65–6, 109, 139–40, 151
urbanisation 59
USAID *see* United States Agency for International Development
USSR *see* Soviet Union, former

Van Ausdal, S. 87, 89
Via Campesina 208–9
Vietnam 31, 57, 68; anti-colonial struggles 116; Asian crisis 153; dam construction 199; development aid 35; Poverty Reduction Strategy Papers 177; radical geography 61; socialism 18, 81, 82; US intervention 115
Vietnam War 104, 107, 111
Vitalis, R. 106
voice 120
'Voices of the Poor' 42, 139
Voluntary Service Overseas (VSO) 36, 171
vulnerability context 181
vulture funds 228

Wade, R. 144
WAIPA *see* World Association of Investment Promotion Agencies
Wales 61, 90
Wallerstein, I. 10, 59, 146, 218
Walling, D. 64
war on terrorism 18, 33, 37, 38–9, 223, 226, 227
Washington consensus 158, 212
Watnick, Morris 77
Watter, Ray 68
Watts, M. J. 62, 64, 88–9, 107, 201
Al Wazir, Intisar 178
WCED *see* Brundtland World Commission on Environment and Development
WDM *see* World Development Movement
WDR *see* World Development Report
WEF *see* World Economic Forum
Weisskopf, W.A. 8, 221
Wellington, J. H. 55
Werbner, R. 126
'the West' concept 75, 76, 95–6, 99–101, 102, 141
Westernisation 87, 101, 129, 154
Westoxication 101
whiteness 12, 232
WHO *see* World Health Organization
Wilks, A. 185
Williams, G. 113
Williams, Raymond 95–6, 112, 221–2
Wills, J. 218
Wisner, Ben 60, 64
Wolfensohn, James 1, 8, 24, 75, 160, 185, 187, 211
women 7, 125–6, 212–14; India 129; marginalisation 156–7; millennium development goals 22; poverty 191; resistance 196, *198*, 199–200, *206*, 209, 217; social capital 161; South Africa 212; 'Third World women' myth 111, 112, 125
Women's Movement (Quebec) 222

World Association of Investment Promotion Agencies (WAIPA) 158, 160
World Bank 7, 24, *146*, 158, 188–91; Afghanistan crisis 38; Bonds Boycott 211, 212; co-optation of NGO agendas 135; conditionality 226; currency regulation 152; debt repayment 23–5, 133; development dissemination 192; 'development knowledge' 184–8; development rhetoric 162; disability issues 92; Enlightenment knowledge 74–5; EU block 35; failure of poverty alleviation 165; Focus on the Global South 145; free trade 73; G8 voting power 84; gender issues 227; globalisation 165–6; HIPCs 154; income data map *9*; income distribution 144; influence on national governments 23; issues ignored by 225; location in Washington 220; market-based loans 22, 23; misleading poverty statistics 20; modernisation theory 79; Mozambique 160; multilateral aid 34; neoliberalism 163, 175, 186, 200, 219; NGO funding 117; paradigm maintenance 228; perception of Third World 113; post-development perspective 87, 88; poverty 7–8, 24, 42, 170, 175–7; primary concern 41–2; privatisation 160; protests against 25, *150*, 206, 207, 209; PRSPs 175, 176, 177, 216, 227–8; social capital 161, 182; South Africa 211, 212; sustainable development 13; terrorism 37; 'Voices of the Poor' 42, 139; website 185, 188; Western influence 84; WTO cooperation 156; *see also* World Development Report
world cities 65–6, 109
World Commission on Environment and Development (WCED) 13
World Development Movement (WDM) 15, 135
World Development Report (WDR) 24, 41–2, 153, 175, 180, 182, 191; issues ignored by 225; knowledge 185, 186, 187; neoliberalism 200
World Economic Forum (WEF) *14*, 143, 180, 211, 214
World Forum for Alternatives 208
World Health Organization (WHO) *21*, 92
World Social Forum (WSF) 143–4, 168, 209, 211, 214–15, 216, 222
World Summit on Sustainable Development (Johannesburg 2002) 1, *2*, 13, 211
World Trade Organisation (WTO) 112, 145–6, 147, 155–6; EU block 35; neoliberalism 163; protests against 25, 146, *150*, 155, 207, 209, 214; rule-making procedures 157
Worseley, P. 102, 117
Wright, Richard 103
WSF *see* World Social Forum
WTO *see* World Trade Organisation

Yapa, Lakshman 67, 71, 85, 86, 169, 188, 221, 231–2
Yemen 39, 227
Yeoh, B. S. A. 45, 46–7

Zaire 114
Zambia 88, 121
Zapatistas 97, 203, 204–5, *205*, 208, 214, 215
Ziegler, Jean 11
Zimbabwe (Rhodesia) 53, 57, 58, 134
Zoubir, Y. H. 225